£10 00

A Specialist Periodical Report

Biosynthesis
Volume 2

A Review of the Literature Published during 1972

Senior Reporter
T. A. Geissman, *Department of Chemistry, University of California at Los Angeles, U.S.A.*

Reporters
T. W. Goodwin, *University of Liverpool*
J. R. Hanson, *University of Sussex*
J. B. Harborne, *University of Reading*
A. Kjær, *Technical University of Denmark, Lyngby, Denmark*
E. Leete, *University of Minnesota, Minneapolis, U.S.A.*
T. Money, *University of British Columbia, Vancouver, Canada*
P. Olesen Larsen, *Technical University of Denmark, Lyngby, Denmark*
H. H. Rees, *University of Liverpool*
M. Tanabe, *Stanford Research Institute, California, U.S.A.*

© Copyright 1973

The Chemical Society
Burlington House, London, W1V 0BN

ISBN: 0 85186 513 5
Library of Congress Catalog No. 72-83455

Printed in Great Britain by Page Bros (Norwich) Ltd, Norwich

Introduction

This second volume on Biosynthesis follows the pattern of Volume 1, with the inclusion of a review of non-alkaloidal nitrogenous compounds and the partition of plant phenolic compounds into the two principal divisions, polyketide-derived and shikimate-derived classes.

As before, there is occasional overlap and repetition, not only with other sections of this volume but with topics reviewed in other volumes of the series. This is not disadvantageous, and an incursion into a related area is usually necessary for maintaining the coherence of a discussion.

The emphasis in this volume will be seen to have become increasingly biochemical. Significant advances are being made in the study of biosynthetic processes in isolated tissues, cell homogenates, and cell-free systems, and much of the work reviewed here deals with the enzymology of the synthetic processes under discussion. It is to be expected that this trend will continue and that future volumes of this series will witness an increasingly rapid dissolution of the chemical–biological interface.

T. A. GEISSMAN

Contents

Chapter 1 Biosynthesis of C_5–C_{20} Terpenoid Compounds 1
By J. R. Hanson

  1 Introduction 1
  2 Mevalonic Acid 1
  3 Hemiterpenoids 2
  4 Monoterpenoids 4
  5 Sesquiterpenoids 7
  6 Diterpenoids 11

Chapter 2 Biosynthesis of Triterpenes, Steroids, and Carotenoids
By H. H. Rees and T. W. Goodwin 16

  1 Introduction 16
  2 Mevalonic Acid 16
  3 Squalene 22
  4 Formation and Cyclization of Squalene 2,3-Oxide 25
  5 Steroids in Vertebrates 27
    Cholesterol Biosynthesis 27
    Further Metabolism of Cholesterol 31
      Steroid Hormones 31
      Bile Acids 44
  6 Triterpenes and Steroids in Higher Plants, Algae, and Fungi 48
    Phytosterol Biosynthesis 48
      Overall Pathway 48
      C-24 Alkylation 55
      General Aspects of Phytosterol Biosynthesis 57
      Further Metabolism of Plant Sterols 57

7 Triterpenes and Steroids in Invertebrates	61
Insects	61
Invertebrates other than Insects	64
Sterol Biosynthesis	64
Metabolism of Sterols	64
8 Carotenoids	65
Prephytoene	65
Stereochemistry of Phytoene	65
Cyclization of Acyclic Carotenes	67
Synthesis in Photosynthetic Bacteria	70

Chapter 3 Non-protein Amino-acids, Cyanogenic Glycosides, and Glucosinolates 71
By A. Kjær and P. Olesen Larsen

1 Introduction	71
2 Non-protein Amino-acids	71
General	71
Serine and Cysteine Derivatives	72
Selenium-containing Amino-acids	79
Homoserine, Homocysteine, Methionine, and Derivatives	79
Chain-lengthening of Amino-acids	82
Branched C_6 and C_7 Amino-acids	82
Imino-Acids	85
Aromatic Amino-acids	87
D-Amino-acids	89
3 Cyanogenic Glycosides	91
General	91
Biosynthesis	92
4 Glucosinolates	95
General	95
Biosynthesis	96
Glucosinolates from Amino-acids, with Preservation of Nitrogen	97
Glucosinolates from Amino-acids, without Preservation of Nitrogen	102

Chapter 4 Biosynthesis of Alkaloids 106
By E. Leete

1 Introduction	106

Contents vii

    2 **Highlights of 1972** 106
       Origin of Vasicine 106
       Formation of Pipecolic acid from D-Lysine 107
       Betanine Biosynthesis 108
       Aberrant Biosynthesis of Unnatural Alkaloids 109
       Biotransformations of Alkaloids 111
       Senecic Acid Biosynthesis 115
    3 **Biosynthesis of Tropic Acid** 115
       Phenylalanine as a Precursor 115
       Tryptophan and Phenylacetic Acid as Precursors 117
       Mechanism of the Rearrangement of the
       Phenylalanine Side-chain 118
    4 **Enzyme Studies** 121
       Nicotiana Species 121
       Ricinus communis 122
       Phalaris tuberosa 124
       Hordeum (Barley) 124
       Mercurialis perennis 124
    5 **Alkaloid Production in Tissue Cultures** 124
    6 **Table of Tracer Work relating to Alkaloid Biosynthesis** 126

Chapter 5 Biosynthesis of Polyketides 183
By T. Money

    1 **Aromatic Polyketides** 183
    2 **Non-aromatic Polyketides** 212

Chapter 6 Biosynthesis of Phenolic Compounds Derived from Shikimate
By J. B. Harborne 215

    1 **Introduction** 215
    2 **Phenols and Phenolic Acids** 216
       Salicylic Acid 216
       3,4-Dihydroxy-2,2-dimethylphenylvaleric Acid 216
       Lack of Repression of the Shikimic Acid Pathway 217
       Hydroxybenzoic Acid Turnover 217
    3 **Phenylpropanoids** 218
       Biosynthesis of Dihydroxycoumarins 218
       Biosynthesis of Lignin 221
       Miscellaneous Phenylpropanoids 223
       Phenylalanine Ammonia Lyase (PAL) 224
       Phenolases 226

4 Flavonoid Biosynthesis		227
	Enzymology of Biosynthesis	227
	Naturally Occurring Intermediates	228
	Glycosylation and Acylation	231
	Anthocyanin Synthesis	232
	Proanthocyanidin Synthesis	233
	Flavonoid Turnover in Higher Plants	235
	Flavonoid Degradation by Micro-organisms	235
5 Isoflavonoids		236
	Biosynthesis of Hydroxyphaseollin	236
6 Neoflavonoids		237
	Biosynthesis of Inophyllide	237
7 Quinones		238
	Biosynthesis of Lawsone	238
	Anthraquinone Biosynthesis	238
	Perinaphthenones	239

Chapter 7 Stable Isotopes in Biosynthetic Studies 241
By M. Tanabe

1 Introduction		241
2 Nuclear Magnetic Resonance Methods		241
	Proton Satellite Method	241
	Griseofulvin	242
	Fusaric Acid	243
	Variotin	245
	Piericidin A	245
	Mollisin	246
	Sepedonin	247
	Direct Carbon Magnetic Resonance Method	252
	Radicinin	252
	Sepedonin	255
	Asperlin	256
	Cephalosporin C	259
	Prodigiosin	260
	Sterigmatocystin	263
	Avenaciolide	267
	Ochratoxin A	270
	Aureothin	272
	Antibiotic X-537A	273
	Virescenosides	275
	Pyrrolnitrin	275

Protoporphyrin-IX	278
Vitamin B_{12}	281
Shanorellin	287
Palmitoleic Acid	288
3 Mass Spectrometry	**288**
General	288
Patulin	288
Deuterium Labelling	290
Vindoline	290
Lincomycin	292
Mycarose	293
Vitamin K_2 and Ubiquinone Q-8	294
Protoporphyrin-IX	295
Nitrogen-15 Labelling	296
Gliotoxin	296
Oxygen-18 Labelling	298
β-Nitropropionic Acid	298
Stipitatonic Acid	298
Carbon-13 Labelling	299
Phlebiarubrone	299
Author Index	**300**

1
Biosynthesis of C_5—C_{20} Terpenoid Compounds

BY J. R. HANSON

1 Introduction

This chapter for 1972 follows the pattern of last year's Report.[1] During the year a number of reviews have appeared which discuss different aspects of terpenoid biosynthesis.[2-5]

2 Mevalonic Acid

The enzyme 3-hydroxy-3-methylglutaryl coenzyme A reductase catalyses the two-step reduction of 3-hydroxy-3-methylglutaryl coenzyme A (HMG-CoA)

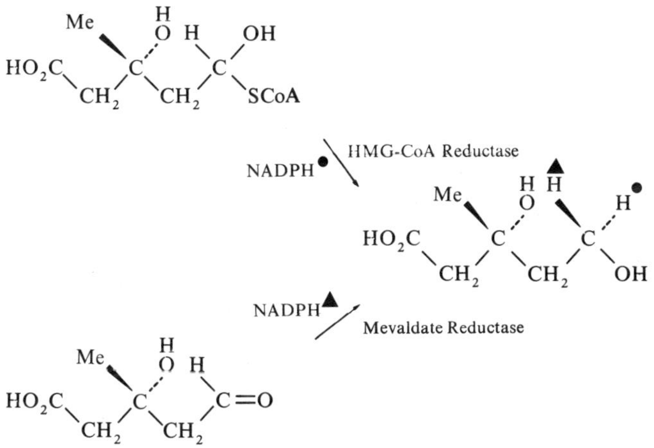

Scheme 1

---

[1] J. R. Hanson, in 'Biosynthesis', ed. T. A. Geissman, (Specialist Periodical Reports), The Chemical Society, London, 1972, vol. 1, p. 41.
[2] G. P. Moss, in 'Terpenoids and Steroids', ed. K. H. Overton, (Specialist Periodical Reports), The Chemical Society, London, 1972, vol. 2, p. 197.
[3] D. V. Banthorpe and B. V. Charlwood, in 'Chemistry of Terpenes and Terpenoids', ed. A. A. Newman, Academic Press, London, 1972, p. 337.
[4] T. K. Devon and A. I. Scott, 'Handbook of Naturally Occurring Compounds', Academic Press, London, 1972, vol. 2.
[5] W. D. Loomis and R. Croteau, in 'Recent Advances in Phytochemistry', ed. V. C. Runeckles, Academic Press, New York, 1972, vol. 6.

to mevalonic acid using NADPH. The hemithioacetal addition compound of mevaldic acid and coenzyme A is a possible intermediate. Both steps involve hydrogen transfer from the 4a (4R) position of NADPH. The mevaldic acid–coenzyme A hemithioacetal addition compound is a good substrate for HMG-CoA reductase. The hydrogen atom which is transferred in this step has been shown to appear at the 5-*pro-S*-position in the resulting mevalonic acid.[6] This is in contrast to the stereochemistry of mevaldate reductase in which a 5-*pro-R* hydrogen atom is introduced. (see Scheme 1)

Cell-free extracts and acetone powder preparations from *Agave americana* have been shown to phosphorylate mevalonic acid to give phosphomevalonic acid and thence pyrophosphomevalonic acid at an optimum pH of 7.0. Glutathione and mercaptoethanol enhance the activity of these preparations.[7] Tracer studies with tissue cultures derived from *Tanacetum vulgare* have revealed[8] the formation of phosphomevalonic acid, pyrophosphomevalonic acid, isopentenyl pyrophosphate, dimethylallyl pyrophosphate, and the incorporation of mevalonic acid into monoterpenes. However, the monoterpene components of the tissue culture differed from those of the whole plant, sabinene being formed rather than isothujone.

Artificial substrates have been studied[9] in systems that form squalene and sterols. Incubation of *trans*-3-methyl[1,1-$^3H_2$]pent-2-enyl pyrophosphate and [1-^{14}C]isopentenyl pyrophosphate with rat liver homogenates gave 1-methylsqualene and 1,24-dimethylsqualene. However, the sterol fraction contained only 27-methyl-lanosterol and 27-methylcholesterol, indicating greater selectivity in either the epoxidation or cyclization stages of this biosynthesis. The substrate specificity of farnesyl pyrophosphate synthetase from pumpkin fruit has also been studied[10] with artificial allylic pyrophosphates. *trans*-3-Methylundec-2-enyl-, *trans*-3-methyldodec-2-enyl-, and *trans*-3-methyltetradec-2-enyl-pyrophosphates were assayed in the enzymatic reaction with isopentenyl pyrophosphate. The tetradecenyl pyrophosphate was inactive. Replacement of the methyl group by ethyl, as in *trans*-3-ethylhept-2-enyl-, *trans*-3-ethyloct-2-enyl-, and *trans*-3-ethyldec-2-enyl-pyrophosphates, gave reactive substrates. However, chain branching in the alkyl residues gave inactive substrates. 3-Ethylbut-3-enyl pyrophosphate also acts[11] as a substrate for farnesyl pyrophosphate synthetase with dimethylallyl pyrophosphate or geranyl pyrophosphate as the starter unit to afford homologues of farnesyl pyrophosphate.

3 Hemiterpenoids

Tryptophan, alanine, and mevalonic acid have been established[12] in earlier

---

[6] A. S. Beadle, K. A. Munday, and D. C. Wilton, *European J. Biochem.*, 1972, **28**, 155; *F.E.B.S. Letters*, 1972, **28**, 13.
[7] E. Garcia-Peregrin, M. D. Suarez, M. C. Aragon, and F. Mayor, *Phytochemistry*, 1972, **11**, 2495.
[8] D. V. Banthorpe and A. Wirz-Justice, *J.C.S. Perkin I*, 1972, 1769.
[9] A. Polito, G. Popjak, and T. Parker, *J. Biol. Chem.*, 1972, **247**, 3464.
[10] T. Nishino, K. Ogura, and S. Seto, *J. Amer. Chem. Soc.*, 1972, **94**, 6849.
[11] K. Ogura, T. Koyama, and S. Seto, *J.C.S. Chem. Comm.*, 1972, 881.
[12] A. J. Birch and K. R. Farrar, *J. Chem. Soc.*, 1963, 4277.

work as biosynthetic precursors of echinulin (1). Cyclo-L-alanyl-L-tryptophanyl (2) is a further intermediate[13] whose isoprenylation has been studied.[14] A cell-free system has been prepared from *Aspergillus amstelodami* which catalyses the transfer of one isoprene unit from dimethylallyl pyrophosphate to the cyclo-L-alanyl-L-tryptophanyl moiety. An enzyme system has been partially purified[15] from cell-free extracts of *E. coli* which catalyses the synthesis of N^6-(3,3-dimethylallyl)adenosine in transfer RNA.

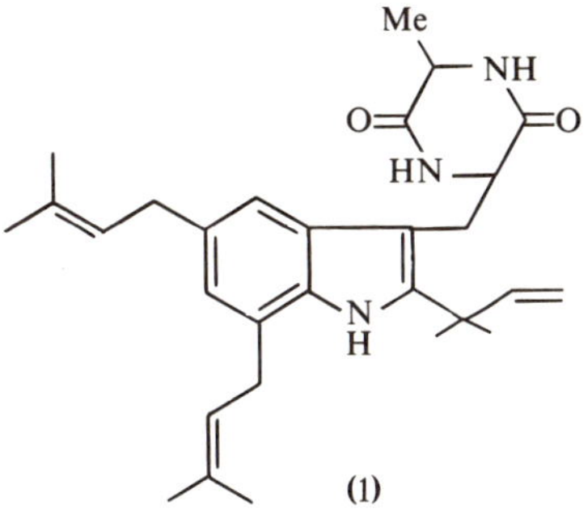

(1)

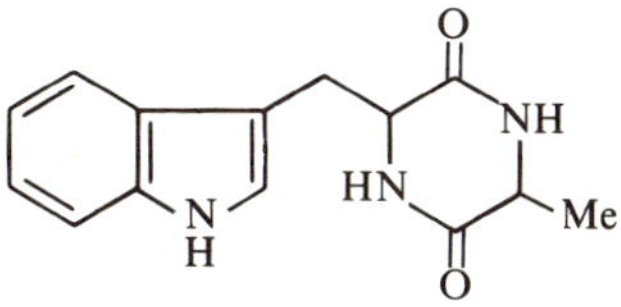

(2)

Further studies have been reported on the biosynthesis of the hemiterpenoid furanocoumarins. A coumarin with a dimethylallyl group adjacent to a hydroxy-group figures in most of the biogenetic speculation concerning the origin of the furan ring in furanocoumarins. Demethyl[1'-^{14}C]suberosin (3) is incorporated[16] by the fruits of *Angelica archangelica* into the linear furanocoumarins such as bergapten (6), imperatorin (7), and isoimperatorin (8). Degradation of the bergapten established that incorporation occurred without randomization. Just as marmesin (4) acts as a precursor of psoralen (5), bergapten (6), and xanthotoxin (9),[17] so it also acts as a precursor of rutaretin (10) in *Ruta graveolens*.[18] 7-Hydroxyumbelliferone is the most effective general precursor of the coumarin portion of these furanocoumarins.

[13] G. P. Slater, J. C. Macdonald, and R. Nakashima, *Biochemistry*, 1970, **9**, 2886.
[14] C. M. Allen, *Biochemistry*, 1972, **11**, 2154.
[15] N. Rosenbaum and M. L. Gefter, *J. Biol. Chem.*, 1972, **247**, 5675.
[16] D. E. Games and D. H. James, *Phytochemistry*, 1972, **11**, 868.
[17] G. Caporale, F. Dall'Acqua, and S. Marciani, *Z. Naturforsch.*, 1972, **27b**, 871.
[18] F. Dall'Acqua, A. Capozzi, S. Marciani, and G. Caporale, *Z. Naturforsch.*, 1972, **27b**, 813.

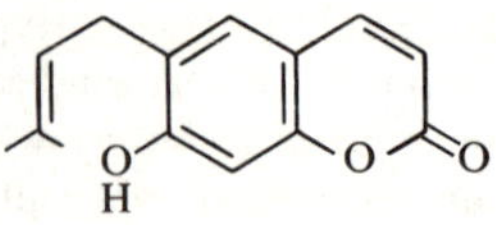

(3)

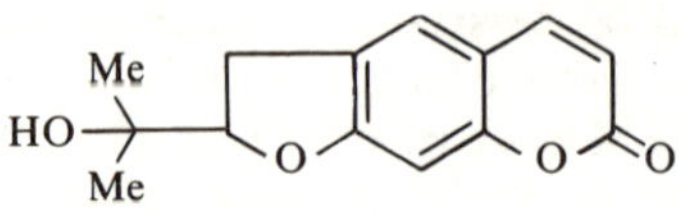

(4)

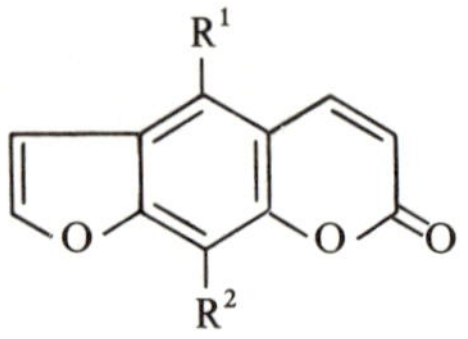

(5) $R^1 = R^2 = H$
(6) $R^1 = OMe; R^2 = H$
(7) $R^1 = H; R^2 = OCH_2CH=CMe_2$
(8) $R^1 = OCH_2CH=CMe_2; R^2 = H$
(9) $R^1 = H; R^2 = OMe$

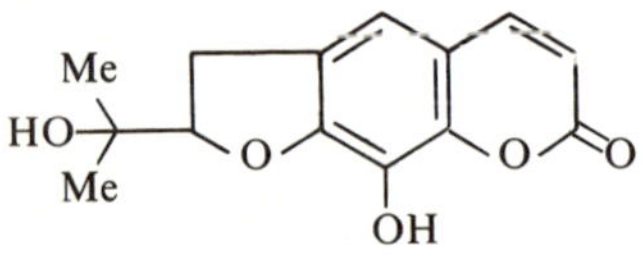

(10)

A thorough review has appeared[19] on the biosynthesis of the ergot and associated alkaloids.

4 Monoterpenoids

A comprehensive and critical review, covering the literature to April 1971, on the biosynthesis and metabolism of the monoterpenes has appeared.[20]

Mevalonic acid is often incorporated selectively into the second isoprene unit of the monoterpenes. This is attributed to compartmentalization effects in the biosynthesis of dimethylallyl pyrophosphate as a starter unit. 3,3-Dimethylacrylic acid has been suggested[21] as an alternative precursor. However, a study of its incorporation into (+)-pulegone (11) by *Mentha pulegium* indicated extensive randomization of the label and hence decomposition and resynthesis.[22] Further evidence for compartmentalization effects and the presence of an endogenous dimethylallyl pyrophosphate pool participating in monoterpene biosynthesis in vegetative tissue has come[23] from the study of the incorporation of [^{14}C]carbon dioxide and [^{14}C]glucose into the monoterpenes of peppermint, *Mentha piperita*. The pulegone (11), derived from $^{14}CO_2$ after

[19] R. Thomas and R. A. Bassett, *Progr. Phytochemistry*, 1972, **3**, 47.
[20] D. V. Banthorpe, B. V. Charlwood, and M. J. O. Francis, *Chem. Rev.*, 1972, **72**, 115.
[21] W. Sandermann and H. Stockmann, *Naturwiss.*, 1956, **43**, 580; *Chem. Ber.*, 1958, **91**, 930.
[22] D. V. Banthorpe, B. V. Charlwood, and M. R. Young, *J.C.S. Perkin I*, 1972, 1532.
[23] R. Croteau, A. J. Burbolt, and W. D. Loomis, *Phytochemistry*, 1972, **11**, 2459.

different time intervals, was degraded and 90% of the radioactivity was found in the second isoprene unit (*i.e.* derived directly from IPP). Sucrose co-administered with [2-^{14}C]mevalonic acid to peppermint cuttings enhances[24] the incorporation of mevalonate into monoterpenes, indicative of energy requirements for this biosynthesis.

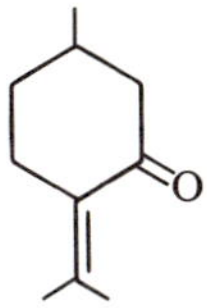

(11)

The isolation and properties of a monoterpene reductase from rose petals have been described.[25] Geraniol and nerol were reduced by a solubilized enzyme preparation to give citronellol. The co-factor requirement was filled only by NADPH and the system had an optimum pH of 8. It was inhibited by *p*-chloromercuriophenylsulphonic acid, suggesting the presence of an S—H group near the active site.

Salvia officinalis has been shown[26] to specifically incorporate [2-^{14}C]geraniol into (−)-camphor (12) and (−)-borneol (13) (0.5 × 10^{-3}% and 3.3 × 10^{-3}%) such that the tracer is incorporated into C-2 of both monoterpenes. This work contains a salutary discussion of the dangers of relying solely on g.l.c. purification of terpenes for radioactive tracer studies.

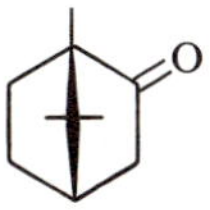

(12) 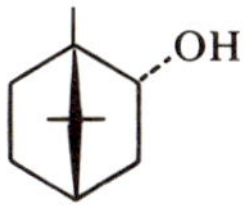   (13)

(14) 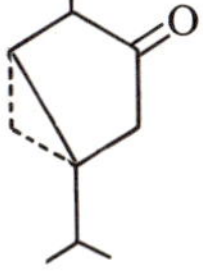   (15)

[24] R. Croteau, A. J. Burbolt, and W. D. Loomis, *Phytochemistry*, 1972, **11**, 2937.
[25] P. J. Dunphy and C. Allcock, *Phytochemistry*, 1972, **11**, 1887.
[26] A. R. Battersby, D. G. Laing, and R. Ramage, *J. C. S. Perkin I*, 1972, 2743.

Monoterpenes are probably metabolically labile in higher plants. Feeding of [^{14}C]-labelled p-menth-1-en-8-ol (α-terpineol) (14) or trans-thujan-3-one (15) to *Tanacetum vulgare* or geraniol to *Artemisia annua* led[27] to a significant uptake into the carotenoids with a labelling pattern that suggested an incorporation of possibly undegraded C_{10} units or dimethylallyl pyrophosphate fragments.

Changes in the monoterpene composition of *Mentha aquatica* have been produced by gene substitution.[28, 29] The oxidation of pulegone (11) to menthofuran (16) is controlled by a single gene that is not completely dominant, whereas the reduction to menthone involves a different set. Changes in favour of low menthofuran composition have been produced in *Mentha aquatica* by gene substitution.

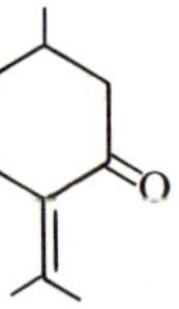

(11)

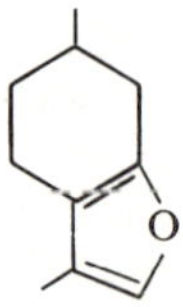

(16)

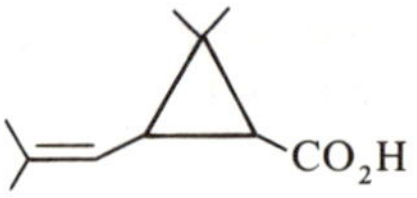

(17)

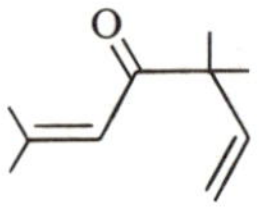

(18)

The cyclopropane monoterpenoid chrysanthemic acid (17) has an unusual linkage of isoprene units and there has been considerable speculation about a possible biogenetic relationship with irregular terpenes involving cleavage of this ring.[30] The mevalonate labelling pattern of artemisia ketone (18) by *Artemisia annua* has been described[31] with results that suggested the intervention of a chrysanthemyl ion or its biogenetic equivalent. However, sodium [*carboxy*-^{14}C]chrysanthemate and the corresponding chrysanthemyl phosphates were not incorporated[32] into artemisia ketone by *Santolina chamaecyparissus*. A number of model studies on the formation of this ion have been reported[33] to afford substances such as artemisia alcohol and yomogi alcohol.

[27] D. V. Banthorpe, H. J. Doonan, and A. Wirz-Justice, *J.C.S. Perkin I*, 1972, 1764.
[28] F. W. Hefendehl and M. J. Murray, *Phytochemistry*, 1972, **11**, 189.
[29] M. J. Murray and F. W. Hefendehl, *Phytochemistry*, 1972, **11**, 2469.
[30] J. R. Hanson, *Perfumery Essent. Oil Record*, 1967, **58**, 787.
[31] D. V. Banthorpe and B. V. Charlwood, *Nature*, 1971, **231**, 285.
[32] L. Crombie, P. A. Firth, R. P. Houghton, D. A. Whiting, and D. K. Woods, *J.C.S. Perkin I*, 1972, 642.
[33] C. D. Poulter, *J. Amer. Chem. Soc.*, 1972, **94**, 5515.

Biosynthesis of C_5–C_{20} Terpenoid Compounds

The incorporation of $^{14}CO_2$ into nepetalactone at different time intervals has been reported.[34] A cell-free preparation has been made[35] from *Vinca rosea* that will methylate loganic acid, which is an intermediate in the biosynthesis of the secoiridoids and the indole alkaloids (see Chapter 4).

5 Sesquiterpenoids

The predominant isomer of farnesol is the 2,6-*trans,trans*-isomer, which is well known as a precursor of the steroids. However, the 2-*cis*-6-*trans*-isomer has often been suggested as a possible precursor of sesquiterpenoid substances. A soluble enzyme preparation has been reported[36] from *Pinus radiata* seedlings which condenses isopentenyl pyrophosphate with geraniol to give both the 2-*cis*- and 2-*trans*-isomers of farnesol. A fungal system from *Helminthosporium sativum* has been shown[37] to isomerize the *trans*-2,3 double bond of epoxyfarnesol and farnesol to a *cis* double bond with the exchange of one hydrogen atom at C-1. The same fungus converts[38] the epoxide into 10,11-dihydroxyfarnesol, 10,11-dihydroxyfarnesic acid, and 9,10-dihydroxygeranyl acetone.

(19) R = Me
(20) R = H

The biosynthesis of the insect juvenile hormone (19) continues to present incorporation problems. The acid 10-epoxy-7-ethyl-3,11-dimethyltrideca-2,6-dienoic acid (20) acts[39] as a substrate for the hormone in the giant silk moth, *Hyalophora cecropia*. L-Methionine gave the ester methyl group. However, it did not contribute to the carbon skeleton whilst farnesol, farnesol pyrophosphate, propionate, and mevalonate were apparently not utilized for the biosynthesis of the hormone under the conditions of these experiments. There was a very low incorporation of [2-^{14}C]acetate into juvenile hormone.

The fungal metabolite siccanin (21) contains a sesquiterpenoid fragment and a fragment derived from orsellinic acid. Previous reports had described a cell-free system from *Helminthosporium siccans* for the synthesis of *trans*-γ-monocyclofarnesol (22). The formation of siccanochromene-A (25) has now been studied[41] by the incubation of cell-free systems from *Helminthosporium siccans*

---

[34] E. D. Mitchell, M. Downing, and G. R. Griffith, *Phytochemistry*, 1972, **11**, 3193.
[35] K. M. Madyastha, R. Guarnaccia, and C. J. Coscia, *Biochem. J.*, 1972, **128**, 34p.
[36] G. Jacob, E. Cardemil, L. Chayet, R. Telliz, R. Pont-Lezica, and O. Cori, *Phytochemistry*, 1972, **11**, 1683.
[37] Y. Suzuki and S. Marumo, *Tetrahedron Letters*, 1972, 5101.
[38] Y. Suzuki and S. Marumo, *Tetrahedron Letters*, 1972, 1887.
[39] M. Metzler, D. Meyer, K. H. Dahm, H. Roller, and J. B. Siddall, *Z. Naturforsch.*, 1972, **27b**, 321.
[40] M. Metzler, K. H. Dahm, D. Meyer, and H. Roller, *Z. Naturforsch.*, 1971, **26b**, 1270.
[41] K. T. Suzuki and S. Nozoe, *J.C.S. Chem. Comm.*, 1972, 1166.

with mevalonolactone and orsellinic acid, and *trans,trans*-farnesyl pyrophosphate and orsellinic acid. Orsellinic acid was found to be the only suitable aromatic co-substrate for the enzyme preparation. Presiccanochromenic acid (23) and siccanochromenic acid (24) were also shown to be precursors of siccanochromene-A.

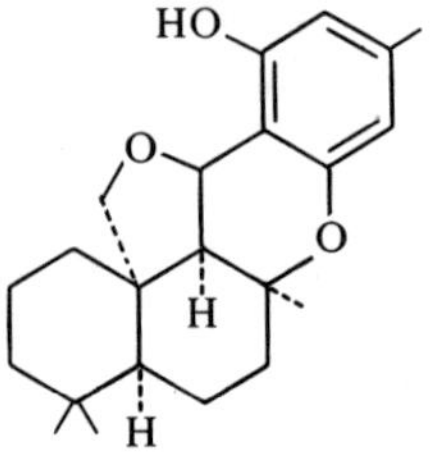

(21)

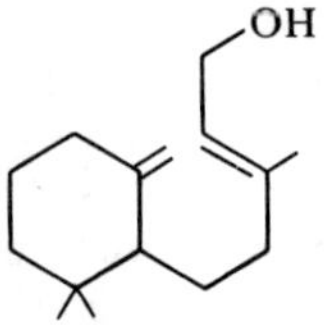

(22)

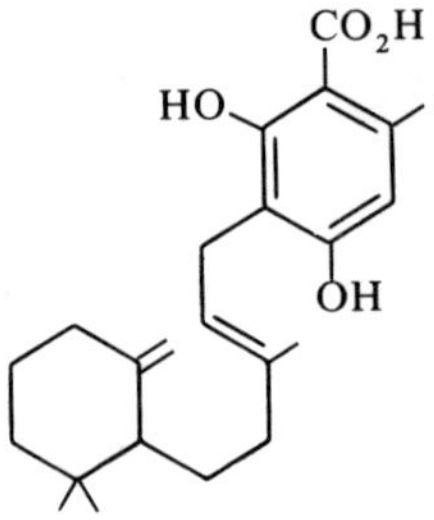

(23)

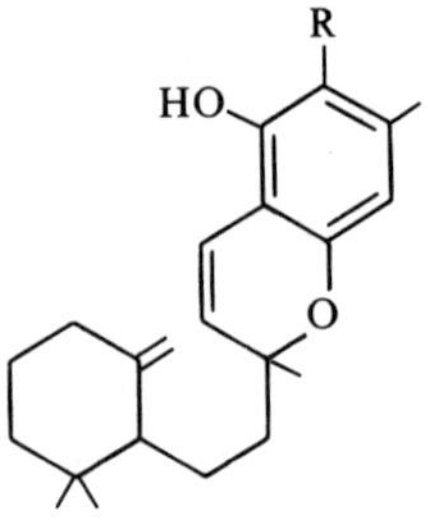

(24) R = CO_2H
(25) R = H

Earlier experiments[42] on the biosynthesis of tutin (28) with [2-^{14}C]mevalonic acid and [2,2-^{3}H$_2$,2-^{14}C]mevalonic acid had indicated its sesquiterpenoid nature and had suggested that it was formed *via* copaborneol (27). Specifically tritiated copaborneol has now been shown[43] to be converted into tutin (28) by *Coriaria japonica* without randomization of the label. The initial step in the cyclization to form copaborneol has been regarded as the cyclization of the electrophilic C-1 of farnesyl pyrophosphate with the distal double bond to form a germacrane cation (26). In order for cyclization to proceed further the reactive centre must then be transferred from C-11 to C-1. One proposal involved a double 1,2-hydride shift from C-10 to C-11 and C-1 to C-10. However, this

---

[42] M. Biollaz and D. Arigoni, *Chem. Comm.*, 1969, 633; A. Corbella, P. Gariboldi, G. Jommi, and C. Scholastico, *Chem. Comm.*, 1969, 634.
[43] K. W. Turnbull, W. Acklin, D. Arigoni, A. Corbella, P. Gariboldi, and G. Jommi, *J.C.S. Chem. Comm.*, 1972, 598.

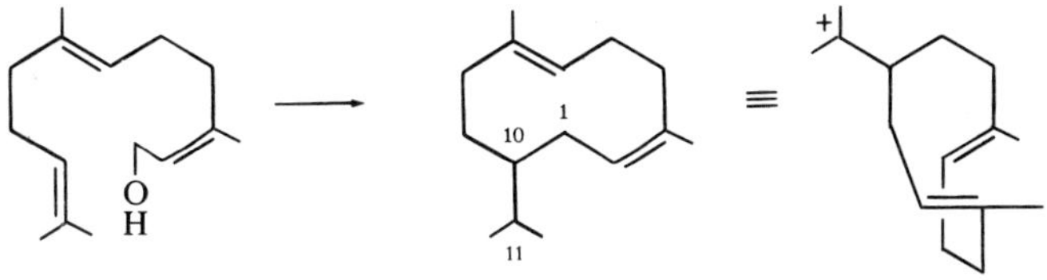

(26)

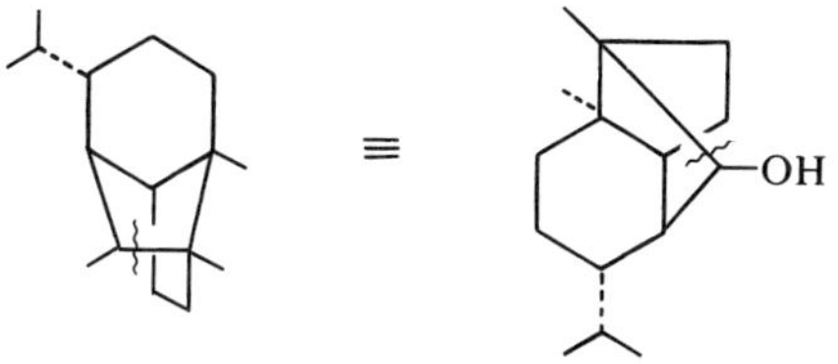

(27)

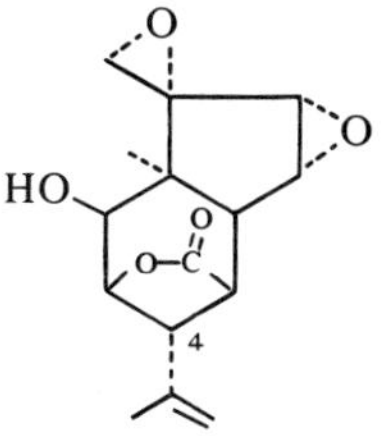

(28)

proposal has been eliminated[44] from tutin biosynthesis by the retention of a (4R)-[4-^{3}H]mevalonoid hydrogen at C-10: this carbon atom becomes C-4 in tutin (28).

The biosynthesis of caryophyllene (29) from [2-^{14}C]mevalonic acid in *Mentha piperita* has been studied.[45] Degradation of the caryophyllene revealed a preferential labelling of those portions of the molecule derived directly from

(29)

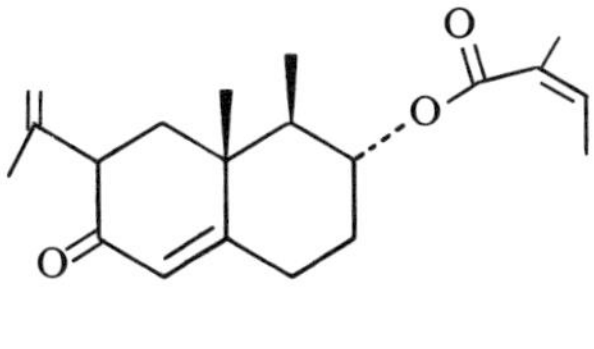

(30)

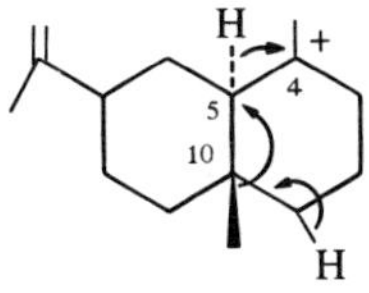

(31)

[44] A. Corbella, P. Gariboldi, and G. Jommi, *J.C.S. Chem. Comm.*, 1972, 600.
[45] R. Croteau and W. D. Loomis, *Phytochemistry*, 1972, **11**, 1055.

isopentenyl pyrophosphate. Petasin (30), formed in *Petasites hybridus*, possesses the rearranged eremophilane skeleton. The proposal has been made that it is formed *via* a β-germacrene, although there are also alternative suggestions involving a spiro-intermediate.[46] The biosynthesis has been studied[47] with [2-^{14}C]- and (4R)-[4-^{3}H]-mevalonic acid. The labelling pattern was consistent with the postulated mode of biosynthesis from *trans,trans*-farnesyl pyrophosphate with a methyl migration from C-10 to C-5 and a hydrogen shift from C-5 to C-4α [see (31)].

There have been further reports from a number of laboratories concerning the biosynthesis of the trichothecane group of fungal metabolites. α-Bisabolol, γ-bisabolene, and monocyclofarnesol are not precursors of trichothecin (34).[48] However, the hydrocarbon trichodiene (33), which is probably the key intermediate in this biosynthesis, was converted[49] into trichothecolone (35), 12,13-epoxytrichothec-9-ene, and trichodiol by *Trichothecium roseum*. Degradation of trichothecolone (35), biosynthesized from (4R)-[4-^{3}H,2-^{14}C]mevalonic acid, had located[50] a (4R)-[4-^{3}H] label at C-10, suggesting that the farnesyl chain was folded in the manner shown in (32). An independent degradation has now[51] been reported, locating a [2-^{14}C] label at C-8 in accord with this scheme. Full papers have now appeared on the formation of the trichothecane nucleus,[52] on the biosynthesis of helicobasidin,[53] and on the biosynthesis of mycophenolic acid.[54]

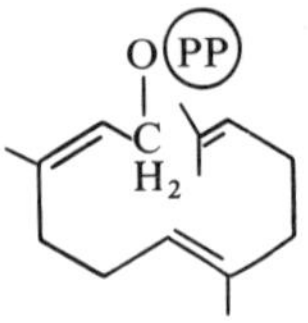

(32)

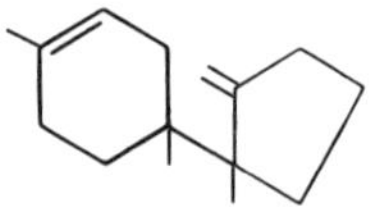

(33)

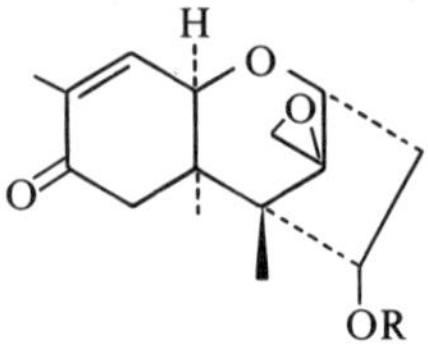

(34) R = OCOCH=CHMe
(35) R = H

---

[46] D. J. Dunham and R. G. Lawton, *J. Amer. Chem. Soc.*, 1971, **93**, 2075.
[47] C. J. W. Brooks and R. B. Keates, *Phytochemistry*, 1972, **11**, 3235.
[48] J. M. Forrester and T. Money, *Canad. J. Chem.*, 1972, **50**, 3310.
[49] Y. Machida and S. Nozoe, *Tetrahedron Letters*, 1972, 1969.
[50] B. Achilladelis, P. M. Adams, and J. R. Hanson, *Chem. Comm.*, 1970, 511.
[51] Y. Machida and S. Nozoe, *Tetrahedron*, 1972, **28**, 5113.
[52] B. A. Achilladelis, P. M. Adams, and J. R. Hanson, *J.C.S. Perkin I*, 1972, 1425.
[53] P. M. Adams and J. R. Hanson, *J.C.S. Perkin I*, 1972, 586.
[54] L. Canonica, W. Kroszszynski, B. M. Ranzi, B. Rindone, E. Santaniallo, and C. Scholastico, *J.C.S. Perkin I*, 1972, 2639.

Biosynthesis of C_5—C_{20} Terpenoid Compounds

The stereochemistry of the formation of the double bonds in abscisic acid (36), biosynthesized by avocado fruit, has been studied[55] using $(2R)$-[2-^{3}H]-, $(2S)$-[2-^{3}H]-, and $(5S)$-[5-^{3}H]-mevalonates. The anticipated stereochemistry of the hydrogen atoms derived from C-2 and C-5 of mevalonate is shown in (37). The C-3' and C-4 hydrogen atoms of abscisic acid (36) were derived from a 2-pro-R-mevalonoid hydrogen atom. The hydrogen atom at C-5 of abscisic acid is derived from a 5-pro-S-mevalonoid hydrogen. The presence of some label at positions 3' and 4 when $(2S)$-[2-^{3}H]mevalonic acid was the precursor was attributed to the action of isopentenyl isomerase.

(36)

(37)

The acid-catalysed cyclization of nerolidol to the bisabolenes and thence to cedrene has been studied[56] as a possible model for the biosynthesis. The same authors have also discussed[57] the possible intervention of spiranic intermediates in the biosynthesis of cedrene.

6 Diterpenoids

The biosynthesis of diterpenoid compounds has been reviewed.[58] The manool:manoyl oxide group of diterpenoids could arise by the cyclization of a geranyllinalool in which even the oxide-ring formation might be concerted with the initial cyclization. However, (−)-labda-8,13-dien-15-[^{3}H]ol pyrophosphate (copalol pyrophosphate) (38) was specifically incorporated[59] into (−)-13-epimanoyl oxide (olearyl oxide) (39), indicating that the geraniol:linalool type of allylic isomerization could take place at the bicyclic level. A number of model studies in the cationic rearrangements and cyclizations related to the biosynthesis of the bicyclic and tricyclic diterpenoids have been reported[60] in the past few years. A biogenetically patterned synthesis of deoxytaondiol methyl ether from manool and toluquinol 4-methyl ether has been described.[61]

The biosynthesis of the virescenosides [D-altropyranosides of diterpenoid aglycones such as virescenol A (40)] from [1-^{13}C]- and [2-^{13}C]-acetate has

---

[55] B. V. Milborrow, *Biochem. J.*, 1972, **128**, 1135.
[56] N. H. Andersen and D. D. Syrdal, *Tetrahedron Letters*, 1972, 2455.
[57] D. D. Syrdal and N. H. Andersen, *Tetrahedron Letters*, 1972, 899.
[58] J. R. Hanson, in 'Progress in the Chemistry of Organic Natural Products', ed. W. Herz, H. Grisebach, and G. W. Kirby, Springer-Verlag, Vienna, 1972, vol. 29, p. 395.
[59] J. R. Hanson and A. F. White, *Phytochemistry*, 1972, **11**, 703.
[60] T. McCreadie and K. H. Overton, *J. Chem. Soc. (C)*, 1971, 312; S. F. Hall and A. C. Oehlschlager, *Tetrahedron*, 1972, **28**, 3155.
[61] A. G. Gonzalez and J. D. Martin, *Tetrahedron Letters*, 1972, 2259.

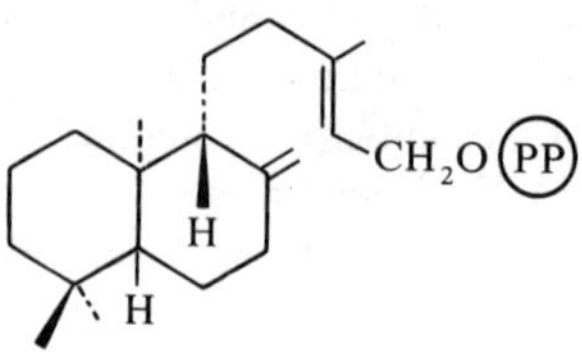

(38) 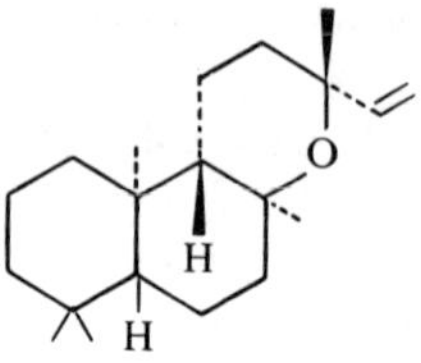   (39)

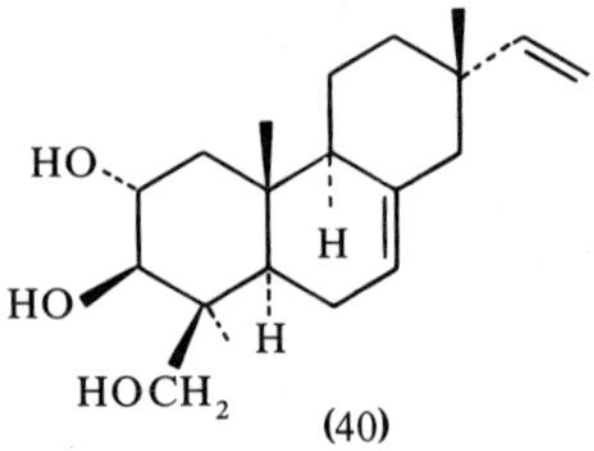

(40) 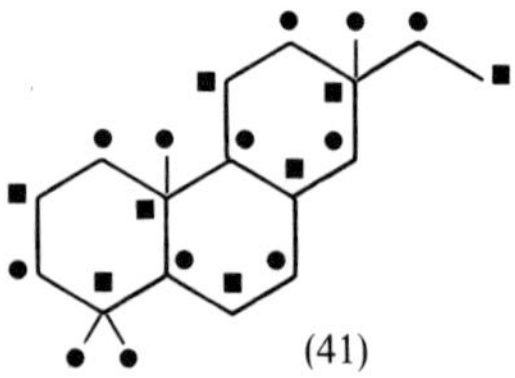   (41)

■ = [1-^{13}C]Acetate
● = [2-^{13}C]Acetate

been studied[62] by ^{13}C n.m.r. spectroscopy. Enrichment of the atoms was observed as shown (41), in accord with currently accepted proposals.

A number of aspects of kaurene and gibberellin biosynthesis have been discussed during the year. The incorporation of [3'-^{2}H$_3$]mevalonic acid into (−)-kaurene (42) by a cell-free system from *Gibberella fujikuroi* has been studied.[63] The labelling pattern excludes a (−)-pimaradiene (43) from the biosynthesis. The previously reported low, but specific, incorporation of

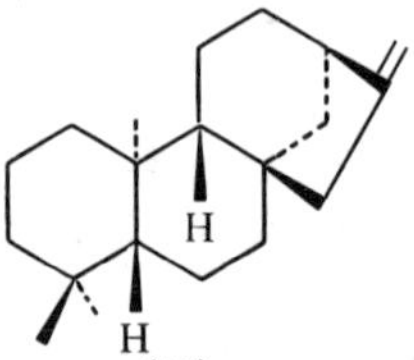

(42) 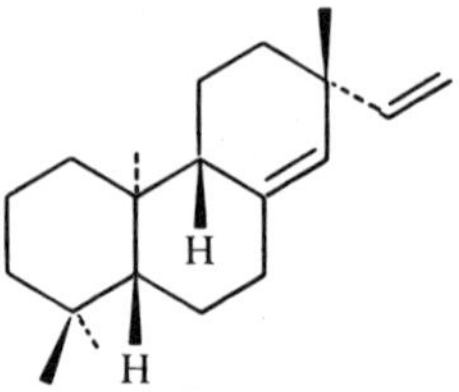   (43)

(−)-pimaradiene was therefore a microbiological transformation. Some evidence has been presented[64] for a kaurene–protein complex in extracts from immature pea, *Pisum sativum*. This renders the lipophilic diterpene accessible to oxidation by mixed function oxidases in the microsomal fraction. Some evidence[65] has also been found for this in *Gibberella fujikuroi*. The sequence of

[62] J. Polonsky, Z. Baskevitch, N. Cagnoli-Bellavita, P. Ceccherelli, B. L. Buckwalter, and E. Wenkert, *J. Amer. Chem. Soc.*, 1972, **94**, 4369.
[63] R. Evans and J. R. Hanson, *J.C.S. Perkin I*, 1972, 2382.
[64] T. C. Moore, S. A. Barlow, and R. C. Coolbaugh, *Phytochemistry*, 1972, **11**, 3225.
[65] R. Evans, J. R. Hanson, and A. M. Holtom, unpublished work.

Biosynthesis of C_5—C_{20} Terpenoid Compounds

oxidation on ring B in kaurene:gibberellin biosynthesis leading to 7β-hydroxy-(−)-kaurenoic acid (44) has been described[66] in a full paper. A cell-free system from the endosperm of *Cucurbita pepo* has been prepared[67] which converts

(44)   (45)

[2-^{14}C]mevalonate into (−)-kaurene (42) and thence into 7β-hydroxy-(−)-kaurenoic acid (44) and the gibberellin A_{12} aldehyde (45). The separation of μg quantities of enzymatic products can be achieved using g.l.c.–mass spectra. In this instance the high degree of labelling obtained with a cell-free system when dilution of the label by endogenous substrates was avoided, was put to advantage and the ^{14}C incorporation was determined by mass spectrometry.[68]

In the conversion of the C_{20} gibberellins into the C_{19} gibberellins, the carbon atom that is lost lies on the same face of the molecule as the lactone ring. Furthermore, there is no loss of nuclear hydrogen atoms during these stages. There has been some discussion as to the oxidation level of this step. Gibberellin A_{13} (46) was not incorporated[69] into gibberellic acid. However, gibberellin A_{13} anhydride (47) was specifically incorporated[70] into gibberellin A_4/A_7 and gibberellic acid (48) in a biosynthetically unusual anhydride–lactone conversion. Thus the loss of this carbon atom, at first sight apparently similar to the loss of the steroid methyl groups, in fact follows a different scheme.

The metabolism of gibberellin A_1 (49) in germinating bean seeds has been studied.[71] The primary product was gibberellin A_8 (50) as its glucoside. There was not a detectable conversion of the gibberellin A_1 into gibberellic acid (48). It had previously been shown[72] that irradiation of segments from etiolated barley leaves and wheat leaves with red light resulted in an increase in the levels of endogenous gibberellin-like substances. Red light apparently enhanced[73] the conversion of gibberellin A_9 into other gibberellin-like substances in homogenates of etiolated barley leaves.

[66] J. R. Hanson, J. Hawker, and A. F. White, *J.C.S. Perkin I*, 1972, 1892.
[67] J. E. Graebe, D. H. Bowen, and J. MacMillan, *Planta*, 1972, **102**, 261.
[68] D. H. Bowen, J. MacMillan, and J. E. Graebe, *Phytochemistry*, 1972, **11**, 2253.
[69] B. E. Cross, K. Norton, and J. C. Stewart, *J. Chem. Soc. (C)*, 1968, 1054.
[70] J. R. Hanson and J. Hawker, *Tetrahedron Letters*, 1972, 4299.
[71] R. Nadeau and L. Rappaport, *Phytochemistry*, 1972, **11**, 1611.
[72] D. M. Reid, J. B. Clememts, and D. J. Carr, *Nature*, 1968, **217**, 580; P. F. Wareing and B. R. Loveys, *Planta*, 1971, **98**, 109.
[73] D. M. Reid, M. S. Tuning, R. C. Durley, and I. D. Railton, *Planta*, 1972, **108**, 67.

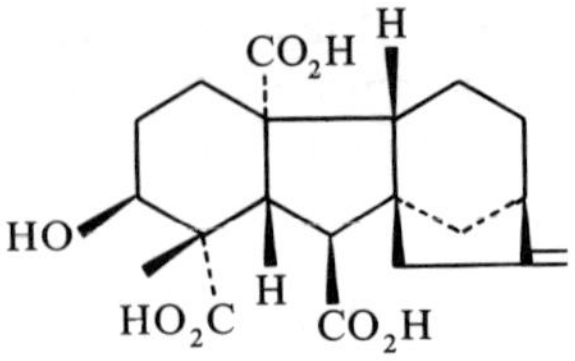

(46) 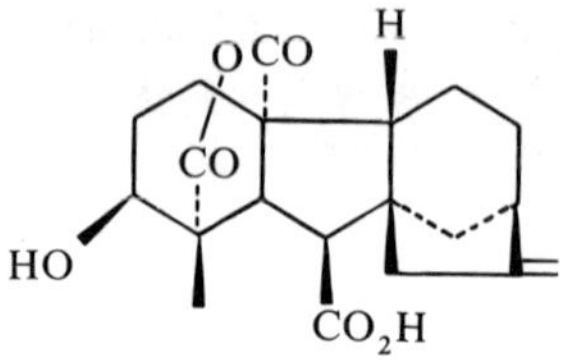 (47)

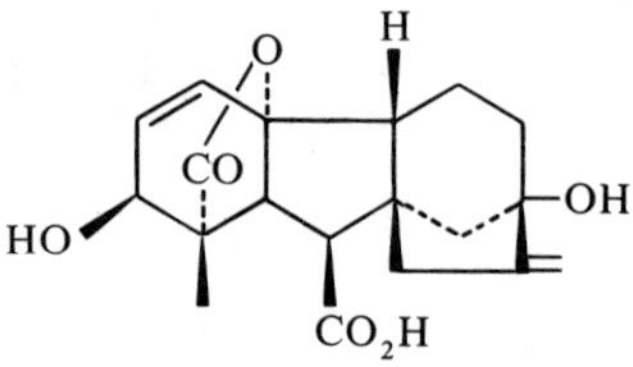

(48)

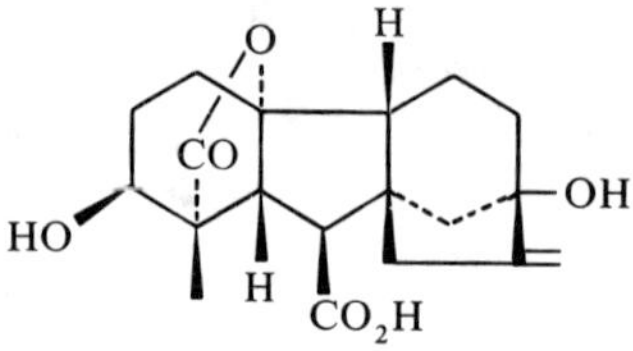

(49) 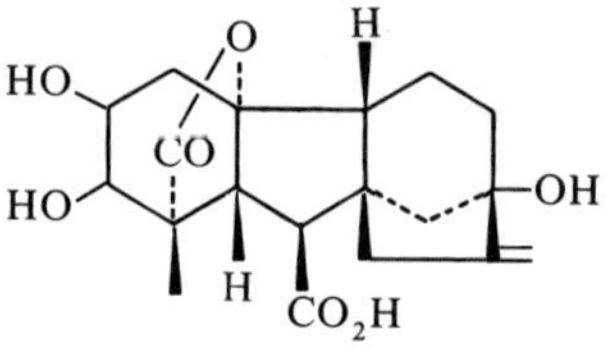 (50)

The introduction of fluorine into the steroid hormones often enhances their biological activity and hence there has been interest in the biological activity of fluorogibberellins.[74] Fluorogibberellic acid and fluorogibberellin A_9 (52) were produced by a fermentation of *Gibberella fujikuroi* to which the fluorogibberellin A_{12} aldehyde (51) had been added.[75] Some interesting gibbane metabolites have been isolated[76] from the microbiological transformation of

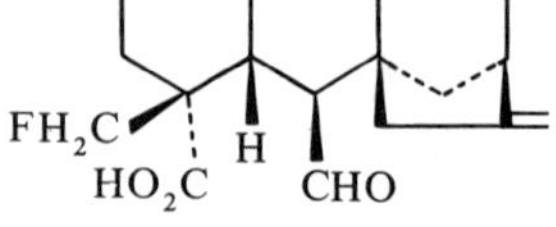

(51) 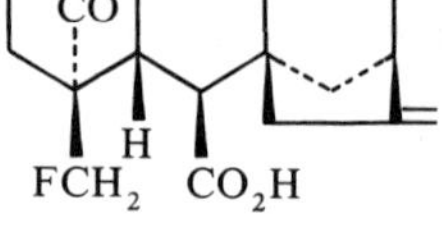 (52)

[74] J. L. Stoddart, *Planta*, 1972, **107**, 81.
[75] J. H. Bateson and B. E. Cross, *J.C.S. Chem. Comm.*, 1972, 649.
[76] H. J. Bakker, P. R. Jefferies, and J. R. Knox, *Tetrahedron Letters*, 1972, 2723.

(−)-kaura-2,16-dien-19-ol (53) by *Gibberella fujikuroi*, including the epoxide (54) and the γ-lactone (55). These were shown to be metabolites by labelling studies.

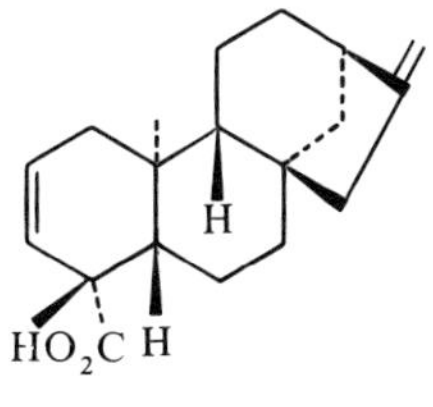

(53)

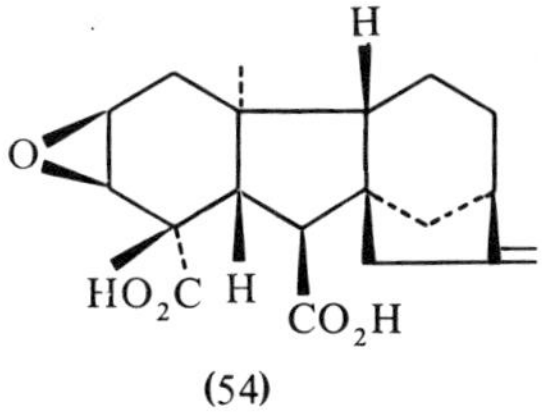

(54)

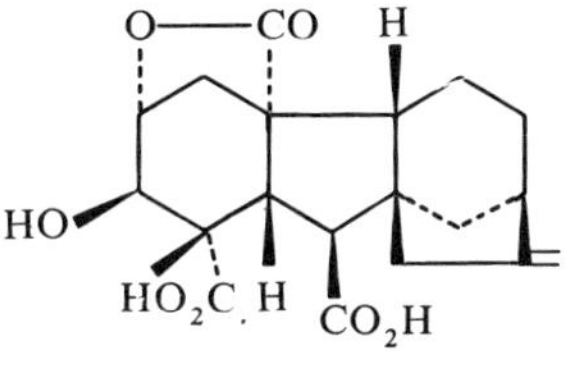

(55)

Beyerene (56) and beyeren-19-ol (57) have been incorporated[77] into the diterpenoids of *Beyeria leschenaultii*. Both serve as precursors of beyerol (58), 17,19-dihydroxybeyer-15-en-3-one, and the 3,4-secobeyerene acid (59), but only beyerene is incorporated into 6β,17-dihydroxybeyer-15-en-3-one. The labelling pattern of the seco-acid (59) suggested that the ring-opening reaction was stereoselective for the C-4 axial methyl group of beyerene.

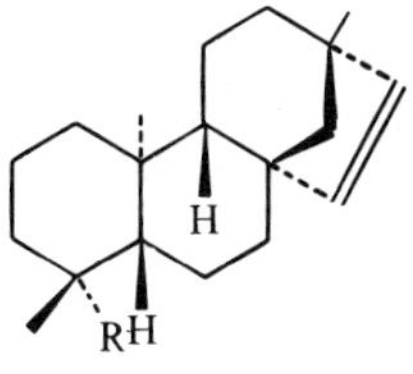

(56) R = Me
(57) R = CH_2OH

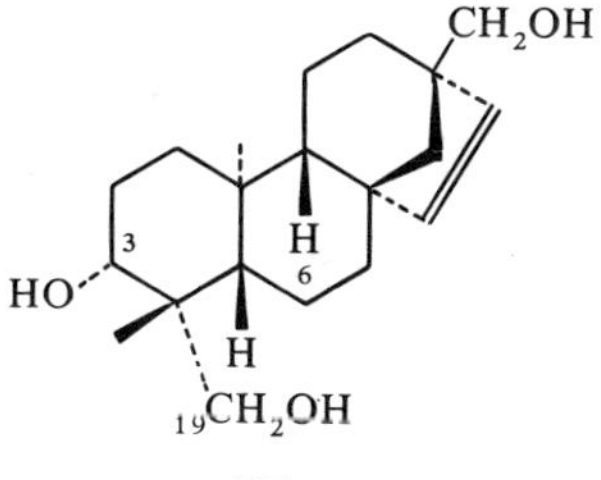

(58)

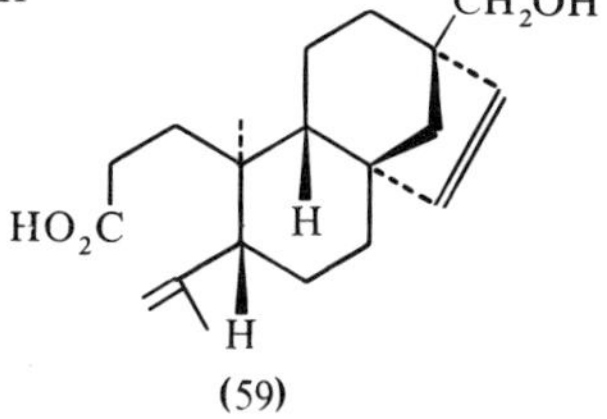

(59)

[77] H. J. Bakker, E. L. Ghisalberti, and P. R. Jefferies, *Phytochemistry*, 1972, **11**, 2221.

2
Biosynthesis of Triterpenes, Steroids, and Carotenoids

BY H. H. REES AND T. W. GOODWIN

1 Introduction

The literature on the biosynthesis of triterpenes, steroids, and carotenoids for 1970 and 1971 has been reviewed[1] in Volume 1. The reader is, therefore, referred to this for much basic information regarding biosynthetic pathways. This Report attempts to survey material related to the biosynthesis of triterpenes, steroids, and carotenoids published during 1972. Since much of the work which has been reported during that period has centred on enzymic aspects, this chapter of necessity reflects such emphasis. It is inevitable that there will be some overlap with chapters in another volume[2] of the Specialist Periodical Reports series. Reviews centred on enzymic[3,4] as well as stereochemical and mechanistic[5] aspects of sterol biosynthesis have appeared.

2 Mevalonic Acid

Numerous publications have appeared recently on the control of cholesterol biosynthesis. For an account of early work on this topic, the reader is referred to a comprehensive review.[6] Substantial acetoacetyl-CoA thiolase [(1)→(2)][7] and β-hydroxy-β-methylglutaryl-CoA (HMG-CoA) synthase [E.C.4.1.3.5; (2)→(3)][8] activities have been reported in the cytoplasmic cell fraction of avian liver. Two cytosolic forms (I and II) of the HMG-CoA synthase, which have been obtained in homogeneous form, have distinctly different electrophoretic and chromatographic properties but similar kinetic properties.[8] Synthase I is composed of a single polypeptide chain of approximately 52 000 daltons,

---

[1] H. H. Rees and T. W. Goodwin, in 'Biosynthesis', ed. T. A. Geissman (Specialist Periodical Reports), The Chemical Society, London, 1972, vol. 1, p. 59.
[2] G. P. Moss, in 'Terpenoids and Steroids', ed. K. H. Overton (Specialist Periodical Reports), The Chemical Society, London, 1972, vol. 2, p. 197, and 1973, vol. 3, p. 245.
[3] J. L. Gaylor, in 'Advances in Lipid Research', ed. R. Paoletti and D. Kritchevsky, Academic Press, New York and London, 1972, vol. 10, p. 89; J. L. Gaylor, in 'M.T.P. International Review of Science, Biochemistry Series, Vol. 4, Biochemistry of Lipids', ed. T. W. Goodwin, in the press.
[4] R. G. Dennick, *Steroids Lipids Res.*, 1972, **3**, 236.
[5] L. J. Mulheirn and P. J. Ramm, *Chem. Soc. Rev.*, 1972, **1**, 259.
[6] M. D. Siperstein, in 'Current Topics in Cellular Regulation', ed. R. L. Horecker and E. R. Stadtman, Academic Press, New York, 1970, vol. 2, p. 65.
[7] T. Sugiyama, K. Clinkenbeard, J. Moss, and M. D. Lane, *Fed. Proc.*, 1972, **31**, 475.
[8] T. Sugiyama, K. Clinkenbeard, J. Moss, and M. D. Lane, *Biochem. Biophys. Res. Comm.*, 1972, **48**, 255.

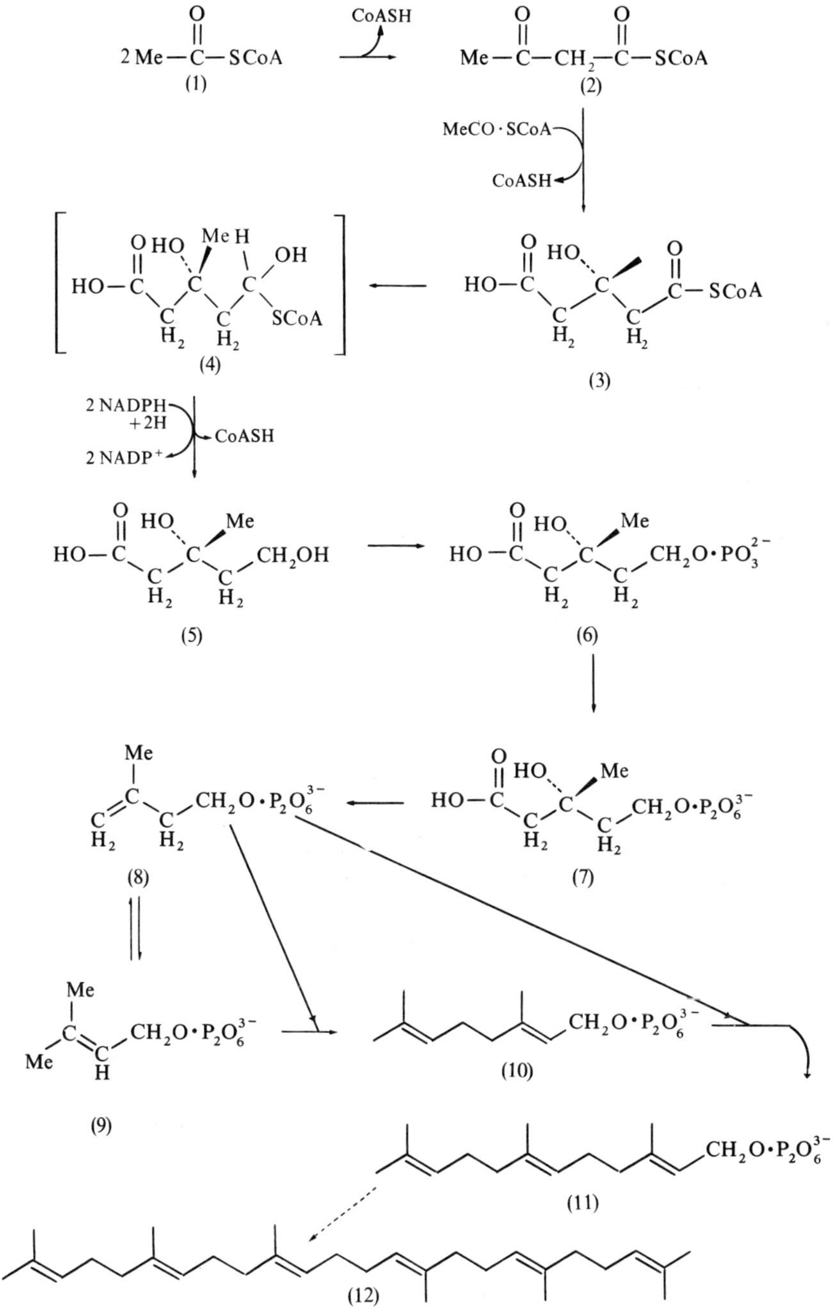

Biosynthesis of Triterpenes, Steroids, and Carotenoids

whereas synthase II consists of two identical or similar polypeptide chains of about 55 000 daltons. On the basis of their cytoplasmic localization and the fact that cholesterol feeding has a marked negative feed-back effect on cytosolic HMG-CoA synthase activity, it has been suggested that one or both of these enzymes carry out the committed step in hepatic cholesterol synthesis. This inhibition of HMG-CoA synthase by cholesterol feeding agrees with previous observations.[9] HMG-CoA synthase has also been purified from baker's yeast and the molecular weight estimated as 130 000.[10] Results of extensive inhibition studies were compatible with a sequential mechanism involving a modified enzyme, tentatively identified as a stable acetyl-enzyme, arising from the reaction with acetyl-CoA in the first step of a ping-pong reaction.

HMG-CoA reductase [mevalonate:NADP oxidoreductase,E.C.1.1.1.34; (3) → (5)], which is considered to be the major rate-controlling enzyme in cholesterol biosynthesis, has been extensively investigated. Cholesterol feeding does not depress HMG-CoA reductase activity by feed-back inhibition but possibly by repression of synthesis of the enzyme.[11] The enzyme is located almost entirely in the microsomal fraction of liver and two reports suggest that 95 % of the microsomal enzyme activity is located in the rough endoplasmic reticulum. However, a recent report indicates that over 80 % of the activity is found in the fractions of smooth membranes composed of smooth endoplasmic reticulum, Golgi apparatus, and plasma membrane.[12] HMG-CoA reductase activity of rat liver microsomal fraction varies diurnally: in rats fed *ad libitum* and subjected to alternate 12 h periods of light and darkness, the activity is maximal during the middle of the dark period and minimal during the light period. The environmental lighting probably establishes the feeding habits of the rat, but is not itself the cause of the rhythm.[13] Further clarification of the situation established[14] that two stimuli are responsible for the rising phase of this diurnal variation. The first begins to act in anticipation of feeding and at a time determined by the time at which the rat habitually feeds. The nature of this stimulus is obscure. The second is the consumption of food, but not necessarily the absorption of a nutrient from the diet. The presence of the first stimulus means that interpretation of much of the earlier work should be re-assessed.

Kinetic studies[15] and measurements of the turnover rate of HMG-CoA reductase using cycloheximide[16] are in agreement with previous studies[17]

---

[9] L. W. White and H. Rudney, *Biochemistry*, 1970, **9**, 2725; K. S. Shepherd and R. Booth, *Biochem. J.*, 1971, **125**, 39P.
[10] B. Middleton and P. K. Tubbs, *Biochem. J.*, 1972, **126**, 27; B. Middleton, *Biochem. J., ibid.,* p.35.
[11] D. J. McNamara and V. W. Rodwell, in 'Biochemical Regulatory Mechanisms in Eukaryotic Cells', ed. E. Kun and S. Grisolia, Wiley-Interscience, New York, 1972, p. 205.
[12] S. Goldfarb, *F.E.B.S. Letters*, 1972, **24**, 153.
[13] R. Booth, K. W. Gregory, and C. Z. Smith, *Biochem. J.*, 1972, **130**, 72P; J. Huber, B. Hamprecht, O.-A. Müller, and W. Guder, *Z. physiol. Chem.*, 1972, **353**, 307.
[14] K. W. Gregory, C. Z. Smith, and R. Booth, *Biochem. J.*, 1972, **130**, 1163.
[15] R. E. Dugan, L. L. Slakey, A. Briedis, and J. W. Porter, *Arch. Biochem. Biophys.*, 1972, **152**, 21.
[16] P. A. Edwards and R. G. Gould, *J. Biol. Chem.*, 1972, **247**, 1520.
[17] M. Higgins, T. Kawachi, and H. Rudney, *Biochem. Biophys. Res. Comm.*, 1971, **45**, 138.

indicating that the circadian rhythm is due solely to changes in the rate of synthesis of the enzyme and not to the rate of degradation. The HMG-CoA reductase peak activity in rats kept under a controlled light (6 a.m.—6 p.m.)–dark (6 p.m.—6 a.m.) cycle has been resolved[18] into two distinct peaks, one at about midnight, with a rapid 33—45% decrease in activity from midnight to about 12.30 a.m., with a further rise to a peak at about 1.45 a.m. Both these increases in reductase activity are prevented by cycloheximide. Evidence was also presented suggesting that cholesterol feeding and fasting may regulate HMG-CoA reductase through different mechanisms. It has been reported that the diurnal rhythm of HMG-CoA reductase in rat liver is not affected by adrenalectomy,[19] indicating that the rhythm is controlled neither by corticosterone nor by catecholamines. Dietary cholestyramine causes an increase in hepatic HMG-CoA reductase activity.[20]

Studies on the development pattern of rat hepatic HMG-CoA reductase levels indicate that the rise which follows weaning from the low reductase levels of suckling to adult daytime levels involves processes other than protein synthesis, and may include relief of *in vivo* inhibition of the enzyme.[21] The supernatant fraction of suckling rat liver contains such an inhibitor of HMG-CoA reductase which may possibly originate in the milk. Whereas in rat liver the HMG-CoA reductase activity is located exclusively in the microsomal fraction, in the intestinal mucosa the enzymic activity is apparently present in both mitochondrial and microsomal fractions.[22] The properties of the microsomal HMG-CoA reductase enzymes from both tissues have also been compared. Previous *in vitro* evidence for the existence of a circadian rhythm in the rate of hepatic cholesterol synthesis in the rat has been confirmed *in vivo* by measuring the conversion of both [^{14}C]acetate and ^3H_2O into non-saponifiable material, which in control experiments contained approximately 70% cholesterol.[23] The circadian changes observed were very similar to changes in the activity of hepatic microsomal HMG-CoA reductase.

Evidence was also presented indicating that food intake is responsible for the major increase in cholesterol biosynthesis during the circadian rhythm. *In vitro* experiments designed to determine whether HMG-CoA reductase is the rate-limiting enzyme in the diurnal variation in the rate of cholesterol synthesis from acetate indicated that another enzyme acting before the reductase may be varying diurnally and that the quantity of the enzyme is also probably markedly affecting the rate of cholesterol synthesis.[15] After 1 day of fasting, the activity of HMG-CoA reductase still changes diurnally whereas cholesterol synthesis from acetate does not. Under such fasting conditions the levels of the enzymes

---

[18] D. J. Shapiro and V. W. Rodwell, *Biochemistry*, 1972, **11**, 1042.
[19] J. Huber and B. Hamprecht, *Z. physiol. Chem.*, 1972, **353**, 313.
[20] S. Goldfarb and H. C. Pitot, *J. Lipid Res.*, 1972, **13**, 797.
[21] D. J. McNamara, F. W. Quackenbush, and V. W. Rodwell, *J. Biol. Chem.*, 1972, **247**, 5805.
[22] S. Shefer, S. Hauser, V. Lapar, and E. H. Mosbach, *J. Lipid Res.*, 1972, **13**, 402.
[23] P. A. Edwards, H. Muroya, and R. G. Gould, *J. Lipid Res.*, 1972, **13**, 396.

catalysing reactions between mevalonate phosphorylation and squalene synthesis decline to values such that a combination of these enzyme-catalysed steps may become rate-limiting for cholesterol synthesis.[24] HMG-CoA reductase is the only enzyme involved in the conversion (4) → (12), which varies diurnally. In contrast to its effect on HMG-CoA reductase,[19] adrenalectomy abolished the diurnal variation in hepatic cholesterol synthesis and caused the synthesis to remain at a uniformly high level.[25] The possibility was considered that corticosterone may play an essential role in the daily rhythm of cholesterogenesis.

Evidence has been obtained indicating that the inhibition of sterol biosynthesis in yeast cell-free extracts by bile acids (as measured by acetate incorporation into non-saponifiable material) is not a non-specific action owing to their detergent action, and that the site of this inhibition is the reduction of HMG-CoA (3) to mevalonate (5).[26] Adenosine 3':5'-cyclic monophosphate (cyclic AMP) markedly decreases incorporation of acetate into fatty acid and cholesterol in rat liver slices, indicating that it may be involved in regulating acetyl-CoA incorporation into these compounds in a specific manner in mammalian liver.[27] Although the mechanism of this inhibition is obscure, cyclic AMP is known to cause accumulation of a 4β-methyl-4α-carboxylic sterol in a rat liver cell-free system.[28] There is a deletion of the cyclic AMP control mechanism of fatty acid and cholesterol biosynthesis in rat hepatoma slices.[29] Results of *in vitro* and *in vivo* experiments indicate that exogenous 3-hydroxy-3-methylglutaric acid (HMGH) can regulate hepatic cholesterol synthesis by reduction of HMG-CoA reductase activity.[30] The true physiological significance of this phenomenon cannot be assessed at present. HMGH can be formed in liver from HMG-CoA by hydrolysis catalysed by HMG-CoA hydrolase (E.C.3.1.2.5). This hydrolase activity increases in rats fasted for 72 h or fed for 14 days on a cholesterol-supplemented diet.[31] This would be expected not only to reduce the available pool of HMG-CoA for mevalonate formation, but HMGH released would also competitively inhibit HMG-CoA reductase.[30] However, the primary factor(s) responsible for observed changes in HMG-CoA reductase in cholesterol-fed and fasted animals is yet unexplained. It has been suggested that light has an inhibitory action on sterologenesis in human skin, possibly by affecting acetate activation.[32] ATP administration to starved rats stimulates hepatic biosynthesis of sterols at a pre-mevalonate site.[33] The inter-relationship

[24] L. L. Slakey, M. C. Craig, E. Beytia, A. Briedis, D. H. Feldbruegge, R. E. Dugan, A. A. Qureshi, C. Subbarayan, and J. W. Porter, *J. Biol. Chem.*, 1972, **247**, 3014.
[25] P. E. Hickman, B. J. Horton, and J. R. Sabine, *Proc. Austral. Biochem. Soc.*, 1971, **4**, 6; *J. Lipid Res.*, 1972, **13**, 17.
[26] H. Hatanaka, A. Kawaguchi, S. Hayakawa, and H. Katsuki, *Biochim. Biophys. Acta*, 1972, **270**, 397.
[27] L. A. Bricker and G. S. Levey, *J. Biol. Chem.*, 1972, **247**, 4914.
[28] D. P. Bloxham and M. Akhtar, *Biochem. J.*, 1971, **123**, 275.
[29] L. A. Bricker and G. S. Levey, *Biochem. Biophys. Res. Comm.*, 1972, **48**, 362.
[30] Z. H. Beg and P. J. Lupien, *Biochim. Biophys. Acta*, 1972, **260**, 439.
[31] M. Saleemuddin and M. Siddiqui, *Lipids*, 1972, **7**, 630.
[32] H. S. Black, J. D. Smith, B. J. Cumbus, and W. B. Lo, *Experientia*, 1972, **28**, 1023.
[33] G. S. Rao and T. Ramasarma, *Biochem. Biophys. Res. Comm.*, 1972, **49**, 225.

Biosynthesis of Triterpenes, Steroids, and Carotenoids

between cholesterol and ubiquinone biosynthesis has also been studied.[34] (−)-Hydroxycitrate, a potent inhibitor of citrate cleavage enzyme, inhibits cholesterol synthesis in rat liver,[35] presumably by stimulation of fatty acid synthesis, thereby reducing the amount of acetyl-CoA available for mevalonate formation.[36]

Both reductive steps in the conversion of HMG-CoA (3) into mevalonic acid (5) catalysed by microsomal rat liver HMG-CoA reductase occur by direct hydrogen transfer from the 4R position of NADPH;[37] this stereochemistry agrees with that previously reported for the corresponding yeast enzyme. The hemithioacetal addition compound (13) of 3R mevaldic acid and coenzyme A

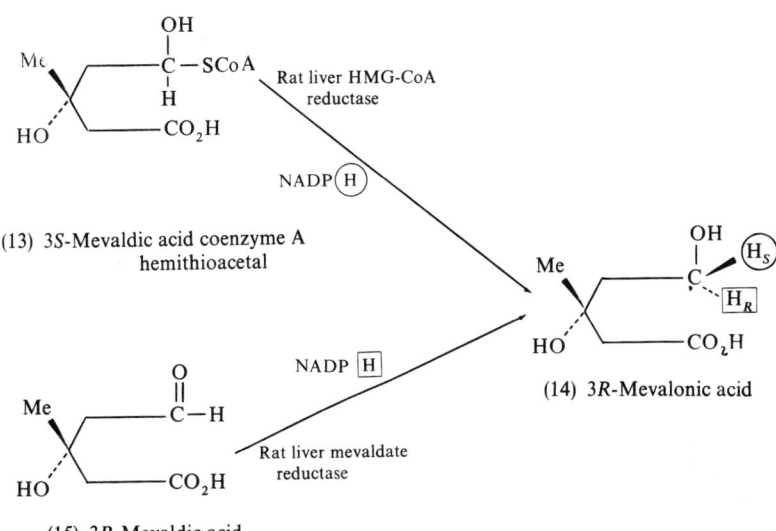

is also a good substrate for the rat liver HMG-CoA reductase, the reduction involving hydride transfer to the 5-pro-S position of mevalonic acid (14).[38] It is interesting that the stereochemistry of this reduction catalysed by rat liver HMG-CoA reductase is opposite to that observed for the reduction of 3R-mevaldic acid (15) by the mevaldate reductase of the liver cytosol, which catalyses hydride transfer to the 5-pro-R position of mevalonic acid (14).[39]

---

[34] T. Ramasarma, *Biochem. J.*, 1972, **128**, 12P.
[35] H. Brunengraber, J. R. Sabine, M. Boutry, and J. M. Lowenstein, *Arch. Biochem. Biophys.*, 1972, **150**, 392.
[36] C. Barth, J. Hackenschmidt, H. Ullmann, and K. Decker, *F.E.B.S. Letters*, 1972, **22**, 343.
[37] A. S. Beedle, K. A. Munday, and D. C. Wilton, *European J. Biochem.*, 1972, **28**, 151.
[38] A. S. Beedle, K. A. Munday, and D. C. Wilton, *F.E.B.S. Letters*, 1972, **28**, 13.
[39] C. Donninger and G. Popják, *Proc. Roy. Soc.*, 1966, **B163**, 465.

3 Squalene

Mevalonate kinase [E.C.2.7.1.36; (5) → (6)] preparations from several plants are inhibited by geranyl, geranylgeranyl, farnesyl, and phytyl pyrophosphates, the most potent inhibitors being the latter two compounds.[40] This enzyme may, therefore, be a control point of isoprenoid biosynthesis in plants. A cell-free system for conversion of mevalonic acid (5) into mevalonic acid 5-pyrophosphate (7) has been prepared from the sapogenin-producing plant *Agave americana*.[41]

Isomerization of isopentenyl pyrophosphate (8) into dimethyllallyl pyrophosphate (9), as well as each chain-lengthening step, involves elimination of a proton from C-4 of the original mevalonic acid. Studies on the incorporation of $(4R)$-[4-^{3}H$_1$]- and $(4S)$-[4-^{3}H$_1$]-mevalonic acids into a variety of isoprenoids led to the acceptance of the general rule that isoprenoids which retain the 4-*pro-R* proton of mevalonic acid are biogenetically *trans*, whereas those which retain the 4-*pro-S* proton are biogenetically *cis*.[42] However, recent evidence suggests that some modification of this generalization may be necessary, especially when applied to shorter-chain compounds. Cell-free extracts from *Pinus radiata* and *Citrus sinensis* have been obtained which synthesize the

(16) $CH_2O \cdot P_2O_6^{3-}$

(17) $CH_2O \cdot P_2O_6^{3-}$

pyrophosphates of geraniol (10), nerol (16), 2,6-*trans,trans*-farnesol (11), and 2-*cis*-6-*trans*-farnesol (17) from mevalonic acid.[43, 44] When $(4R)$-[2-^{14}C,4-^{3}H$_1$]- and $(4S)$-[2-^{14}C,4-^{3}H$_1$]-mevalonic acids were incorporated by such extracts into these pyrophosphates as well as isopentenyl pyrophosphate (8) and dimethylallyl pyrophosphate (9), the 4-*pro-R* hydrogen was retained and the 4-*pro-S* proton was eliminated in each case. Since this enzyme system apparently does not catalyse direct *trans–cis* isomerization of exogenous C_{10} or C_{15} prenyl pyrophosphates, two alternative mechanisms were considered.[44] One proposal considers that the enzyme preparation contains two different forms

---

[40] J. C. Gray and R. G. O. Kekwick, *Biochim. Biophys. Acta,* 1972, **279**, 290.
[41] E. Garcia-Peregrin, M. D. Suárez, M. C. Aragón, and F. Mayor, *Phytochemistry,* 1972, **11**, 2495.
[42] F. W. Hemming, in 'Natural Substances Formed Biologically from Mevalonic Acid', Biochemical Society Symposium No. 29, ed. T. W. Goodwin, Academic Press, London and New York, 1970, p. 105; R. Bentley, 'Molecular Asymmetry in Biology', Academic Press, New York, 1970, vol. 2, p. 316.
[43] G. Jacob, E. Cardemil, L. Chayet, R. Tellez, R. Pont-Lezica, and O. Cori, *Phytochemistry,* 1972, **11**, 1683.
[44] E. Jedlicki, G. Jacob, F. Faini, and O. Cori, *Arch. Biochem. Biophys.,* 1972, **152**, 590.

Biosynthesis of Triterpenes, Steroids, and Carotenoids

of prenyl transferase, catalysing formation of *trans*- or *cis*-isoprenoids, respectively. Alternatively, the existence of a single prenyl transferase was suggested with rotation of the hydrophobic residue originating from dimethylallyl pyrophosphate (9) occurring on the enzyme immediately following the initial condensation of (8) and (9), yielding the *cis* α-residue in neryl pyrophosphate (16). Formation of (17) could be explained in an analogous manner. Similarly, removal of the 4-*pro-S* proton has been observed during formation of geraniol and nerol, together with their respective glycosides, in higher plants.[45] However, the absence of *trans–cis* isomerization has not been demonstrated in this system. There is now evidence that such *trans–cis* isomerization of geraniol[46] and farnesol[47] occurs *via* the corresponding aldehydic intermediate.

Details of the elegant elucidation of the stereochemistry of hydrogen addition at C-4 of isopentenyl pyrophosphate (8) during its isomerization to dimethylallyl pyrophosphate (9) have been published.[48] Addition of hydrogen to the double bond of (8) is from the 3*re*,4*re* face, so that the stereochemistry of isomerization (2-*pro-R* elimination, 3*re*,4*re* addition) is consistent with a concerted mechanism, whereby electrons from the C—H bond of the 2*pro-R* hydrogen form the new double bond and release (from the opposite side of C-3) electrons for formation of the new C—H bond at C-4. Studies on the substrate specificity of farnesyl pyrophosphate synthetase from pig liver have shown that it has a relatively broad specificity towards the allylic pyrophosphate but a stringent requirement for the non-allylic pyrophosphate. For example, only the ethyl homologue (18; R^1 = Et) acts as substrate for the enzyme in place of isopentenyl pyrophosphate, to give farnesyl pyrophosphate homologues; other homologues (18; R^1 = H, Pr^n, or Bu^n) were inactive.[49] Similar studies on the substrate specificity of farnesyl pyrophosphate synthetase from pumpkin fruit with regard to the allylic pyrophosphate revealed that disubstitution at C-3 is

(18) $CH_2O \cdot P_2O_6^{3-}$, R^1

(19) $CT_2 \cdot OP_2O_6^{3-}$, T ≡ ³H

(20) R^1 = Et, R^2 = Me
(21) R^1 = Me, R^2 = Et
(22) R^1 = Pr^n, R^2 = Me
(23) R^1 = Me, R^2 = Pr^n
$OP_2O_6^{3-}$

---

[45] D. V. Banthorpe, G. N. J. Le Patourel, and M. J. O. Francis, *Biochem. J.*, 1972, **130**, 1045.
[46] Dunphy, quoted in reference 45.
[47] Y. Suzuki and S. Marumo, *Tetrahedron Letters*, 1972, 5101.
[48] J. W. Cornforth, K. Clifford, R. Mallaby, and G. T. Phillips, *Proc. Roy. Soc.*, 1972, **B182**, 277.
[49] K. Ogura, T. Koyama, and S. Seto, *J.C.S. Chem. Comm.*, 1972, 881.

essential, regardless of chain length of the alkyl group, but branching of the latter is unfavourable.[50] Similarly, incubation of *trans*-3-methyl[1-^{3}H$_2$]pent-2-enyl pyrophosphate (19) and [1-^{14}C]isopentenyl pyrophosphate with a 10 000 × *g* supernatant of rat liver homogenate gave not only [^{14}C]squalene but also [^{3}H,^{14}C]-1-methylsqualene and [^{3}H,^{14}C]-1,24-dimethylsqualene.[51] Gas–liquid chromatographic evidence was also obtained for the presence of 27-methyl-lanosterol and 27-methylcholesterol, which suggests that 1-methylsqualene is epoxidized enzymically only at its isopropylidene end. Also in agreement with the above observations was the finding that 12-methyl- and 12′-methyl-farnesyl pyrophosphates [(20) and (21)] undergo dimeric condensation catalysed by pig liver microsomal enzyme to give squalene homologues, whereas the ethyl derivatives (22) and (23) did not condense.[52]

The intermediacy of presqualene pyrophosphate (24) in the conversion of farnesyl pyrophosphate (11) into squalene (12) has been established in yeast and rat liver.[1] Formation of (24) from mevalonic acid by a cell-free system from

$^{3-}$O$_6$P$_2$OCH$_2$ ... (24)

bramble tissue cultures in the absence of NADH,[53] coupled with the conversion of (24) into squalene by a cell-free preparation from *Pisum sativum*,[54] establish a similar role of presqualene pyrophosphate (24) in higher plants. Particulate yeast squalene synthetase which has been solubilized and purified to near homogeneity has been resolved into 'protomeric' and 'polymeric' units of different molecular weights.[55] The larger form catalyses the conversion of farnesyl pyrophosphate (11) into squalene (12), whereas the smaller form only catalyses the formation of presqualene pyrophosphate (24); both forms of the enzyme are interconvertible. The molecular weight of the smaller form (450 000 daltons) is in close agreement with that previously reported[56] for the soluble squalene synthetase. In contrast, attempts to solubilize and purify the hog liver squalene synthetase have been unsuccessful. However, some degree

---

[50] T. Nishimoto, K. Ogura, and S. Seto, *J. Amer. Chem. Soc.*, 1972, **94**, 6849.
[51] A. Polito, G. Popják, and T. Parker, *J. Biol. Chem.*, 1972, **247**, 3464.
[52] K. Ogura, T. Koyama, and S. Seto, *J. Amer. Chem. Soc.*, 1972, **94**, 307.
[53] R. Heintz, P. Benveniste, W. H. Robinson, and R. M. Coates, *Biochem. Biophys. Res. Comm.*, 1972, **49**, 1547.
[54] G. H. Beastall, H. H. Rees, and T. W. Goodwin, *F.E.B.S. Letters*, 1972, **28**, 243.
[55] A. A. Qureshi, E. D. Beytia, and J. W. Porter, *Biochem. Biophys. Res. Comm.*, 1972, **48**, 1123.
[56] J. Schechter and K. Bloch, *J. Biol. Chem.*, 1971, **246**, 7690.

of concentration of the enzyme in microsomal subfragments produced by sonication has been achieved.[57] Kinetic analysis using such particles indicates that squalene formation from farnesyl pyrophosphate occurs by a ping-pong mechanism, with irreversible steps separating the addition of the first from the addition of the second molecule of farnesyl pyrophosphate to the enzyme, and addition of farnesyl pyrophosphate from the addition of NADPH to the enzyme. Sterol carrier protein (SCP) from liver has been shown to bind presqualene pyrophosphate (24) but not the water-soluble farnesyl pyrophosphate (11), and to stimulate the enzymic conversion of farnesyl pyrophosphate into both presqualene pyrophosphate and squalene.[58] Similar SCP activity was also observed in yeast.

The development of the activities of all the enzymes catalysing conversion of mevalonic acid into squalene during early stages of pea seed germination has been studied.[59] The results suggest that, after 16 h germination, prenyltransferase may be the rate-limiting enzyme in squalene synthesis, although it is not known whether the enzyme activities measured *in vitro* reflect the relative activities of the enzymes under physiological conditions. The antibiotic bacitracin is a potent inhibitor of squalene and sterol biosynthesis from mevalonic acid (5), isopentenyl pyrophosphate (8), or farnesyl pyrophosphate (11), presumably by a mechanism involving complex formation between bacitracin, a bivalent cation, and the isoprenyl pyrophosphate.[60] It is interesting that the bacterium *Halobacterium cutirubrum* is capable of effecting partial saturation and desaturation of squalene.[61]

4 Formation and Cyclization of Squalene 2,3-Oxide

Further studies have been reported[62] on the rat liver microsomal enzyme squalene epoxidase, which catalyses epoxidation of squalene (12) to squalene 2,3-oxide (25). For epoxidase activity there is a requirement for oxygen, NADPH, microsomes, a heat-labile cytoplasmic protein, and a phospholipid. The system is also slightly stimulated by FAD, which might be a component of the electron-transport chain for oxygen activation. The heat-labile supernatant protein has a molecular weight of 44000, but does not bind either squalene, 10,11-dihydrosqualene (which can also act as substrate for the enzyme), or squalene 2,3-oxide, which suggests that it does not have carrier activity analogous to the reported sterol carrier proteins. There is also no evidence at present that this supernatant acts catalytically.

Work on the cyclization of squalene 2,3-oxide (25) has been discussed.[1] During sterol biosynthesis, cyclization of (25) is believed to yield the cation (26).

---

[57] R. E. Dugan and J. W. Porter, *Arch. Biochem. Biophys.*, 1972, **152**, 28.
[58] H. C. Rilling, *Biochem. Biophys. Res. Comm.*, 1972, **46**, 470.
[59] T. R. Green and D. J. Baisted, *Biochem. J.*, 1972, **130**, 983.
[60] K. J. Stone and J. L. Strominger, *Proc. Nat. Acad. Sci. U.S.A.*, 1972, **69**, 1287.
[61] S. C. Kushwaha, E. L. Pugh, J. K. G. Kramer, and M. Kates, *Biochim. Biophys. Acta*, 1972, **260**, 492; J. K. G. Kramer, S. C. Kushwaha, and M. Kates, *Biochim. Biophys. Acta*, 1972, **270**, 103.
[62] H.-H. Tai and K. Bloch, *J. Biol. Chem.*, 1972, **247**, 3767.

However, it is difficult to rationalize, on a chemical basis alone, the stabilization of cation (26) to prosterols such as fusidic acid (27) which have the Z-geometry at $\Delta^{17(20)}$. Direct stabilization of (26) by elimination of the 17β-proton would give, according to Cornforth's[63] hypothesis, the incorrect geometry about the 17(20) double bond. The possibility that cation (26) is stabilized by formation

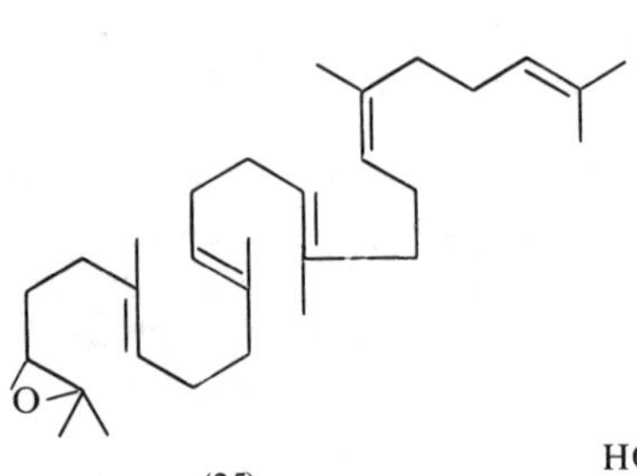

(25)

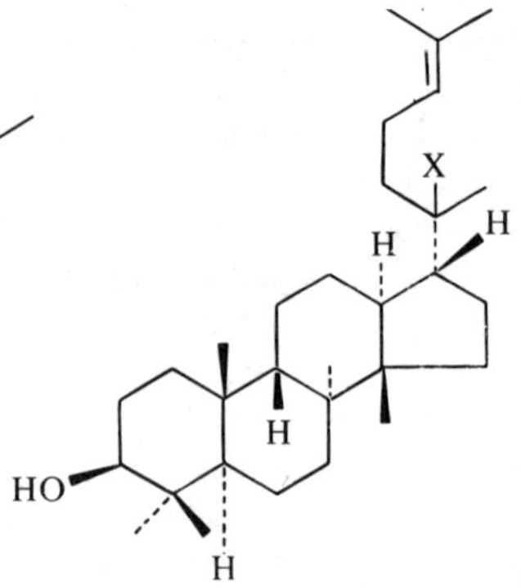

(26) X = Positive charge or nucleophilic group

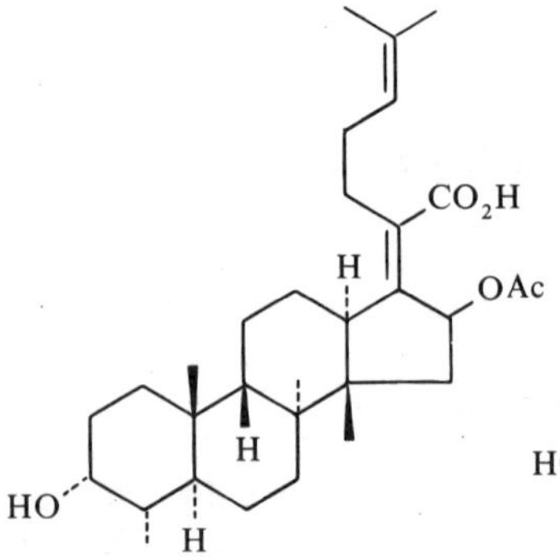

(27)

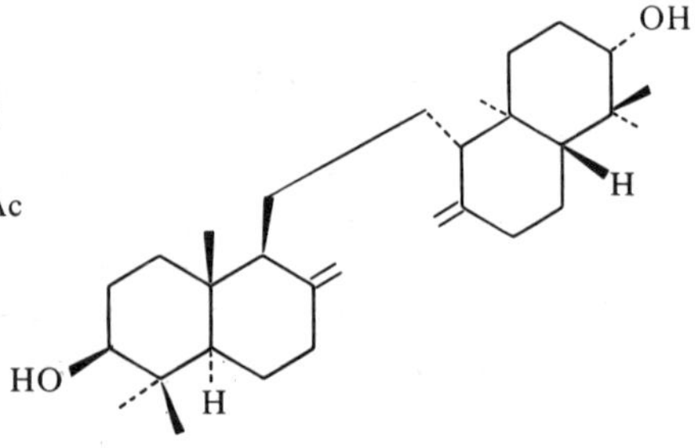

(28)

of a $\Delta^{20(22)}$-intermediate, which could then be enzymically isomerized to (27), has been explored.[64] Degradation of fusidic acid (27) samples biosynthesized from (2R)-[2-^{14}C,2-^{3}H$_1$]- and (2S)-[2-^{14}C,2-^{3}H$_1$]-mevalonic acids, which label structure (26) at C-22, clearly showed that there was no loss of label from C-22 with either substrate, and that the C-2 hydrogens of mevalonic acid are incorporated with retention of configuration into C-22 of fusidic acid. It therefore seems that enzymic participation must be invoked in the stabilization of

[63] J. W. Cornforth, *Angew Chem. Internat. Edn.*, 1968, **7**, 903.
[64] E. Caspi, R. C. Ebersole, W. O. Godtfredsen, and S. Vangedal, *J.C.S. Chem. Comm.*, 1972, 1191.

structure (26). Chemical biogenetic-type total syntheses of 24,25-dihydro-lanosterol, 24,25-dihydro-$\Delta^{13(17)}$-protosterol, isoeuphenol, (−)-isotirucallol, porkeol, tetrahymanol, δ-amyrin, β-amyrin, and germanicol have been reported.[65]

The properties of a soluble enzyme system from plants which catalyses cyclization of squalene 2(3),22(23)-diepoxide into α-onocerin (28) have been reported.[66] The enzyme requires both substrate epoxide groups for recognition and cyclization of the substrate. Additional inhibitors of squalene 2,3-oxide cyclase have been described,[67] and a comparison of the effects of numerous hypocholesterolaemic agents on squalene metabolism in rat liver and the protozoan *Tetrahymena pyriformis* has been reported.[68]

5 Steroids in Vertebrates

Cholesterol Biosynthesis.—Some work on cholesterol biosynthesis which appeared during the early part of 1972 was included in Volume 1 of this series[1] in an attempt to present a more complete picture. The reader is also referred to this work for much background information regarding this section. In agreement with previous observations,[69] it has been demonstrated that cholesterol (30) biosynthesis from [2-^{14}C]mevalonate, [^{14}C]squalene, or [^{14}C]lanosterol is inhibited by carbon monoxide.[70] Since lanosterol (29) and

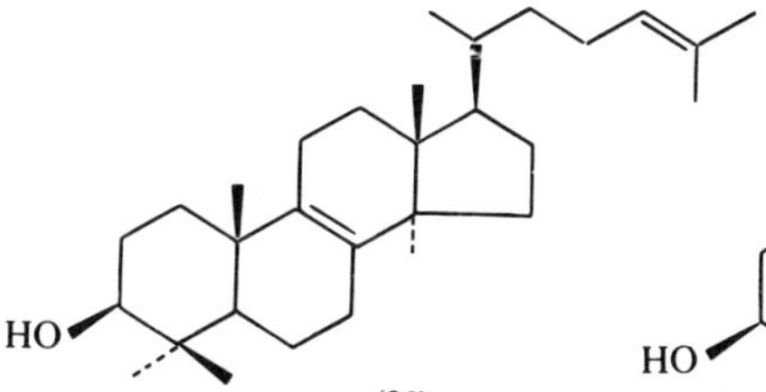

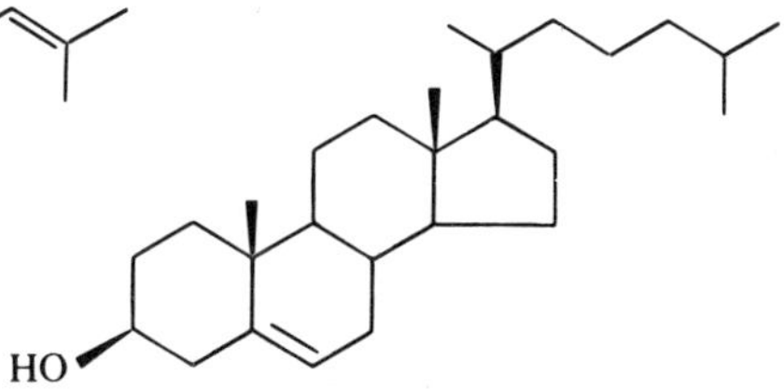

its 24,25-dihydro-derivative accumulate exclusively during cholesterol biosynthesis from [2-^{14}C]mevalonic acid in the presence of CO, the latter probably influences an early step during oxidative elimination of the 14α-methyl group of lanosterol. It is difficult to reconcile indirect evidence, that reduction of the

---

[65] E. E. van Tamelen and R. J. Anderson, *J. Amer. Chem. Soc.*, 1972, **94**, 8225; E. E. van Tamelen, R. A. Holton, R. E. Hopla, and W. E. Konz, *ibid.*, p. 8228; E. E. van Tamelen, M. P. Seiler, and W. Wierenga, *ibid.*, p. 9229.
[66] M. G. Rowan and P. D. G. Dean, *Phytochemistry*, 1972, **11**, 3111.
[67] S. D. Atkin, B. Morgan, K. H. Baggaley, and J. Green, *Biochem. J.*, 1972, **130**, 153; T. J. Douglas and L. G. Paleg, *Plant Physiol.*, 1972, **49**, 417.
[68] J. D. Sipe and C. E. Holmlund, *Biochim. Biophys. Acta*, 1972, **280**, 145.
[69] F. Wada, K. Kirata, and Y. Sakamoto, *J. Biochem. (Japan)*, 1969, **65**, 171.
[70] G. F. Gibbons and K. A. Mitropoulos, *Biochem. J.*, 1972, **127**, 315.

hepatic concentration of cytochrome P-450 does not affect overall hepatic biogenesis of cholesterol,[71] with the above results.

Whereas the C-4 methyl groups of lanosterol (29) are removed by oxidation to carboxylic acid derivatives and eliminated as carbon dioxide,[1,72] it has now been demonstrated that removal of the C-14 methyl group occurs by oxidation

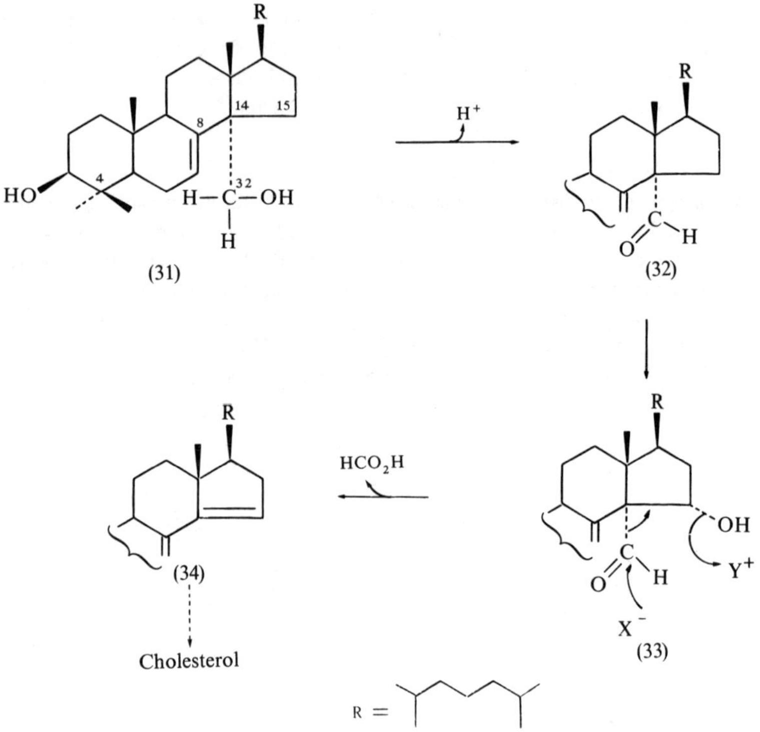

Scheme 1

to the aldehyde and elimination as formaldehyde.[72] Formation of formic acid from [32-^{3}H]lanost-7-ene-3β,32-diol (31) could be correlated with formation of the product of C-14 demethylation, 4,4-dimethylcholesta-7,14-dien-3β-ol (34), from [3α-^{3}H]lanost-7-ene-3β,32-diol. A mechanism (Scheme 1) has been suggested[72] for the overall microsomal reaction, which requires NADPH and

[71] S. D. Atkin, E. D. Palmer, P. D. English, B. Morgan, M. A. Cawthorne, and J. Green, *Biochem. J.*, 1972, **28**, 237.
[72] K. Alexander, M. Akhtar, R. B. Boar, J. F. McGhie, and D. H. R. Barton, *J. C. S. Chem. Comm.*, 1972, 383.

oxygen. It is interesting that the C-19 methyl group of androgens is also removed as formic acid during oestrogen biosynthesis.[73] The overall enzymic activity of demethylation of C-4 methyl sterols (R—CH_3 → RH + CO_2) can be referred to as methylsterol demethylase. This encompasses three enzymes:[74] (i) methylsterol oxidase (a mixed function oxidase), R(3β-OH)—CH_3 → R(3β-OH)—CO_2H, (ii) decarboxylase, R(3β-OH)CO_2H → R(3=O)H + CO_2, and (iii) 3-ketosteroid reductase, R(3=O) → R(3β-OH). Evidence has been obtained for the presence of an endogenous, NAD^+-dependent cytochrome b_5-reducing system in rat liver microsomes,[75] which appears to obtain reducing equivalents via a microsomal alcohol dehydrogenase. The rate of generation of reducing equivalents by this endogenous NAD^+-dependent system is slow, but is substantially faster than the rate of methylsterol oxidase, which is the mixed function oxidase of demethylase. The terminal oxidase may be cytochrome P-450[76] and it has been suggested[75] that methylsterol oxidase of liver microsomes is an unusual NADH-dependent mixed-function oxidase system.

Cholesta-7,9-dien-3β-ol can be enzymically converted into cholesterol via cholesta-8,14-dien-3β-ol.[77] However, the physiological significance of this observation is uncertain. The intermediary role of a number of oxygenated sterols in the conversion of cholest-7-en-3β-ol into cholesta-5,7-dien-3β-ol by rat liver homogenates has been investigated.[78] 7α,8α-Epoxycholestan-3β-ol, cholestane-3β,7β,8α-triol, cholest-8-ene-3β,7ξ-diols, cholestane-3β,8α-diol-7-one, and cholest-8(14)-ene-3β,7α-diol, but not cholestane-3β,7α,8α-triol, were efficiently transformed into cholesterol aerobically. However, they are not normal intermediates in cholesterol biosynthesis, since anaerobically they were transformed into cholest-7-en-3β-ol.

It is still difficult to reconcile completely the work of Ritter and Dempsey with that of Scallen's group on a non-catalytic carrier protein (sterol carrier protein or SCP) which is involved in the conversion of squalene into cholesterol by liver microsomes. Whereas the former workers have isolated[79] a heat-stable protein which activates the microsomal enzymic steps for conversion of squalene into cholesterol, Scallen's group have reported[80] a heat-labile protein with SCP properties. It has been reported[81] that one of the two major apo-high-density lipoprotein peptides, apo-LP-gln II, can substitute specifically in

---

[73] S. J. M. Skinner and M. Akhtar, *Biochem. J.*, 1969, **114**, 75.
[74] A. D. Rahimtula and J. L. Gaylor, *J. Biol. Chem.*, 1972, **247**, 9.
[75] M. M. Bechtold, C. V. Delwiche, K. Comai, and J. L. Gaylor, *J. Biol. Chem.*, 1972. **247**, 7650.
[76] J. L. Gaylor, N. J. Moir, H. E. Seifried, and C. R. E. Jefcoate, *J. Biol. Chem.*, 1970, **245**, 5511; K. Comai and J. L. Gaylor, *Fed. Proc.*, 1972, **31**, 484.
[77] M. Akhtar, C. W. Freeman, A. D. Rahimtula, and D. C. Wilton, *Biochem. J.*, 1972, **129**, 225.
[78] A. Fiecchi, M. G. Kienle, A. Scala, G. Galli, R. Paoletti, and E. G. Paoletti, *J. Biol. Chem.*, 1972, **247**, 5898.
[79] M. C. Ritter and M. E. Dempsey, *Biochem. Biophys. Res. Comm.*, 1970, **38**, 921; M. C. Ritter and M. E. Dempsey, *J. Biol. Chem.* 1971, **246**, 1536.
[80] T. J. Scallen, M. W. Schuster, and A. K. Dhar, *J. Biol. Chem.*, 1971, **246**, 224; T. J. Scallen, M. W. Schuster, A. K. Dhar, and H. B. Skrdlant, *Lipids*, 1971, **6**, 162.
[81] M. E. Dempsey, M. C. Ritter, and S. E. Lux, *Fed. Proc.*, 1972, **31**, 430.

cholesterol biosynthesis for substrate binding and enzyme activation functions of the heat-stable SCP. However, Scallen's group claim[82] that Ritter and Dempsey's SCP does not originate from native SCP, but originates from breakdown of a protein fraction which has no SCP activity prior to heating. The heat-labile SCP has been purified from rat liver 105000 × g supernatant[82] and binds (by hydrophobic forces) not only water-insoluble precursors of cholesterol but also lipid components of lipoprotein, i.e. cholesterol, cholesterol ester, phospholipid, and triglyceride. The amino-acid composition of this SCP also closely resembled the amino-acid composition of the protein component of serum low-density lipoprotein (LDL). Based on these findings it was suggested[82] that SCP simultaneously serves both as a carrier for the enzymic synthesis of cholesterol and as the protein component of LDL, specifying LDL assembly, so that enzymic synthesis by the liver of lipid components (e.g. cholesterol) of LDL may coincide with the assembly of LDL.

It is interesting that ^{3}H and ^{14}C are incorporated from L-[methyl-^{3}H,^{14}C]-methionine into cholesterol and 5α-cholest-7-en-3β-ol in normal and tumorous rats.[83] The exact mechanism of this incorporation is obscure at present. There has been an increased interest in the biosynthesis[84] and metabolism[85] of cholesterol in brain tissue. This area has also been reviewed recently.[86] The primary pathway of sterol biosynthesis in adult rat brain seems to be via Δ^{24}-intermediates.[87] It is interesting that the conversion of squalene into sterols by microsomal fractions from brains of immature rats requires the 100000 × g supernatant fraction from liver, the corresponding supernatant fraction from brain being inactive.[88]

Several systems for formation and hydrolysis of cholesteryl esters in rat liver are known. Microsomes contain an acyl-CoA:cholesterol acyltransferase, which requires coenzyme A and ATP for fatty acid activation,[89] and operates at neutral pH. Enzymic transfer of fatty acids from lecithin to cholesterol occurs in the soluble fraction of rat liver.[90] A third enzyme, cholesterol esterase, occurs in rat liver and its main function is probably hydrolytic.[91] Although human liver apparently does not have acyl-CoA:cholesterol acyltransferase activity, it does have a reversible cholesterol esterase (E.C.3.1.1.13) with optimal

---

[82] T. J. Scallen, M. V. Srikantaiah, H. B. Skrdlant, and E. Hansbury, *F.E.B.S. Letters*, 1972, **25**, 227.
[83] J. G. Lloyd-Jones, P. Heidel, B. Yagen, P. J. Doyle, G. H. Friedell, and E. Caspi, *J. Biol. Chem.*, 1972, **247**, 6347.
[84] R. B. Ramsey, R. T. Aexel, J. P. Jones, and H. J. Nicholas, *J. Biol. Chem.*, 1972, **247**, 3471; R. B. Ramsey, J. P. Jones, A. Rios, and H. J. Nicholas, *J. Neurochem.*, 1972, **19**, 101; A. A. Kandutsch and S. E. Saucier, *Biochim. Biophys. Acta*, 1972, **260**, 26.
[85] Y. Eto and K. Suzuki, *J. Neurochem.*, 1972, **19**, 117; M. Spohn and A. N. Davison, *J. Lipid Res.*, 1972, **13**, 563.
[86] R. B. Ramsey and J. Nicholas, *Adv. Lipid Res.*, 1972, **10**, 144.
[87] R. B. Ramsey, R. T. Aexel, and H. J. Nicholas, *J. Biol. Chem.*, 1971, **246**, 6393.
[88] S. N. Shah, *F.E.B.S. Letters*, 1972, **20**, 75.
[89] K. T. Stokke and K. R. Norum, *Biochim. Biophys. Acta*, 1970, **210**, 202.
[90] M. Akiyama, O. Minari, and T. Sakagami, *Biochim. Biophys. Acta*, 1967, **137**, 525.
[91] W. Stoffel and H. Greten, *Z. physiol. Chem.*, 1967, **348**, 1145.

pH of 4.5.[92] Such acid cholesterol esterase in bovine liver is probably lysosomal.[93] The cytosol cholesterol esterase of rat adrenal cortex is stimulated by cyclic AMP, this activation being further enhanced by ATP.[94] Cholesterol esterase has been isolated from pig pancreas powder as a complex of molecular weight greater than 800 000, which by lipid extraction was converted into enzyme sub-units of 15 000—20 000 molecular weight, which were still capable of synthesizing and hydrolysing cholesterol esters.[95] According to another report, the molecular weight of pancreatic juice cholesterol esterase is 65 000—69 000.[96] Evidence was also presented for a direct molecular interaction between cholic acid (or conjugates) and this enzyme, resulting in polymerization of the enzyme protein to yield an apparent molecular weight of approximately 400 000. The polymerized protein is probably the active enzyme. It is interesting that diesters of cholest-5-ene-3β,26-diol have been identified in human atherosclerotic aorta.[97]

Further Metabolism of Cholesterol.—*Steroid Hormones*. Since a complete coverage of the recent literature on the biosynthesis of steroid hormones is impossible, only a few aspects will be discussed. Proceedings of the 3rd International Congress on Hormonal Steroids have been published.[98] Reviews on the biosynthesis of pregnenolone[99] and also biological hydroxylation mechanisms[100,101] have appeared. For work on steroids in non-mammalian vertebrates, the reader is referred to a recent volume.[102]

Until recently the major pathways for conversion of cholesterol (30) into pregnenolone (36; R = H) were considered to involve: cholesterol (30) → (22R)-22-hydroxycholesterol (35; R^1 = H, R^2 = OH) → (22R)-20α,22-dihydroxycholesterol (35; R^1 = R^2 = OH) → pregnenolone (36; R = H) and cholesterol (30) → 20α-hydroxycholesterol (35; R^1 = OH, R^2 = H) → (22R)-20α,22-dihydroxycholesterol (35; R^1 = R^2 = OH). The former pathway is probably quantitatively considerably more important.[103,104] However, there

---

[92] K. T. Stokke, *Biochim. Biophys. Acta*, 1972, **270**, 156.
[93] K. T. Stokke, *Biochim. Biophys. Acta*, 1972, **280**, 329.
[94] E. R. Simpson, W. H. Trzeciak, J. L. McCarthy, C. R. Jefcoate, and G. S. Boyd, *Biochem. J.*, 1972, **129**, 10P.
[95] J. D. Teale, T. Davies, and D. A. Hall, *Biochem. Biophys. Res. Comm.*, 1972, **47**, 234.
[96] J. Hyun, C. R. Treadwell, and G. V. Vahouny, *Arch. Biochem. Biophys.*, 1972, **152**, 233.
[97] J. D. Gilbert, C. J. W. Brooks, and W. A. Harland, *Biochim. Biophys. Acta*, 1972, **270**, 149.
[98] 'Hormonal Steroids', Proceedings of the 3rd International Congress, Hamburg, 1970, ed. V. H. T. James, Excerpta Medica, Amsterdam, 1971.
[99] S. Burstein and M. Gut, *Recent Progr. Hormone Res.*, 1971, **27**, 303.
[100] 'Biological Hydroxylation Mechanisms', Biochemical Society Symposium No. 34, ed. G. S. Boyd and R. M. S. Smellie, Academic Press, London and New York, 1972.
[101] 'Molecular Mechanisms of Oxygen Activation' ed. O. Hayaishi, Academic Press, London and New York, 1972.
[102] 'Steroids in Nonmammalian Vertebrates', ed. D. R. Idler, Academic Press, London and New York, 1972.
[103] S. Burstein, H. Zamoscianyk, H. L. Kimball, N. R. Chaudhuri, and M. Gut, *Steroids*, 1970, **15**, 13; N. K. Chaudhuri, R. Nicholson, H. L. Kimball and M. Gut, *ibid.*, p. 252.
[104] S. Burstein, H. L. Kimball, and M. Gut, *Steroids*, 1970, **15**, 809.

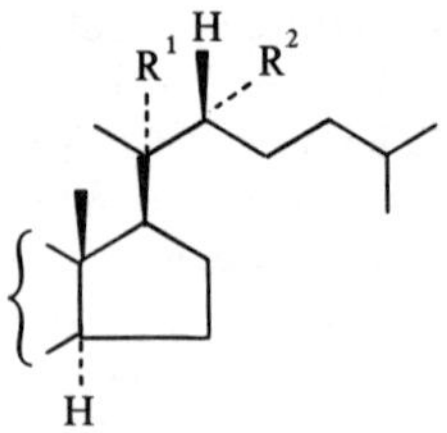

is also evidence that only a small fraction of the pregnenolone formed from cholesterol can be accounted for by these pathways.[104] In an attempt to gain further insight into the process, (20R)-20-t-butyl-pregn-5-ene-3β,20-diol (37), an analogue of 20α-hydroxycholesterol, has been synthesized and its metabolism studied.[105] When injected intravenously into a rabbit, (37) is metabolized to pregnanediol (38; 0.04% yield), whereas pregnenolone (36; R = H) is formed on incubation with adrenal mitochondria (0.5—1.2% yield). It is argued that since C-22 in compound (37) cannot become hydroxylated, being fully substituted, formation of such a free hydroxylated compound is not obligatory in pregnenolone production. It was therefore proposed[105] that conversion of cholesterol into pregnenolone occurs *via* transient intermediate complexes possessing apparent radical or ionic character, and that the transitional side-chain-hydroxylated compounds are not obligatory intermediates, but by-products resulting from competitive reactions of short-lived, reactive species. It has been demonstrated that pregnenolone formation in adrenal mitochondria from rats subjected to stress by ether anaesthesia is increased compared with preparations from quiescent animals.[94, 106] The suggestion was made that the action of stress, mediated by adrenocorticotropin (ACTH) is to increase the amount of cholesterol bound in an active high-spin complex to the side-chain-

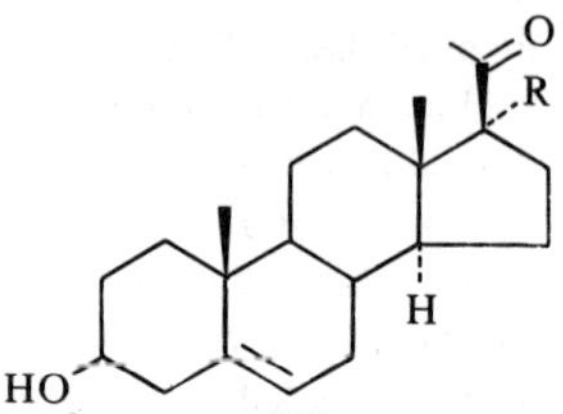

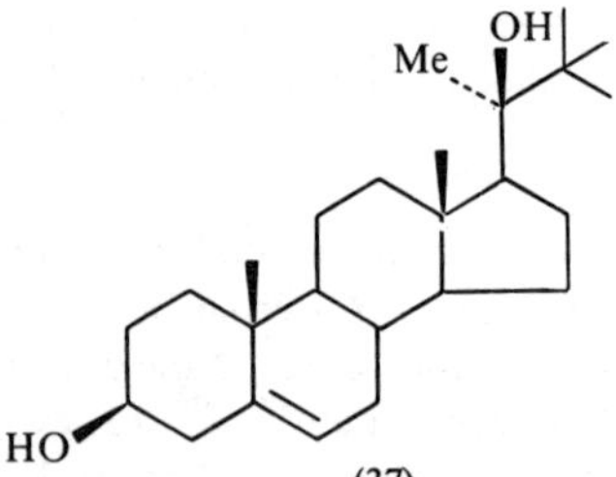

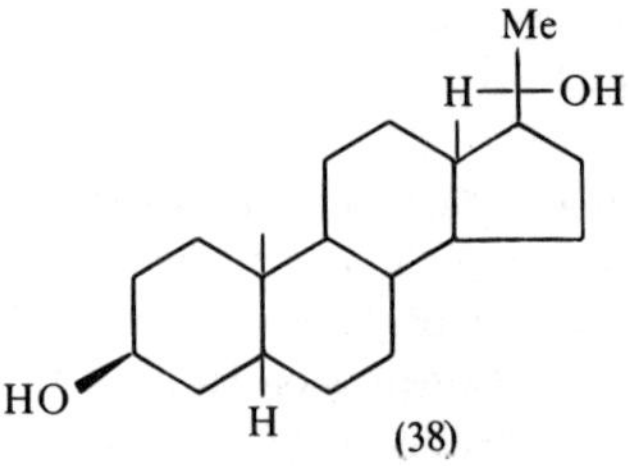

---

[105] B. Luttrell, R. B. Hochberg, W. R. Dixon, P. D. McDonald, and S. Lieberman, *J. Biol. Chem.*, 1972, **247**, 1462.

[106] S. Burstein, N. Co, M. Gut, H. Schleyer, D. Y. Cooper, and O. Rosenthal, *Biochemistry*, 1972, **11**, 573.

Biosynthesis of Triterpenes, Steroids, and Carotenoids

cleavage cytochrome P-450, probably by redistribution of cholesterol within the mitochondria, so that more becomes associated with cytochrome P-450 and thus available for side-chain cleavage. Care should be exercised in interpretation of substrate-induced difference spectra with adrenal haem protein cytochrome P-450 and various substrates since they may drastically vary both with respect to magnitude and affinity in different enzyme preparations, without significant differences in enzymic activities for pregnenolone formation.[106] The liver heat-stable carrier protein (SCP) of Ritter and Dempsey also stimulates adrenal cholesterol synthesis by microsomal enzymes and cholesterol side-chain cleavage by the mitochondrial enzymes. Adrenal mitochondria also contain a heat-stable protein with similar properties, which can also function with liver enzymes catalysing cholesterol biosynthesis.[107] This adrenal activator is presumed to participate not only in cholesterol synthesis and transport in that organ, but also in the initial steps of steroidogenesis. Probable lack of substrate specificity of the side-chain cleavage enzymes is illustrated by conversion of 23,24-dinor-cholest-5-en-3β-ol into the C-20- and C-21-hydroxylated compounds, and of the latter into pregnane and androstane derivatives by bovine adrenal tissues.[108]

17α-Hydroxy-C_{21}-steroids are generally accepted as intermediates for side-chain cleavage in the transformation of pregnenolone (36; R = H) or progesterone (39) into C_{19} steroids, but there are conflicting reports regarding their formation in the rat adrenal gland. No detectable 17α-hydroxy-C_{21}-steroid was formed during metabolism of [^{14}C]pregnenolone to C_{19} metabolites, indicating that, if present, the 17α-hydroxy-C_{21}-intermediates are turned over very rapidly.[109] Stereospecific syntheses of (20S,22R)- and (20S,22S)-17α,20,22-trihydroxycholesterols have been reported.[110]

Although steroid hydroxylations are considered to be catalysed by mixed-function oxidases, it has been suggested that hydroperoxides may possibly be intermediates in certain hydroxylations by a different mechanism. For example, 17α-hydroperoxypregnenolone (36; R = OOH) on incubation with bovine adrenal

(39)   (40)

[107] K. W. Kan and F. Ungar, *Fed. Proc.*, 1972, **31**, 429; K. W. Kan, M. C. Ritter, F. Ungar, and M. E. Dempsey, *Biochem. Biophys. Res. Comm.*, 1972, **48**, 423.
[108] A. D. Tait, *Biochem. J.*, 1972, **128**, 467.
[109] L. Milewich and L. R. Axelrod, *J. Endocrinol.*, 1972, **54**, 515.
[110] R. C. Nickolson and M. Gut, *J. Org. Chem.*, 1972, **37**, 2119.

preparations (either aerobically or anaerobically) yields 17α-hydroxypregnenolone (36; R = OH) by an enzymic process, in addition to certain androgens.[111] Preliminary evidence has also been presented that enol forms of Δ^4-3-oxosteroids may possibly be involved in cytochrome P-450 mediated hydroxylations at C-2 and C-6.[112] Whereas the pterin folic acid causes slight stimulation of steroid 11β- and 21-hydroxylases of duck adrenals, aminopterin inhibits both these processes.[113] The mechanism of these effects is unknown at present. Evidence has been presented that the inhibitory effect of steroid (substrate or product) on succinate-supported hydroxylation in rat adrenal mitochondria is due to an inhibition of reversed electron transport in the region of the first phosphorylation site.[114] From studies with deuterium isotope it has been concluded that the rate-limiting step in the 16α- and 24-hydroxylation of pregnenolone (36; R = H) and 5β-cholestane-3α,7α,12α-triol, respectively, is the splitting of the C—H bond in the substrate, whereas another step is rate-limiting in the 6β-hydroxylation of testosterone and the 26-hydroxylation of 5β-cholestane-3α,7α,12α-triol.[115] Furthermore, a common feature of hydroxylation in which splitting of the C—H bond in the substrate is rate-limiting seems to be a relatively low sensitivity towards carbon monoxide. Studies on rat liver microsomal hydroxylation of cholesterol (30), pregnenolone (36; R = H), and dehydroepiandrosterone (40) have shown the pronounced influence of the side-chain on the patter of hydroxylation of Δ^5-3β-hydroxy-steroids.[116] Whereas cholesterol is essentially only hydroxylated in the 7α-position, dehydroepiandrosterone and pregnenolone undergo more extensive hydroxylation. There is evidence for the existence of two different C-21 hydroxylases, one for 17α-deoxy- and another for 17α-hydroxy-C_{21}-steroids.[117]

A common reaction in the biosynthesis of many steroid hormones is the conversion of the Δ^5-3β-ol grouping into a Δ^4-3-oxo-system by oxidation at C-3 followed by isomerization, i.e. (41) → (43). This can either occur at the

(41)   (42)   (43)

---

[111] L. Tan, H. M. Wang, and P. Falardeau, *Biochim. Biophys. Acta*, 1972, **260**, 731.
[112] A. J. Liston and P. Toft, *Biochim. Biophys. Acta*, 1972, **273**, 52.
[113] J.-G. Lehoux, A. G. Fazekas, H. Leblanc, A. Chapdelaine, and T. Sandor, *J. Steroid Biochem.*, 1972, 3, 773.
[114] L. A. Sauer, *Arch. Biochem. Biophys.*, 1972, **149**, 42.
[115] I. Björkhem, *European J. Biochem.*, 1972, **27**, 354.
[116] G. Johansson, *European J. Biochem.*, 1971, **21**, 68; H. Danielsson and G. Johansson, *F.E.B.S. Letters* 1972, **25**, 329.
[117] F. W. Kahnt and R. Neher, *Acta Endocrinol.*, 1972, **70**, 315.

pregnenolone (36; R = H) stage or later. Comparison of the distribution of the NAD$^+$-dependent 3β-hydroxy-steroid dehydrogenase [(41) → (42)] and 3-oxo-steroid $\Delta^4 \to \Delta^5$ isomerase [E.C.5.3.3.1; (42) → (43)] enzymes, required for conversion of pregnenolone (41) into progesterone (43) in adrenal cortex, with that of marker enzymes showed that these two enzymes are firmly associated with both mitochondrial and microsomal fractions.[118] Rat liver microsomal fraction contains small amounts of 3β-hydroxy-Δ^5-C_{19}- and -C_{21}-steroid oxidoreductase, the enzyme(s) being apparently different from the hepatic 3β-hydroxy-Δ^5-C_{27}-steroid oxidoreductase involved in the biosynthesis of bile acids, since the two enzyme activities were affected differently by age and by a 3β-hydroxy-Δ^5-steroid oxidoreductase inhibitor.[119] Kinetic evidence on calcium-chloride-'solubilized' beef adrenal microsomal 3-oxo-steroid $\Delta^5 \to \Delta^4$ isomerase [(42) → (43)] suggests the existence of only one C_{19} and C_{21} 3-oxo-steroid isomerase in the adrenal glands.[120] Conversion of 17α,21-dihydroxypregn-4-ene-3,20-dione into 3β,17α,21-trihydroxypregn-5-en-20-one, by bovine adrenal cortex microsomes, indicates that the 3β-hydroxy-steroid dehydrogenase $\Delta^5 \to \Delta^4$ isomerase reactions in this tissue are reversible.[121] The effects of solvent on the 3-oxo-steroid $\Delta^5 \to \Delta^4$ isomerase reaction, which involves an intramolecular proton transfer from the C-4β to the C-6β position, have been studied.[122] Circular dichroism studies on the secondary and tertiary structure of the purified isomerase enzyme from *Pseudomonas testosteroni* indicate that it has a very high percentage of β-structure.[123] Inhibition studies on the *Ps. testosteroni* enzyme by various bromo-3-keto-steroids are consistent with the presence of a basic group situated above the β-face of ring A in the enzyme–steroid complex and with an open and readily accessible active site which is tolerant of wide structural variations in the region of the ring A binding loci.[124] The gonadal isomerase is inhibited by the synthetic steroid medrogestone.[125]

The conversion of biologically active steroid hormones into inactive metabolites (usually more polar) is mainly a hepatic process. The most important reactions in the case of 3β-hydroxy-Δ^5- and 3-oxo-Δ^4-steroids are various hydroxylations, conversion of the 3-oxo-Δ^4-structure into a 3β- or 3α-hydroxy-5α- or 5β-structure, and conjugation of hydroxy-groups with sulphuric or glucuronic acids. It is currently believed that the microsomal fraction of liver contains enzymes for hydroxylations and for conversion of 3-oxo-Δ^4-structures into 5α-saturated compounds, whereas the soluble fraction contains enzymes responsible for conversion of 3-oxo-Δ^4-compounds into 5β-saturated compounds. Recent *in vitro* studies have definitely revealed the presence in human liver microsomes of various hydroxylases, 3-oxo-Δ^4-steroid 5α-

[118] R. A. Cowan, J. K. Grant, C. A. Giles, and W. Biddlecombe, *Biochem. J.*, 1971, 126, 12P.
[119] I. Björkhem, K. Einarsson, and J.-Å. Gustafsson, *Steroids*, 1972, 19, 471.
[120] P. Geynet, J. Gallay, and A. Alfsen, *Eurpoean J. Biochem.*, 1972, 31, 464.
[121] M. J. Shapiro, *Steroids*, 1972, 20, 1.
[122] H. Weintraub, E.-E. Baulieu, and A. Alfsen, *Biochim. Biophys. Acta*, 1972, 258, 655.
[123] F. Vincent, H. Weintraub, and A. Alfsen, *F.E.B.S. Letters*, 1972, 22, 319.
[124] J. B. Jones and S. Ship, *Biochim. Biophys. Acta*, 1972, 258, 800.
[125] M. L. Givner and D. Dvornik, *Experientia*, 1972, 28, 1105.

reductases, 3β- and 17β-hydroxy-steroid oxidoreductases, and 3β-hydroxy-Δ^5-steroid oxidoreductase(s).[126] The metabolites of pregnenolone (36; R = H) and corticosterone (44; R^1 = H, R^2β-OH,α-H) in bile of male rats have been examined.[127] It is important to take species[128] and sex[129] differences into account during interpretation of results of steroid hormone metabolism.

Saturation of the Δ^4-bond of Δ^4-3-oxo-steroids catalysed by Δ^4-3-keto-steroid- 5β-reductase involves transfer of the 4-*pro-R* hydrogen of NADPH to the 5β-position of the steroid,[130] whereas the 4-*pro-S* hydrogen of NADPH is transferred to the 5α-position in the reaction catalysed by Δ^4-3-keto-steroid 5α-reductase.[130] This is in agreement with previous results.[131] The Δ^4-3-keto-steroid 5α-oxidoreductase of rat prostate is located partly in the cytoplasm and partly in the nucleus, the latter being present in a nuclear membrane.[132] In agreement with previous results, most of the Δ^4-5α-C_{19}-steroid reductase activity of female rat liver has been found in the microsomal fraction.[133] It is interesting that the 5α-reduced metabolites produced by the soluble fraction when incubated with NADH constituted about 30% and with NADPH only about 6% of the total from nuclear, mitochondrial, microsomal, and soluble fractions, thus possibly suggesting the existence of NADH- and NADPH-specific Δ^4-5α-C_{19}-steroid reductase activities. Similarly, the 3α-hydroxy-steroid dehydrogenase activity in the NADPH system was essentially localized in the soluble fraction, whereas in the NADH system it was almost equally distributed between the microsomal and soluble fractions.[133] Studies on the reduction of the Δ^4-bond of testosterone (45) and two of its glycoside conjugates indicated that, in contrast to (45), the conjugates were not good substrates for the male rat liver microsomal Δ^4-5α-reductase and hydroxylases, while they were better substrates for the soluble Δ^5-5β-reductase.[134] However, it is most interesting that the glycosidic group at C-17 of testosterone (45) affects the stereochemical course of reduction of the Δ^4-bond. For example, incubation of the glycosides with a 20 000 × *g* supernatant of male rat liver homogenate gave only a 5β-metabolite with the sugar still attached, in good yield, whereas free testosterone gave both 5α- and 5β-metabolites.[134] Possible explanations for this phenomenon were also discussed. The microsomal fraction of rat ovary catalyses reduction of 3-oxo-Δ^4-steroids to yield exclusively 3β-hydroxy-5α-oriented metabolites. A soluble enzyme system from the same organ catalyses irreversible epimerization of these metabolites to their 3α-oriented isomers.[135] The mechanism of this epimerization is not yet proven.

[126] I. Björkhem, K. Einarsson, J.-Å. Gustafsson, and A. Somell, *Acta Endocrinol.*, 1972, **71**, 569.
[127] T. Cronholm, H. Eriksson, and J.-Å. Gustafsson, *Steroids*. 1972, **19**, 455.
[128] K. Leung and S. Solomon, *Endocrinol.*, 1972, **91**, 341.
[129] H. Eriksson, J.-Å. Gustafsson, and Å. Pousette, *European J. Biochem.*, 1972, 27. 327; K. Einarsson, J.-Å. Gustafsson, and A. S. Goldman, *European J. Biochem.*, 1972, **31**, 345.
[130] Y. J. Abul-Hajj, *Steroids*, 1972, **20**, 215.
[131] I. Björkhem and H. Danielsson, *European J. Biochem.* 1970, **12**, 80; I. Björkhem, *ibid.* **8**, 345.
[132] R. J. Moore and J. D. Wilson, *J. Biol. Chem.*, 1972, **247**, 958.
[133] N. Arimasa and C. D. Kochakian, *Steroids*, 1972, **19**, 325.
[134] M. Matsui, F. Abe, K. Kimura, and M. Okada, *Chem. and Pharm. Bull*, (*Japan*), 1972, **20**, 1913.
[135] A. Nimrod and H. R. Lindner. *Biochim. Biophys. Acta*, 1972, **261**, 291.

(44)   (45)   (46) R = β-OH, α-H
                        (47) R = O

Studies on steroid metabolism in rats given [1-^{2}H$_2$]ethanol indicate that the coenzyme pool(s) used in different reductions at C-3 is metabolically related to NADH formed in the alcohol dehydrogenase reaction, but since little or no deuterium was found at C-5 and C-20 other coenzyme pools are probably used for the reductions of these positions.[136] The soluble 20α-hydroxy-steroid dehydrogenase from pig testes has been purified to near homogeneity and its properties have been studied.[137] It has a molecular weight of 35 000 and probably no apparent subunit structure. The same enzyme from foetal sheep blood has also been partly purified.[138] It is interesting that when 15α-hydroxyprogesterone is incubated with rabbit liver homogenate, it directs reduction of the C-20-oxo-group towards the 20α-alcohol rather than the 20β-epimer, which is formed when progesterone (39) is used as substrate.[139] 20β-Hydroxy-steroid dehydrogenase from *Streptomyces hydrogenens* stereospecifically reduces the 20-oxo-group only of 20-oxo-21-al steroids to a 20β-hydroxy-structure.[140] The nature of the binding site of this enzyme has also been studied.[141] Extensive studies have been carried out on the effect of different substrate structures on the C-3-[142] and C-20-[143]hydroxy-steroid–NAD$^+$ oxidoreductase activities of cortisone (44; R^1 = OH; R^2 = O) reductase (E.C.1.1.1.53). Steroid biosynthetic pathways in the human adrenal have been reviewed.[144]

Of the several 17β-hydroxy-steroid NAD(P) oxidoreductase (E.C.1.1.1.64) activities which have been separated from rabbit liver cytosol, two have been particularly studied.[145] One of these is a weakly anionic 17β-hydroxy-steroid oxidoreductase of molecular weight 25 000—30 000, both testosterone (45) and

---

[136] T. Cronholm, *European J. Biochem.*, 1972, **27**, 10.
[137] F. Sato, Y. Takagi, and M. Shikita, *J. Biol. Chem.*, 1972, **247**, 815.
[138] R. F. Seamark, D. Herriot, and C. D. Nancarrow, *J. Endocrinol.*, 1972, **52**, xiv.
[139] B. R. Bhavnani, K. N. Shah, and S. Solomon, *Biochemistry*, 1972, **11**, 753.
[140] E. S. Szymanski, C. S. Furfine, and C. F. Hammer, *Steroids*, 1972, **19**, 243.
[141] C. C. Chin and J. C. Warren, *Biochemistry*, 1972, **11**, 2720.
[142] W. Gibb and J. Jeffery, *Biochim. Biophys. Acta*, 1972, **268**, 13; *Biochem. J.*, 1972, **128**, 136P; 1972, **130**, 37P.
[143] I. H. White and J. Jeffery, *European J. Biochem.*, 1972, **25**, 409; *Biochem. J.*, 1972, **129**, 24P.
[144] K. Griffiths and E. H. D. Cameron, *Adv. Steroid Biochem. Pharmacol.*, 1970, **2**, 223.
[145] A. Thaler-Dao, B. Descomps, M. Saintot, and A. Crastes de Paulet, *Biochimie*, 1972, **54**, 83.

17β-oestradiol (46) being substrates; NAD or NADP are coenzymes. The second, a very anionic testosterone NADP oxidoreductase with high affinity for testosterone, exists as isozymes which can be resolved into peptide chains of molecular weight 36000. This enzyme also has high 3α-hydroxy-steroid–NADP oxidoreductase activity. A rat skin microsomal 17β-hydroxy-steroid dehydrogenase has been studied, which utilizes both oestradiol (46) and testosterone (45).[146] A NADP-dependent 17β-hydroxy-steroid dehydrogenase of molecular weight 64000 and broad substrate specificity has been partly purified from mammalian erythrocytes.[147] It is interesting that adenosine 3′:5′-cyclic-monophosphate inhibits the conversion of (45) into androst-4-ene-3,17-dione (48) by testicular 17β-hydroxy-steroid dehydrogenase in the presence of NAD^+ but not when $NADP^+$ was the cofactor.[148]

Earlier studies on biosynthesis of C_{19}-Δ^{16}-steroids have been reviewed.[1,149] Further evidence for the intermediary role of pregn-5-ene-3β,20β-diol (50) in androsta-5,16-dien-3β-ol (52) formation from pregnenolone in boar testis has been furnished.[150] However, there is some evidence for the existence of another

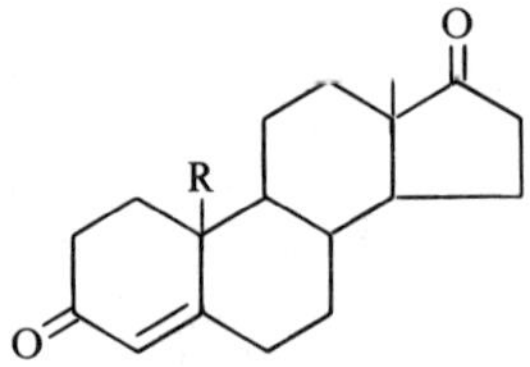

(48) R = Me
(49) R = CHO

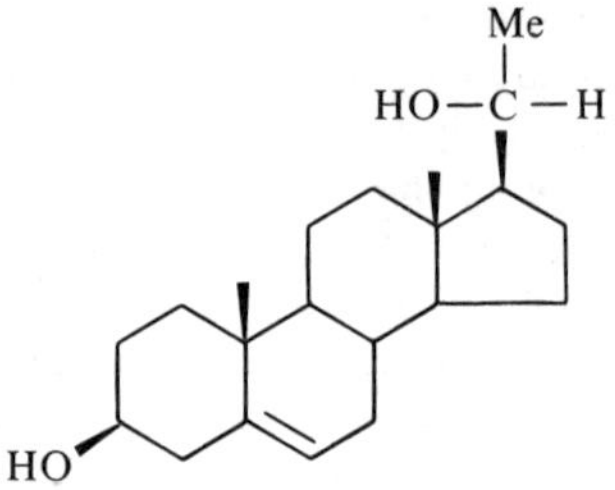

(50)

pathway without the obligatory involvement of (50). Conceivable free-radical pathways for the biosynthesis of C_{19}-Δ^{16}-steroids from pregnenolone have also been discussed.[150] The conversion of androsta-5,16-dien-3β-ol (52) into androsta-4,16-dien-3-one (54) is reversible. Further metabolism of (54) into 5α-androst-16-en-3-one (56), 5α-androst-16-en-3α-ol (55), and 5α-androst-16-en-3β-ol (57) has also been demonstrated *in vitro*.[151] It is interesting that optimal yields of the 3α-alcohol (55) and 3β-alcohol (57) were obtained from the ketone (56), when NADPH and NADH, respectively, were used as cofactors. Based on these and other observations, the pathway shown in Scheme 2 has been

[146] B. P. Davis, E. Rampini, and S. L. Hsia, *J. Biol. Chem.*, 1972, **247**, 1407.
[147] E. Mulder, G. J. M. Lamers-Stahlhofen, and H. J. Van Der Molen, *Biochem. J.*, 1972, **127**, 649; *Biochim. Biophys. Acta*, 1972, **260**, 290.
[148] S. Sulimovici and B. Lunenfeld, *J. Steroid Biochem.*, 1972, **3**, 781.
[149] D. B. Gower, *J. Steroid Biochem.*, 1972, **3**, 45.
[150] K. H. Loke and D. B. Gower, *Biochem. J.*, 1972, **127**, 545.
[151] P. J. Brophy and D. B. Gower, *Biochem. J.*, 1972, **128**, 945.

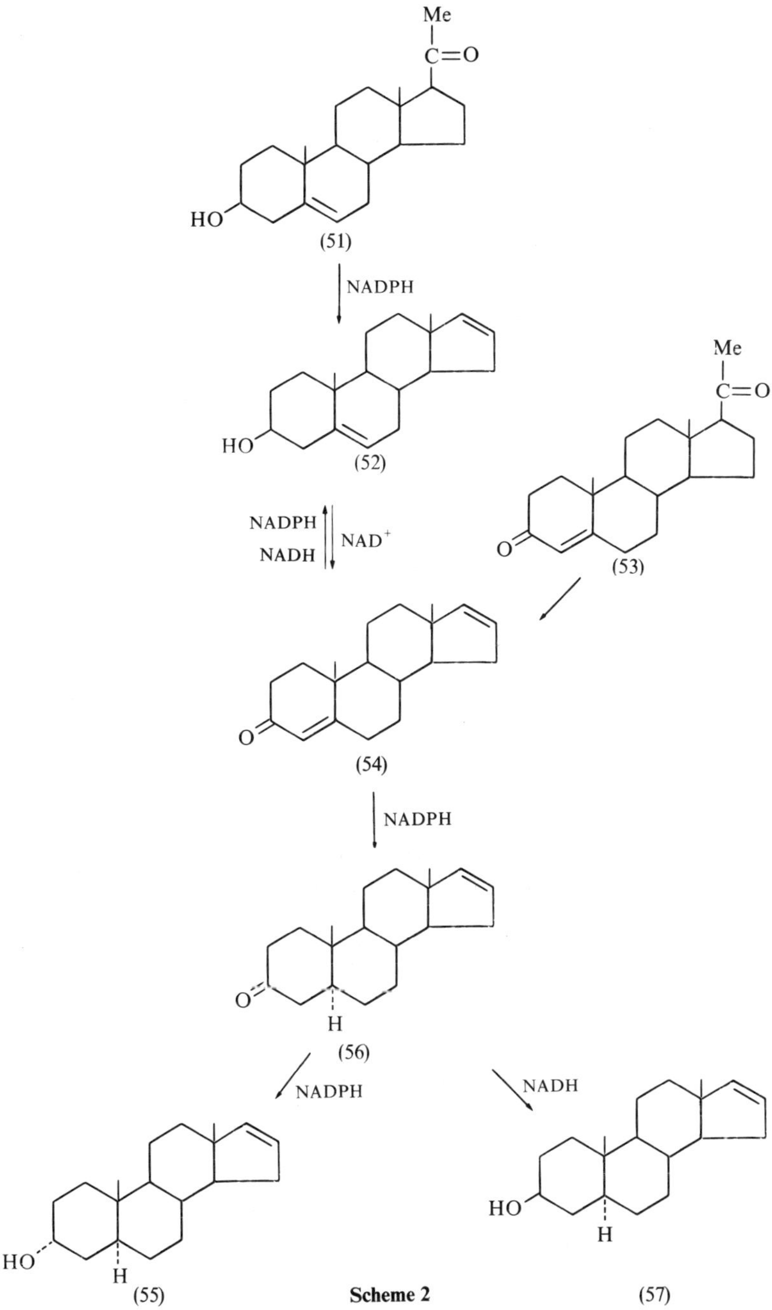

Biosynthesis of Triterpenes, Steroids, and Carotenoids

Scheme 2

proposed[151] for the biosynthesis of C_{19}-Δ^{16}-steroids. *In vivo* experiments using [7α-³H]pregnenolone indicated that the alcohols (55) and (57) were further metabolized to their sulphate conjugates; the 5α-androst-16-en-3β-ol (57) was mainly present in urine as its glucosiduronate.[152] Since C_{21}-steroids are probably not converted into C_{19}-Δ^{16}-steroids in boar salivary glands, formation of C_{19}-Δ^{16}-3-hydroxylated steroids in such glands probably occurs by reduction of Δ^{16}-steroids pre-formed in the testes.[153] The origin of 5α-androst-16-en-3α-ol in man has also been studied.[154] The transformation of C_{19}-Δ^{16}-steroids into 16,17-dihydroxylated metabolites by liver microsomes involves cytochrome P-450.[155]

The true *in vivo* significance of C_{19}-norsteroids in the biosynthesis of oestrogens from androgens is still uncertain; there may possibly be more than one pathway.[156,157] The 1β,2β-hydrogens are removed during aromatization of C_{19}-norsteroids by human placental preparations,[158] which is the same stereochemistry as that previously observed with C_{19}-steroids as substrates. The same stereochemistry has also been observed during aromatization by baboon placental microsomes.[159] Since cyano-ketone (2α-cyano-17β-hydroxy-4,4,17α-trimethyl-androst-5-en-3-one), a potent inhibitor of the 3β-hydroxy-steroid dehydrogenase enzyme, strongly inhibited oestrone (47) and 17β-oestradiol (46) formation from 3β-hydroxyandrost-5-en-17-one (40) but not from androst-4-ene-3,17-dione (48), the 3β-hydroxy-steroid dehydrogenase–isomerase enzyme is probably obligatory for oestrogen biosynthesis from (40).[160] The effect of many inhibitors on the conversion of androst-4-ene-3,17-dione (48) into oestrone (47) by human placental microsomal preparations has been studied.[161] The absence of effects by metyrapone, carbon monoxide, and ethyl isocyanide was interpreted as conclusive evidence that cytochrome P-450 is not involved in the conversion of androgens into oestrogens in the placenta. This is at variance with a previous observation.[156] Androst-4-ene-3,17,19-trione (49), an apparently obligatory intermediate in oestrogen biosynthesis by human placental microsomes, has now also been isolated as a testosterone (45) metabolite in baboon microsomes.[162]

The soluble human placental oestra-3,17β-diol dehydrogenase (E.C.1.1.1.62) has been purified by affinity chromatography on an oestrone-3-linked Sepharose column and its amino-acid composition analysed.[163] Pre-incubation

---

[152] Y. A. Saat, D. B. Gower, F. A. Harrison, and R. B. Heap, *Biochem. J.*, 1972, **129**, 657.
[153] T. Katkov, W. D. Booth, and D. B. Gower, *Biochim. Biophys. Acta*, 1972, **270**, 546.
[154] W. L. Brooksbank, D. A. A. Wilson, and D. A. MacSweeney, *J. Endocrinol.*, 1972, **52**, 239.
[155] C. von Bahr, K. Brandt, and J.-Å. Gustafsson, *F.E.B.S. Letters*, 1972, **25**, 65.
[156] R. A. Meigs and K. J. Ryan, *J. Biol. Chem.*, 1971, **246**, 83;
[157] L. Milewich and L. R. Axelrod, *Fed. Proc.*, 1971, **30**, 311.
[158] J. D. Townsley and H. J. Brodie, *Biochemistry*, 1968, **7**, 33; T. Nambara, T. Anjyo, and H. Hosoda, *Chem. and Pharm. Bull. (Japan)*, 1972, **20**, 853.
[159] L. Milewich and L. R. Axelrod, *J. Endocrinol.*, 1972, **52**, 137.
[160] J. Yates and R. E. Oakey, *Steroids*, 1972, **19**, 119.
[161] J. Chakraborty, R. Hopkins, and D. V. Parke, *Biochem. J.*, 1972, **130**, 19P.
[162] L. Milewich and L. R. Axelrod, *Endocrinol.*, 1972, **91**, 1101.
[163] J. C. Nicolas, M. Pons, B. Descomps, and A. Crastes de Paulet, *F.E.B.S. Letters*, 1972, **23**, 175.

Biosynthesis of Triterpenes, Steroids, and Carotenoids

of oestra-3,17β-diol (46) with rabbit uterus cytosol, which contains oestrogen-binding proteins, reduces its rate of oxidation by placental 17β-hydroxy-steroid dehydrogenase.[164] The cytoplasmic fraction of rat liver[165] and human placenta[166] contain both 17β-hydroxy-steroid oxidoreductase and 'transhydrogenase' activity, the latter catalysing the transfer of hydrogen from the 17β-position of oestra-3,17β-diol (46) to the 17β-position of androst-4-ene-3,17-dione (48). Although the two enzymic activities from liver or placenta could not be separated, some evidence was obtained that the hepatic 'transhydrogenase activity' is identical with the already known 17β-hydroxy-steroid oxidoreductase; however, the characteristics of the two placental enzymic activities provide some evidence for the existence of two distinct enzymes. Similar 'transhydrogenase' activity is present in the microsomal fractions of rat liver and ovary.[167] The cytoplasmic fraction of rat ovaries contains an analogous 'transhydrogenase enzyme', which transfers hydrogen specifically from C-17 of oestradiol (46) to C-20 of progesterone (36; R = H), with formation of 20α-hydroxypregn-4-en-3-one.[168] Similarly, studies on the transfer of tritium from [17α-^{3}H]oestradiol (46) to the 20-oxo-group of cortisone (58) in rat liver slices indicate that such enzymic hydrogen transfer is quanti-

(58)

tatively important only if the concentrations of donor and acceptor steroids differ by more than 10—100-fold, irrespective of the absolute concentrations.[169]

A full account of the further metabolism and conjugation of steroid hormones is beyond the scope of this chapter, but a few remarks might be germane. Oestrogens can be transformed into the corresponding glycosides,[170] sul-

---

[164] E. M. Whitaker and R. E. Oakey, *Steroids*, 1972, **20**, 295.
[165] K. Pollow and B. Pollow, *Z. Naturforsch.*, 1972, **27b**, 981.
[166] K. Pollow and B. Pollow, *Z. physiol. Chem.*, 1972, **353**, 53.
[167] K. Pollow, G. Sokolowski, and B. Pollow, *Z. physiol. Chem.*, 1972, **353**, 1094.
[168] K. Pollow, G. Sokolowski, and B. Pollow, *Z. physiol. Chem.*, 1972, **353**, 43.
[169] M. Wenzel and B. Hieronimus, *Z. physiol. Chem.*, 1972, **353**, 1477.
[170] D. G. Williamson, D. S. Layne, and D. C. Collins, *J. Biol. Chem.*, 1972, **247**, 3286.

Biosynthesis

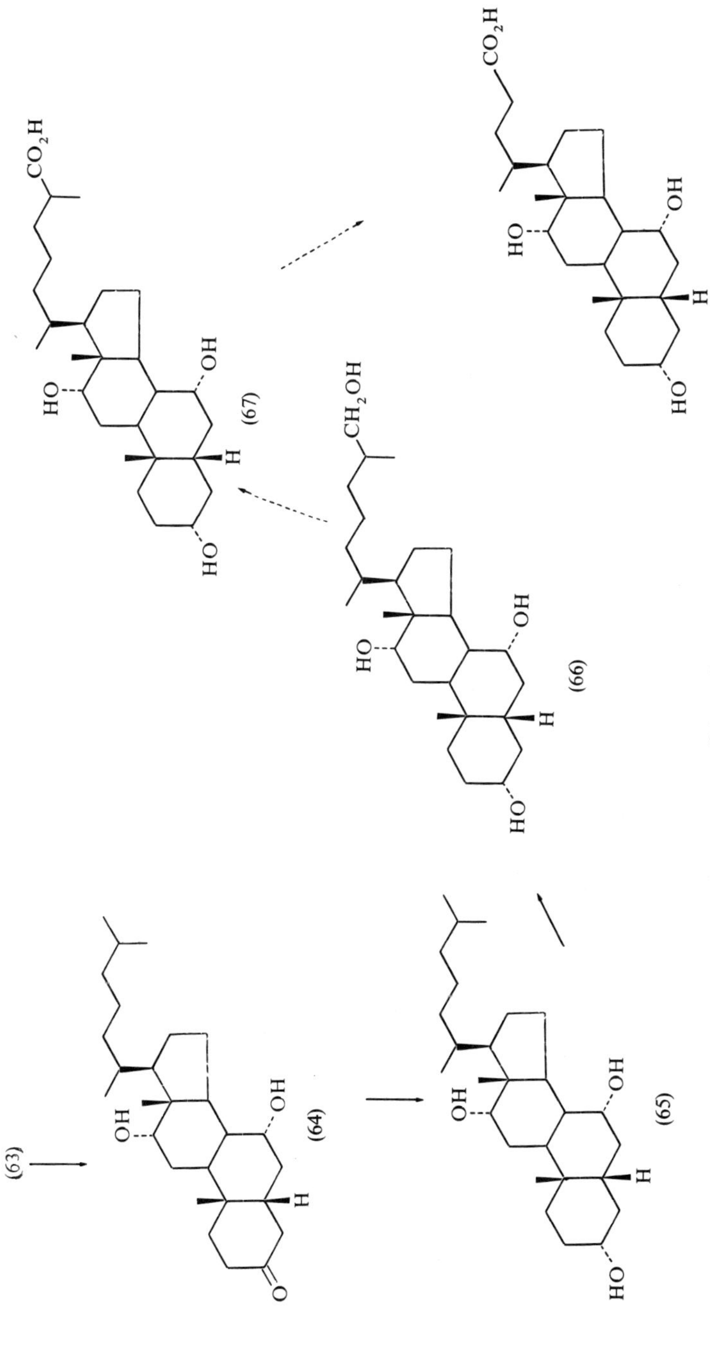

Scheme 3

phates,[171] and methyl ethers,[172] undergo peroxidase-catalysed metabolism,[173] or further hydroxylation.[174] A single enzyme present in the outer mitochondrial membrane of pig intestine catalyses glucuronidation of oestrone (47) and oestra-3,17β-diol (46).[175] Incubation of androst-4-ene-3,17-dione (48) with human foetal liver microsomes revealed the presence of 1β-, 2α-, 2β-, 6α-, 6β-, and 18-hydroxylase activities.[176] An UDP glucuronyl transferase with high specificity for testosterone occurs in rat liver microsomes.[177] Some evidence has also been obtained for the mitochondrial formation of steroid sulphatides from cytidine phosphosulphate steroid complex, which is probably formed from steroid sulphate and cytidine triphosphate.[178] Finally, it might be pertinent to draw attention to the possible formation of artifacts during enzymic hydrolysis of steroid conjugates unless precautions are taken to remove or suppress bacteria.[179]

Bile Acids. Reviews which have appeared[180] on bile acids include two excellent complementary articles covering both the metabolic pathways[181] and enzymic regulation[182] of their biosynthesis. One major pathway for 3α,7α,12α-trihydroxy-5β-cholan-24-oic acid (68) formation in mammalian liver is outlined in Scheme 3, in which modification of the nucleus is completed before removal of the C_3 side-chain unit. However, further evidence has been presented supporting a possible alternative pathway (Scheme 4) suggested[183] for 3α,7α-dihydroxy-5β-cholan-24-oic acid (73) formation, involving side-chain cleavage at an early stage. For example, formation of 3β,7α-dihydroxychol-5-enoic acid (69) from [^{14}C]-mevalonate or -cholesterol has been demonstrated in rats,[184] as well as the efficient conversion of (71) into (73).[185] Conversion of 7α-hydroxycholesterol into (69) has also been reported in the hen.[186] Evidence for the formation of cholest-5-ene-3β,26-diol as an intermediate in the conversion of cholesterol into bile acids by rat liver mitochondria[187] indicates that a slight variant of the pathway shown in Scheme 4 might exist.

[171] J. B. Adams and R. K. Ellyard, *Biochim. Biophys. Acta*, 1972, **260**, 724.
[172] R. Knuppen, P. Ball, O. Haupt, and H. Breuer, *Z. physiol. Chem.*, 1972, **353**, 565; T. Nambara, S. Honma, and K. Kanayama, *Chem. and Pharm. Bull. (Japan)*, 1972, **20**, 2235.
[173] C. R. Lyttle and P. H. Jellinck, *Steroids*, 1972, **20**, 89.
[174] K. Leung, J. Merkatz, and S. Solomon, *Endocrinology*, 1972, **91**, 523.
[175] R. Schumacher, G. S. Rao, M. L. Rao, and H. Breuer, *Z. physiol. Chem.*, 1972, **353**, 1784; G. S. Rao, R. Schumacher, M. L. Rao, and H. Breuer, *ibid.*, p. 1789.
[176] B. P. Lisboa and J.-C. Plasse, *European J. Biochem.*, 1972, **31**, 378.
[177] H. Schriefers, R. Ghraf, and B. Lehnen, *Z. Naturforsch.*, 1972, **27b**, 49.
[178] P. Benes and G. W. Oertel, *J. Steroid Biochem.*, 1972, **3**, 925.
[179] M. E. Manson, L. Nocke-Fink, J.-Å. Gustafsson, and C. H. L. Shackleton, *Clinica Chim. Acta*, 1972, **38**, 45.
[180] K. Yamasaki, *Yonago Acta Medica*, 1971, **15**, 171; B. Croizat, M. Lambiotte, N. Simonet-Thierry, and J. Kande, *Ann. Nutr. Aliment.*, 1971 **25**, 203.
[181] W. H. Elliott and P. M. Hyde, *Amer. J. Med.*, 1971, **51**, 568.
[182] G. S. Boyd and I. W. Percy-Robb, *Amer. J. Med.*, 1971, **51**, 580.
[183] Y. Ayaki and K. Yamasaki, *J. Biochem.*, 1970, **68**, 341.
[184] Y. Ayaki and K. Yamasaki, *J. Biochem.*, 1972, **71**, 85.
[185] S. Ikawa, Y. Ayaki, M. Ogura, and K. Yamasaki *J. Biochem.*, 1972, **71**, 579.
[186] H. Yamasaki and K. Yamasaki, *J. Biochem.*, 1972, **71**, 77.
[187] K. A. Mitropoulos, M. D. Avery, N. B. Myant, and G. F. Gibbons, *Biochem. J.*, 1972, **130**, 363.

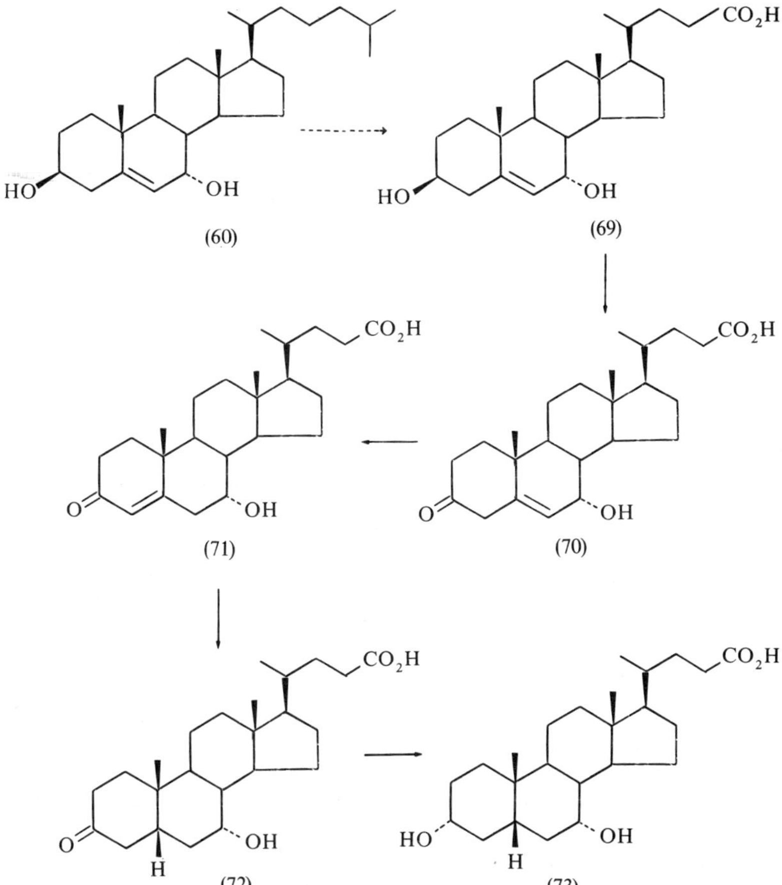

Scheme 4

Several experiments have been reported[188] on the biosynthesis of bile acids in rats administered [1-^{2}H$_2$]ethanol for labelling of the NADH and NADPH pools. The fact that deuterium was present in the 3β-position but not in the 5α-position of 3α-hydroxy-5α-cholanoic acid formed from 3-oxochol-4-en-24-oic acid indicated that different coenzyme pools are utilized in the reductions of the oxo-group and the double bond. This is in agreement with other observations.[136] It is interesting that unidentified unsaturated monohydroxy bile acids have been isolated from patients with cholestatic liver disease.[189]

[188] T. Cronholm, I. Makino, and J. Sjövall, *European J. Biochem.*, 1972, **24**, 507; *ibid.*, 1972, **26**, 251.
[189] G. M. Murphy, F. H. Jansen, and B. H. Billing, *Biochem. J.*, 1972, **129**, 491.

5α Bile acids can be formed by several routes.[1] The conversion of 7α-hydroxycholesterol (60) into 3α,7α,12α-trihydroxy-5α-cholan-24-oic acid in the rat and rabbit suggests that this 5α bile acid is possibly derived from cholesterol by a biosynthetic pathway analogous to that of 3α,7α,12α-trihydroxy-5β-cholan-24-oic acid (68).[190] Circumstantial evidence for such a pathway in Nature is furnished by the isolation of 3α,7α,12α-trihydroxy-5α-cholestan-26-oic acid.[191] However, the possible low specificity of the enzymes is illustrated by the conversion of 5α-cholestan-3β-ol into 3α,7α,12α-trihydroxy-5α-cholan-24-oic acid in the Mongolian Gerbil, although this species does not normally contain 5α bile acids.[192] Another example of such non-specificity may possibly be the conversion of 3α-hydroxy-5β-cholan-24-oic acid into the corresponding 3-oxo- and 6β-hydroxy-derivatives by rat brain cell-free homogenates.[193]

There is a good deal of evidence that cholesterol 7α-hydroxylase [(59) → (60)] is the major rate-limiting enzyme in bile acid biosynthesis.[182] This enzyme shows a diurnal variation,[194, 195] following the same pattern as the incorporation of acetate into cholesterol.[194] The rise in activity of cholesterol 7α-hydroxylase can be prevented by treating rats with inhibitors of protein synthesis, indicating that an increase in enzyme synthesis is responsible for the diurnal rise in the activity of the enzyme.[195] It has been tentatively suggested that cholesterol 7α-hydroxylase is influenced by an unknown substance, possibly a protein or peptide, to a much greater extent than by bile acids;[196] regulation by the latter does not occur after removal of the pituitary gland.[197] Phenobarbital has entirely different effects on this enzymic activity with different strains of rats.[198] It is important to eliminate the aberrant formation of 7-oxocholesterol during *in vitro* studies on 7α-hydroxylase, since this enzyme is strongly inhibited by the keto-compound.[199] Formation of 7α-hydroxycholesterol, together with the non-physiological compounds 7-oxocholesterol, 7β-hydroxycholesterol, and 5α-cholestane-3β,5,6β-triol, is observed on aerobic incubation of cholesterol with NADPH and microsomes prepared from rat livers homogenized in the absence of H_4edta, whereas 7α-hydroxycholesterol is the main product when H_4edta is present in the homogenizing medium.[200] In agreement with previous observations, formation of 7α-hydroxycholesterol was inhibited by carbon monoxide, thus implicating cytochrome P-450 as the terminal oxidase. Indirect evidence indicates that cytochrome P-450 may be distinct from such components involved in drug hydroxylation.[71] As

---

[190] K. Yamasaki, Y. Ayaki, and G. Yamasaki, *J. Biochem.*, 1972, **71**, 927.
[191] K. Okuda, M. G. Horning, and E. C. Horning, *J. Biochem.*, 1972, **71**, 885.
[192] B. W. Nall, L. B. Walsh, E. A. Doisy, and W. H. Elliott, *J. Lipid Res.*, 1972, **13**, 71.
[193] C. W. Martin and H. J. Nicholas, *Steroids*, 1972, **19**, 549.
[194] H. Danielsson, *Steroids*, 1972, **20**, 63.
[195] K. A. Mitropoulos, S. Balasubramaniam, G. F. Gibbons, and B. E. A. Reeves, *F.E.B.S. Letters*, 1972, **27**, 203.
[196] D. Mayer and U. Petrosilius, *Z. physiol. Chem.*, 1972, **353**, 1185.
[197] D. Mayer and A. Voges, *Z. physiol. Chem.*, 1972, **353**, 1187.
[198] S. Shefer, S. Hauser, and E. H. Mosbach, *J. Lipid Res.*, 1972, **13**, 69.
[199] J. Van Cantfort, *Life Sci.*, Part II, 1972, **11**, 773.
[200] K. A. Mitropoulos and S. Balasubramaniam, *Biochem. J.*, 1972, **128**, 1.

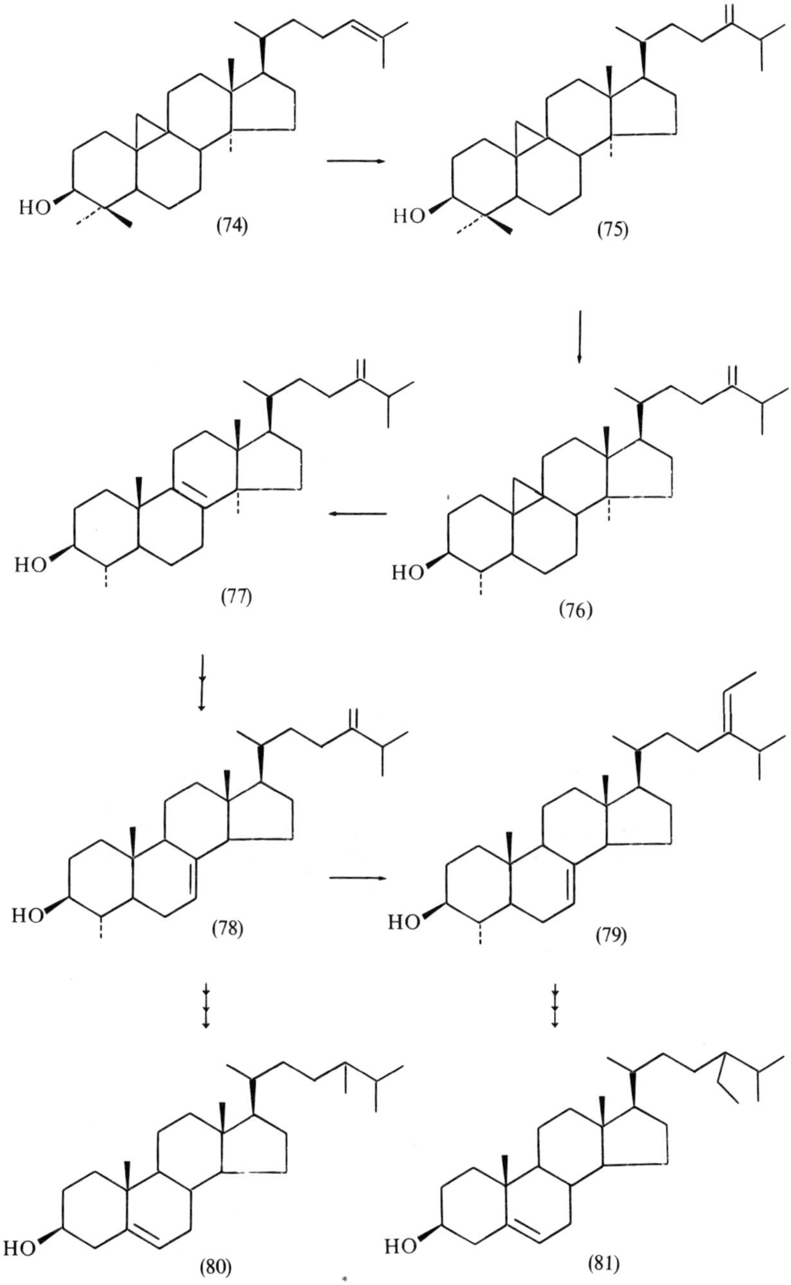

Biosynthesis of Triterpenes, Steroids, and Carotenoids

Scheme 5

expected, administration of the cholesterol biosynthesis inhibitor AY-9944 to rats leads to reduced secretion of bile salts.[201]

6 Triterpenes and Steroids in Higher Plants, Algae, and Fungi

Phytosterol Biosynthesis.—*Overall Pathway*.Biosynthesis of plant sterols has been reviewed.[202] A plausible partial scheme[202] for biosynthesis of some phytosterols in higher plants is shown in Scheme 5. However, the structural features of numerous naturally occurring sterols mean that other diverging pathways must be considered. For example, the isolation[203] of 14α-methyl-9β,19-cyclo-5α-ergost-24(28)-en-3β-ol (82) from *Musa sapientum* may imply that opening of the cyclopropane ring in this species occurs at a late stage; it is also possible that (82) is not further metabolized. An enzyme capable of opening the cyclo-

(82)

propane ring of cycloeucalenol (76; 4α,14α-dimethyl-24-methylene-9β,19β-cyclocholestan-3β-ol) yielding obtusifoliol (77; 4α,14α-dimethyl-24-methylenecholest-8-en-3β-ol), as required by Scheme 5, has been demonstrated in a cell-free system from *Rubus fruticosus*.[204] In contrast, cycloartenol (74; 9β,19β-cyclolanost-24-en-3β-ol) and its 24-methylene derivatives (75) are very poor substrates for this enzyme, thus indicating that *in vivo* opening of the cyclopropane ring occurs at the 4α-methyl-sterol stage,[205] and that lanosterol is probably not formed from (74). A cell-free system has been reported from the endosperm of germinating seed of *Pinus pinea* which transforms mevalonic acid into squalene, cycloartenol (74), and 24-methylenecycloartanol (75).[206] Since no detectable lanosterol was formed, the results suggest that non-photosynthetic tissue of a photosynthetic plant can biosynthesize sterols by the cycloartenol route as opposed to a pathway involving lanosterol.

Since phytosterols contain extra carbon atoms at C-24 and often a Δ^{22}-bond,

---

[201] N. H. Alam and J. Glover, *European J. Biochem.*, 1972, **27**, 413.
[202] L. J. Goad and T. W. Goodwin, in 'Progress in Phytochemistry', ed. L. Reinhold and Y. Liwschitz, Interscience, London, 1972, vol. 3, p. 113.
[203] F. F. Knapp, D. O. Phillips, L. J. Goad, and T. W. Goodwin, *Phytochemistry*, 1972, **11**, 3497.
[204] R. Heintz, P. Benveniste, and T. Bimpson, *Biochem. Biophys. Res. Comm.*, 1972, **46**, 766.
[205] R. Heintz, T. Bimpson, and P. Benveniste, *Biochem. Biophys. Res. Comm.*, 1972, **49**, 820.
[206] H. C. Malhotra and W. R. Nes, *J. Biol. Chem.*, 1972, **247**, 6243.

Biosynthesis of Triterpenes, Steroids, and Carotenoids   49

there are many more plausible sequences for their biosynthesis than in cholesterol formation. This is well illustrated by recent studies with inhibitors in algae. For example, the isolation of numerous sterols from triparanol-treated cultures of *Chlorella emersonii*, coupled with the presence of most of these compounds in untreated cultures, led to proposal of Scheme 6 as a possible pathway for sterol biosynthesis in this alga.[207] All the sterols shown in Scheme 6, except (83), (84), (93), and (94), were identified in *C. emersonii*. Sterols (85), (86), (90), (92), (95), (96), and (97) constitute new compounds in Nature. During biosynthesis of phytosterols in higher plants, removal of the 14α-methyl group usually precedes demethylation of the 4α-methyl group, whereas cholesterol synthesis in animals usually involves removal of the 14α-methyl group after both C-4 *gem*-dimethyl substituents. It is evident from the structures shown in Scheme 6 that the pathway in *C. emersonii*, at least during triparanol inhibition, surprisingly more closely resembles the demethylation pattern observed in animals. Similar studies on inhibition of sterol biosynthesis in another *Chlorella* species (*C. ellipsoidea*) by AY-9944 resulted in isolation of several new sterols and proposal of the pathway shown in Scheme 7 for sterol synthesis in this organism.[208] Sterols (103), (109),[209] (102), (104), (105), (108), (110), and (111) are novel compounds from biological material. Whereas AY-9944 is regarded as a Δ^7-reductase inhibitor in animal tissues, it is evident from examination of the structures in Scheme 7 that this inhibitor is also inhibiting reduction of the Δ^{14}-bond and inhibiting the $\Delta^8 \rightarrow \Delta^7$ isomerase in this alga. It is interesting that the demethylation pattern indicated in Scheme 7 is analogous to that usually found in higher plants. Finally, it should be pointed out that inhibitor studies have their shortcomings and could conceivably cause accumulation of non-physiological compounds, so that final proof of the operation of pathways shown in Schemes 6 and 7 *in vivo* under normal conditions must await further studies.

Extensive structural studies on the minor sterols of yeast (*Saccharomyces cerevisiae*) have been carried out.[210, 211] The pathways shown in Scheme 8 have been suggested for the later stages of ergosterol (119) biosynthesis, based upon these structural studies and feeding experiments with sterols (113)—(117) as substrates.[210] The latter studies indicated that the major pathway to ergosterol involves (113)→(115)→(116)→(119), corresponding to introduction of unsaturation at C-22 prior to C-5, which is introduced prior to reduction of the 24-methylene function. The pathways (113)→(114)→(116)→(119) and (113)→(117)→(118)→(119) also operate to a lesser degree in this organism. Independent structural studies,[211] as well as the observed incorporation of sterol (115)

---

[207] P. J. Doyle, G. W. Patterson, S. R. Dutky, and M. J. Thompson, *Phytochemistry*, 1972, **11**, 1951.
[208] L. G. Dickson and G. W. Patterson, *Lipids*, 1972, **7**, 635.
[209] L. G. Dickson, G. W. Patterson, C. F. Cohen, and S. R. Dutky, *Phytochemistry*, 1972, **11**, 3473.
[210] M. Fryberg, A. C. Oehlschlager, and A. M. Unrau, *Biochem. Biophys. Res. Comm.*, 1972, **48**, 593.
[211] D. H. R. Barton, U. M. Kempe, and D. A. Widdowson, *J. C. S. Perkin I*, 1972, 513.

50

Biosynthesis

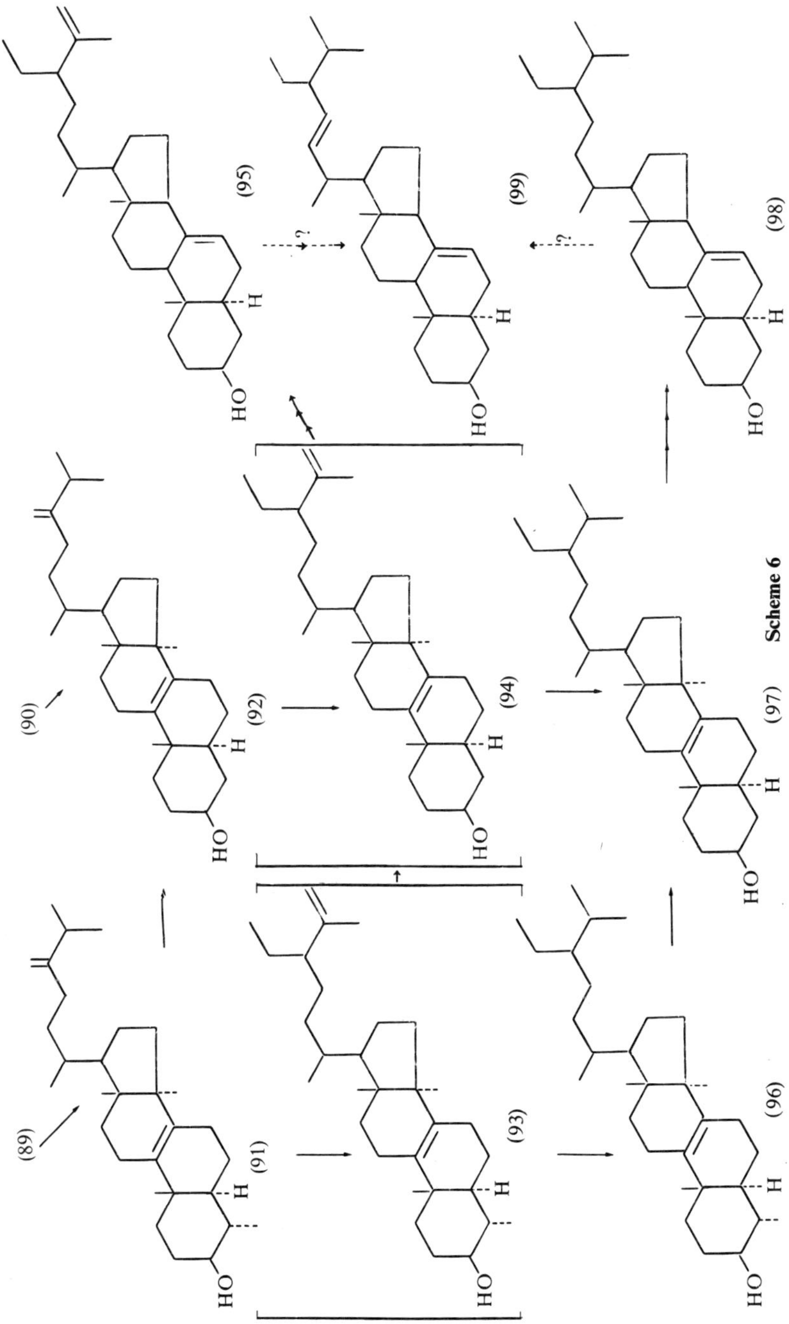

Biosynthesis of Triterpenes, Steroids, and Carotenoids

Scheme 6

52   Biosynthesis

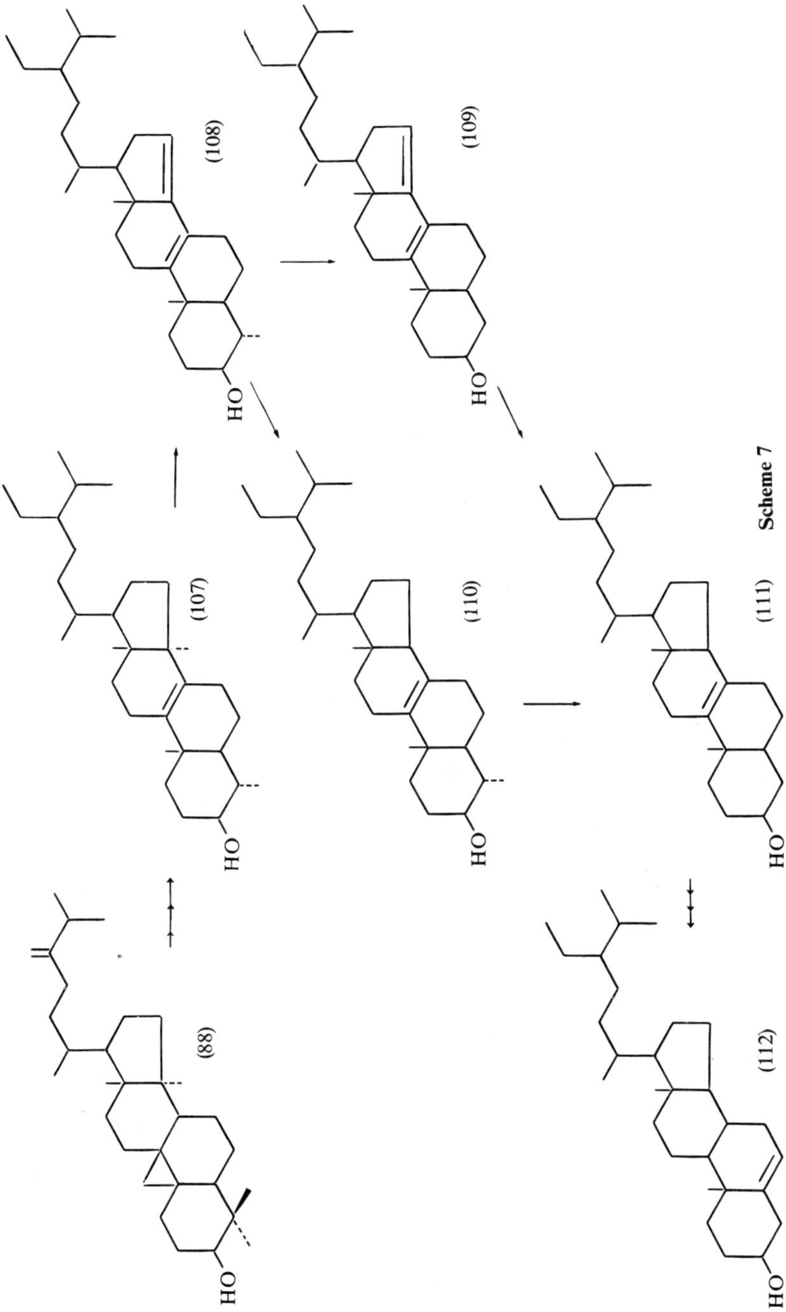

Biosynthesis of Triterpenes, Steroids, and Carotenoids

Scheme 7

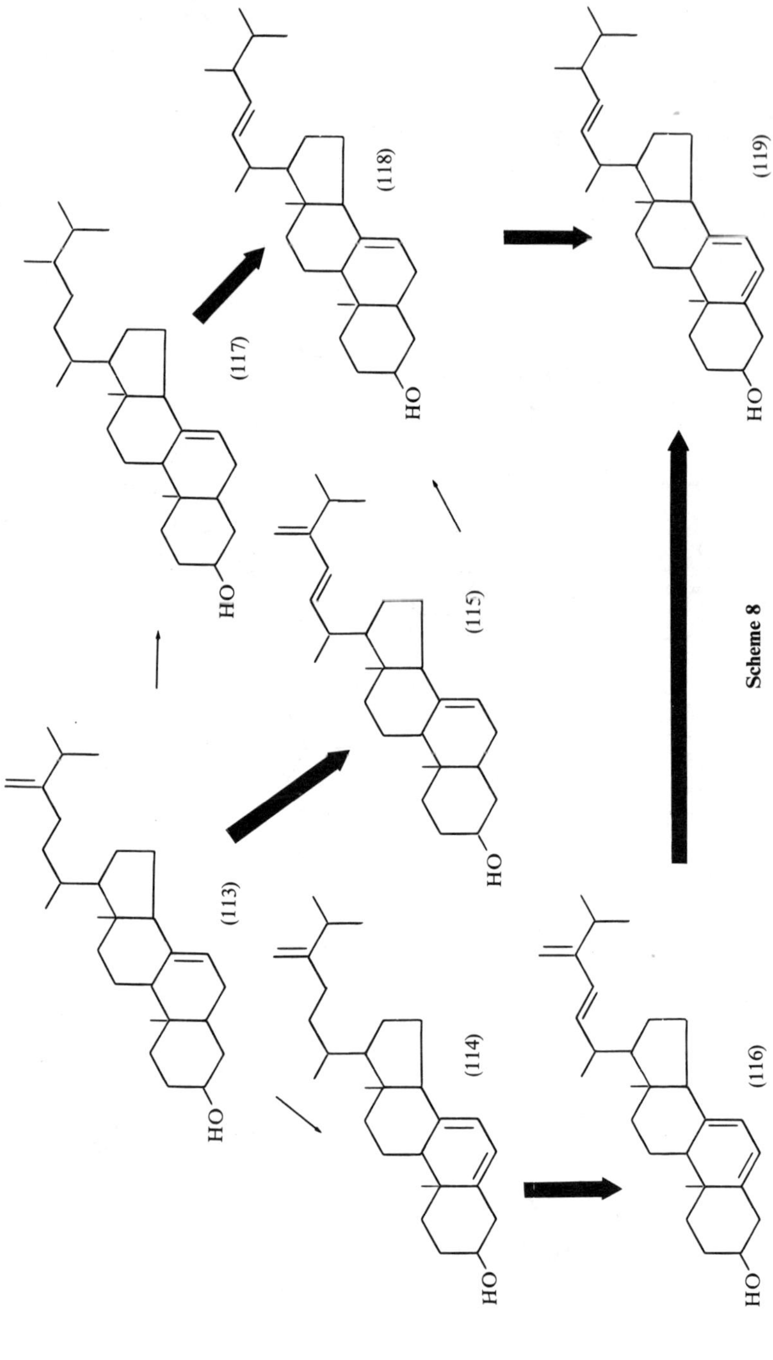

Scheme 8

into compounds (119) and (118), led to the postulation[212] of similar pathways to account for the later stages of ergosterol biosynthesis, but excluding structure (113). 5α-Ergosta-8(9),22-dien-3β-ol has also been proposed as an intermediate in ergosterol biosynthesis, based on its occurrence in wild-type and mutant yeast as well as its conversion into ergosterol in the former.[213] The possible

(120)

involvement of C-5 oxygenated intermediates in the introduction of the Δ^5-bond during ergosterol biosynthesis is still an open question. Evidence that the oxygen function at C-5 does not have to be a hydroxy-group was furnished by the formation of ergosterol at equal rates from ergosta-7,22-diene-3β,5α-diol and 5α,8α-epidioxyergosta-6,22-dien-3β-ol (120).[214] Furthermore, the mechanism shown in Scheme 9 has been proposed to account for the loss of the ^{3}H from [3α-^{3}H₁]ergosta-7,22-diene-3β,5α-diol (121) upon its conversion into ergosterol (122) by a partially-purified yeast enzyme.

Efficient conversion of ergosterol into ergosta-5,7,9(11),22-tetraen-3β-ol in *Mucor rouxii* has been reported.[215]

C-24 *Alkylation.* The alkyl group at C-24 in phytosterols is derived by trans-methylation from methionine. Incorporation of [*methyl*-^{2}H₃]methionine into the alga *Trebouxia* showed that three deuterium atoms were present in the C_{28}-sterol ergost-5-en-3β-ol, whereas five deuterium atoms were present in the C_{29}-sterols poriferasterol [(24R)-24-ethylcholesta-5,22-dien-3β-ol] and clionasterol [(24S)-24-ethylcholest-5-en-3β-ol].[216] Therefore, the C-24 alkylation reactions in *Trebouxia* are similar to those previously reported in *Chlorella*[217] and do not involve $\Delta^{24(28)}$-intermediates. It is interesting that the composition of the sterol mixtures from cultures grown on normal and deuterium-labelled methionine was different, indicating the operation of a pronounced deuterium

---

[212] D. H. R. Barton, E. J. Davies, U. M. Kempe, J. F. McGarrity and D. A. Widdowson, *J. C. S. Perkin I*, 1972, 1231.
[213] L. W. Parks, F. T. Bond, E. D. Thompson, and P. R. Starr, *J. Lipid Res.*, 1972, **13**, 311.
[214] R. W. Topham and J. L. Gaylor, *Biochem. Biophys. Res. Comm.*, 1972, **47**, 180.
[215] L. Atherton, J. M. Duncan, and S. Safe, *J. C. S. Chem. Comm.*, 1972, 882.
[216] L. J. Goad, F. F. Knapp, J. L. Lenton, and T. W. Goodwin, *Biochem. J.*, 1972, **129**, 219.
[217] Y. Tomita, A. Uomori, and H. Minato, *Phytochemistry*, 1970, **9**, 555.

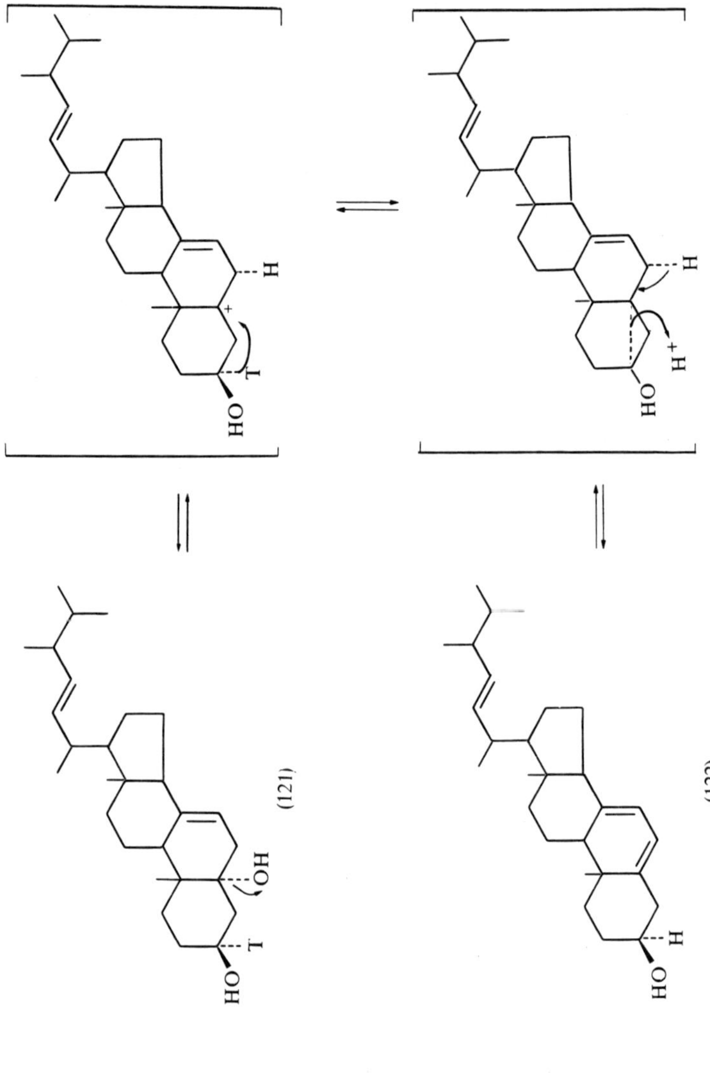

(121)

(122)

Scheme 9

isotope effect. The deuterium-labelling results were compatible with several possible biosynthetic routes. The observed conversion of 31-norcyclolaudenol (4α,14α-dimethyl-9β,19-cyclo-5α-ergost-25-en-3β-ol) into ergost-5-en-3β-ol by the organism adds some support to the suggestion that one product of the first alkylation reaction may be a Δ^{25}-compound, which can then be converted into the C_{28}-sterol in *Trebouxia*. Incorporation of $(4R)$-[2-^{14}C, 4-^{3}H$_1$]mevalonic acid into sitosterol in *Larix decidua* demonstrated[218] that, during alkylation, the C-24 hydrogen of the Δ^{24}-precursor is eliminated, whereas this hydrogen is retained in 28-isofucosterol [stigmasta-5(Z),24(28)-dien-3β-ol], probably at C-25. These results implicate the possible involvement of a Δ^{24}-sterol[1] in sitosterol biosynthesis and indicate that if 28-isofucosterol is on the major pathway to sitosterol in *Larix decidua*, it must be isomerized to a Δ^{24}-structure. However, the presence of ^{3}H at C-25 in poriferasterol biosynthesized from $(4R)$-[2-^{14}C,4-^{3}H$_1$]mevalonic acid in *Ochromonas malhamensis* is consistent with the involvement of a 24-ethylidene-sterol in the C-24 alkylation sequence in this organism.[219] The preceding account, therefore, illustrates that the same alkylation mechanism may not operate in different organisms. A microsomal system from *Rubus fruticosus* has been shown to biosynthesize 24-methylene-cycloartanol (75) from [^{3}H]squalene 2,3-oxide and *S*-adenosylmethionine.[220]

General Aspects of Phytosterol Biosynthesis. Plant sterols occur in a number of forms: free sterol, esterified sterol, steryl glucoside, and acylated steryl glucoside, as well as the recently recognized[221] bound, water-soluble form. A uridine diphosphate glucose:sterol glucosyltransferase catalyses formation of steryl glucosides,[222] which can be further transformed into their acylated derivatives.[223] Water-soluble complexes of sterols represent 0.1—1 % by weight of the total sterols in the higher plant *Zea mays*.[224] The composition of this water-soluble fraction is different from that of the lipid-soluble fraction; cholesterol, sitosterol, and 28-isofucosterol are preferentially complexed compared with campesterol and stigmasterol. The significance of the occurrence of such complexes in plants is still obscure.

It has been suggested that the observed increase in specific radioactivity of sterols biosynthesized from [^{14}C]acetate by plant tissues grown under far-red light, as opposed to dark conditions, may be a phytochrome-mediated process.[225]

Further Metabolism of Plant Sterols.—Cholesterol in plants can undergo further transformations either with or without side-chain cleavage. The latter include formation of the C_{27}-sapogenins, C_{27}-alkaloids, and of the C_{27}-insect-

[218] P. J. Randall, H. H. Rees, and T. W. Goodwin, *J. C. S. Chem. Comm.*, 1972, 1295.
[219] A. R. H. Smith, L. J. Goad, and T. W. Goodwin, *Phytochemistry*, 1972, 11, 2775.
[220] R. Heintz and P. Benveniste, *Compt. rend.*, 1972, 274, D, 947.
[221] C. Anding, R. D. Brandt, G. Ourisson, R. J. Price, and M. Rohmer, *Proc. Roy. Soc.*, 1972, B180, 115.
[222] T. W. Esders and R. J. Light, *J. Biol. Chem.*, 1972, 247, 7494.
[223] Z. Wojciechowski, *Acta Biochim. Polon.*, 1972, 19, 43.
[224] M. Rohmer, G. Ourisson, and R. D. Brandt, *European J. Biochem.*, 1972, 31, 172.
[225] M. A. Hartmann, P. Benveniste, and F. Durst, *Phytochemistry*, 1972, 11, 3003.

moulting hormones. Side-chain cleavage of cholesterol and other plant sterols occurs, yielding pregnenolone and then progesterone, which are precursors of cardenolides, bufadienolides, and C_{21}-alkaloids.

The biosynthesis of cardenolides, bufadienolides, and steroid sapogenins has been reviewed.[226] In confirmation of other results, labelled acetate, mevalonate, and cholesterol were efficiently incorporated into several sapogenins in *Trigonella foenumgraecum*, and into diosgenin (123) in germinating seeds of *Balanites aegyptiaca*; pregnenolone was not incorporated into (123).[227]

(123)

(124)

(125)

(126) R = O
(127) R = α-OH, β-H

(128)

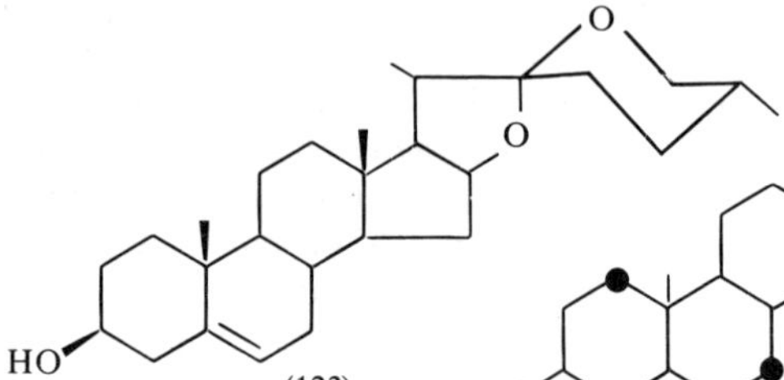

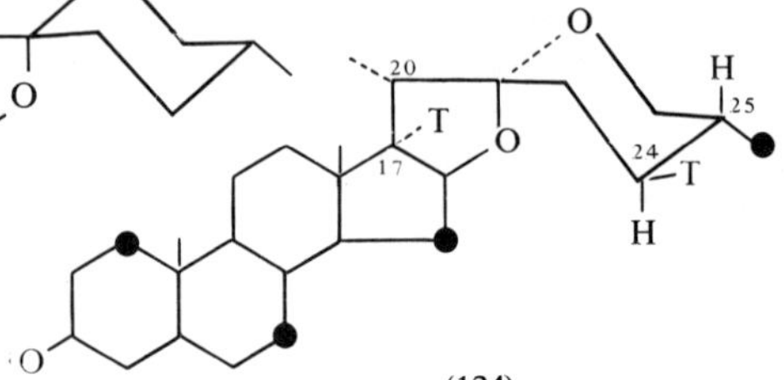

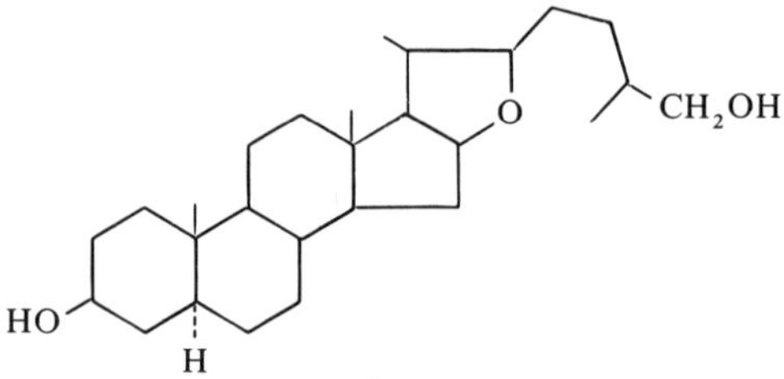

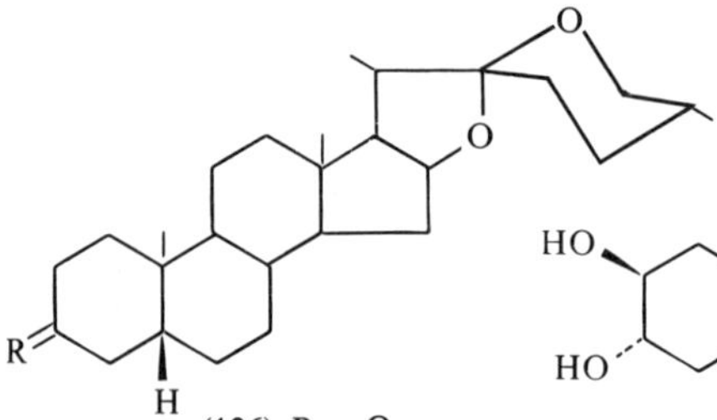

[226] R. Tschesche, *Proc. Roy. Soc.*, 1972, **B180**, 187.

[227] R. Hardman and E. A. Sofowora, *Planta Med.*, 1971, **20**, 193; R. Hardman and F. R. Y. Fazli, *ibid.*, 1972, **21**, 188.

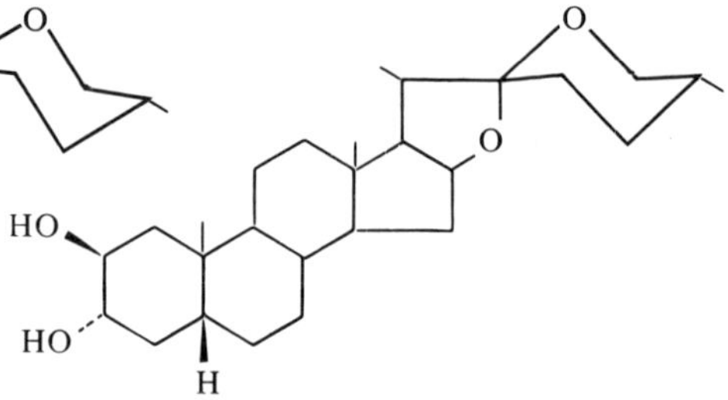

Incorporation of 4R-[2-^{14}C,4-^{3}H$_1$]mevalonic acid into tigogenin (124) by *Digitals lanata* and elucidation of the stereochemistry of the ^{3}H atom at C-24 revealed that during conversion of a Δ^{24}-precursor into (124), the incoming hydrogen assumes the 24-*pro-R* position.[228] It therefore follows from the structure of (124) that the overall process of Δ^{24}-bond reduction occurs with *trans* stereochemistry. This is completely in contrast to the stereochemistry of Δ^{24}-bond reduction during cholesterol biosynthesis, which involves addition of a 24-*pro-S* hydrogen and an overall *cis*-addition of hydrogens at C-24 and C-25.[1, 229] The observed incorporation of 5α-[^{3}H]furostan-3β,26-diol (125) into tigogenin by *Digitalis lanata* plants is consistent with the hypothesis that the biosynthesis of the latter occurs *via* a pathway such as cholesterol→26-hydroxycholesterol (or 26-hydroxycholestanol)→16β,26-dihydroxycholestanol →dihydrotigogenin→tigogenin.[230] Results of experiments utilizing various labelled sapogenins indicated that during the biosynthesis of the 3α-hydroxylated steroidal sapogenin yonogenin (128), smilagenone (126) is an important intermediate, and that smilagenin (127) is a probable precursor in the biosynthesis of 3α-hydroxylated steroidal sapogenins.[231] The biosynthesis of steroidal sapogenins of the Dioscoreaceae has been briefly reviewed.[232]

In agreement with previous results, [^{3}H]cholest-4-en-3-one was not incorporated into the insect-moulting hormones ecdysone (129; R = H) and ecdysterone (129; R = OH) in the plant *Sesuvium portulacastrum*.[233] These results do not rule out the possibility that a 3-keto-Δ^4-steroid could be involved in formation of the A/B-*cis* ring-junction at a later stage, *e.g.* after insertion of some hydroxy-groups. However, results of incorporation of [4-^{14}C,3α-^{3}H]cholesterol into ecdysterone (129; R = OH) in *Taxus baccata* indicated that biosynthesis of this compound does not involve obligatory oxidation at C-3 during the formation of the A/B-*cis* ring-junction.[234]

The fact that cholesterol side-chain cleavage enzyme activity could be demonstrated in *Digitalis purpurea* seedlings but not in tissue cultures may explain why the latter cannot biosynthesize cardenolide glycosides.[235] Leaf homogenates of various cardenolide-producing plants metabolized pregnenolone (36; R = H) into progesterone (39) and 5α-pregnane-3,20-dione (130), as well as progesterone into 5α-pregnane-3,20-dione and small amounts of 5α-pregnan-3β-ol-20-one (131); no 5β-metabolites could be detected.[236] The following sequence seems to occur: pregnenolone→progesterone→5α-

---

[228] L. Canonica, F. Ronchetti, and G. Russo, *J.C.S. Chem. Comm.,* 1972, 1309.
[229] D. J. Duchamp, C. G. Chidester, J. A. F. Wickramasinghe, E. Caspi, and B. Yagen, *J. Amer. Chem. Soc.,* 1971, 93, 6283.
[230] L. Canonica, F. Ronchetti, and G. Russo, *Phytochemistry,* 1972, 11, 243.
[231] K. Takeda, H. Minato, A. Shimaoka, and T. Nagasaki, *J.C.S. Perkin I,* 1972, 957.
[232] K. Takeda, in 'Progress in Phytochemistry', ed. L. Reinhold and Y. Liwschitz, Interscience, London, 1972, vol. 3, p. 287.
[233] A. T. Sipahimalani, A. Banerji, and M. S. Chadha, *J.C.S. Chem. Comm.,* 1972, 692.
[234] J. G. Lloyd-Jones, H. H. Rees, and T. W. Goodwin, *Phytochemistry,* 1973, 12, 569.
[235] H. Pilgrim, *Phytochemistry,* 1972, 11, 1725.
[236] S. J. Stohs and M. M. El-Olemy, *Phytochemistry,* 1972, 11, 2409.

pregnane-3,20-dione→5α-pregnan-3β-ol-20-one. However, the biochemical significance of these 5α-metabolites in the plants is obscure. *Dioscorea deltoidea* tissue cultures possess 20β-hydroxy-steroid dehydrogenase activity, as demonstrated by metabolism of progesterone (36; R = H) into 5α-pregnan-3β-ol-20-one (131) and 5α-pregnane-3β,20β-diol (132),[237]

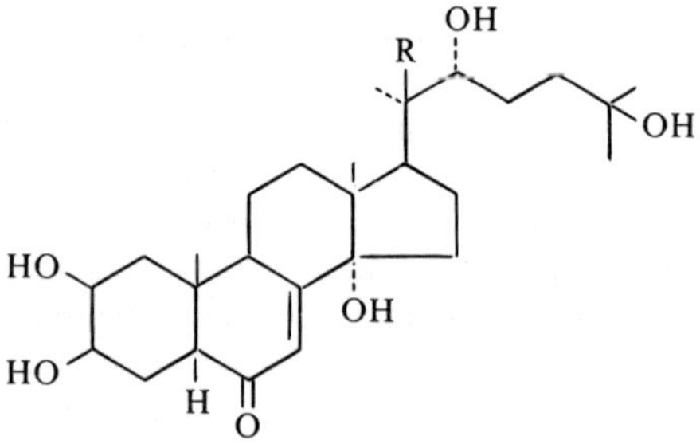

(129)

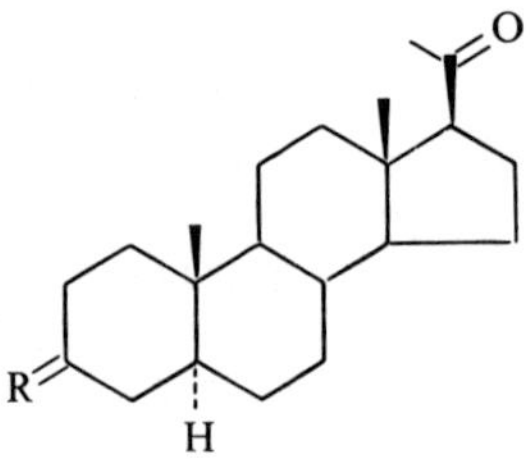

(130) R = O
(131) R = β-OH, α-H

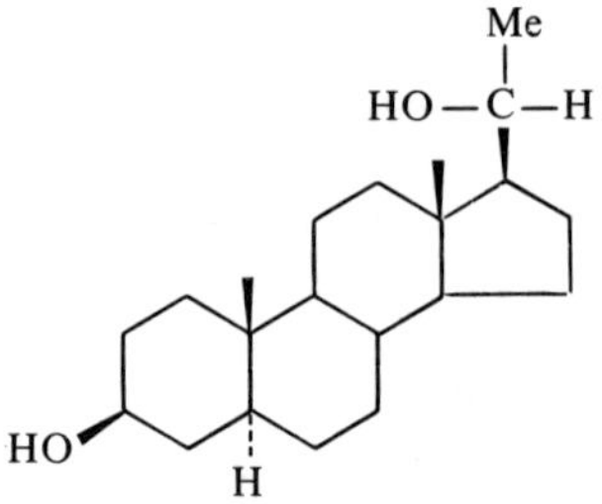

(132)

but the function of such enzymic activity in a plant is uncertain at present. The presence of 17β-hydroxy-steroid dehydrogenase activity has also been demonstrated in such cultures.[238]

Further studies have been carried out on the role of oestrogen-like compounds in plants.[239] The hypothesis that ergosterol peroxide is an artifact arising by attack of 'singlet oxygen' on ergosterol has been further substantiated.[240] A few aspects of the biosynthesis of the cucurbitanes, a class of oxygenated tetracyclic triterpenes, have been reviewed.[241]

---

[237] S. J. Stohs and M. M. El-Olemy, *Phytochemistry*, 1972, **11**, 1397.
[238] S. J. Stohs and M. M. El-Olemy, *Lloydia*, 1972, **35**, 81.
[239] J. Kopcewicz, *Z. Pflanzenphysiol.*, 1972, **67**, 373; *New Phytol.*, 1972, **71**, 129; J. Kopcewicz and A. Chromiński, *Experientia*, 1972, **28**, 603.
[240] J. Arditti, R. Ernst, M. H. Fisch, and B. H. Flick, *J.C.S. Chem. Comm.*, 1972, 1217.
[241] D. Lavie and E. Glotter, *Fortschr. Chem. org. Naturstoffe*, 1971, **29**, 307.
R. D. Goodfellow and G. C. K. Liu, *J. Insect Physiol.*, 1972, **18**, 95.

7 Triterpenes and Steroids in Invertebrates

Insects.—Terpenoid metabolism in insects will be considered separately from other invertebrates. Although insects do not possess the complete machinery required for biosynthesis of sterols from small molecules, it has been reported that a *Sarcophaga bullata* homogenate converts squalene into a compound with the chromatographic properties of squalene 2,3-oxide.[242] However, this report warrants substantiation. The major insect juvenile hormone (133; R = Me) becomes labelled from [^{14}C]acetate and also from L-[^{3}H or

(133)

^{14}C]methionine, which exclusively labels the ester methyl group.[243] The free acid (133; R = H) also serves as precursor,[244] but no incorporation of mevalonic acid or farnesol has been observed so far. Metabolism of juvenile hormone appears to occur by carboxylesterase-catalysed cleavage of the ester methyl group, followed by hydration of the epoxide ring.[245]

Since insects do not synthesize sterols, they are dependent upon a dietary supply of these essential compounds. Phytophagous insects transform the dietary C_{28}- and C_{29}-sterols into C_{27}-sterols (usually cholesterol) before they are used for various functions, including structural components of membranes and moulting-hormone precursors. This dealkylation process has been reviewed,[246] and it appears that fucosterol (135) is an intermediate in the conversion of sitosterol (134) into cholesterol. Scheme 10 has been proposed as a main dealkylation pathway in *Bombyx mori*, based on the efficient incorporation of [^{3}H]fucosterol 24,28-epoxide (136) into cholesterol (138) and also trapping of (136) as a probable intermediate in the conversion of fucosterol (135) into (138).[247] It is interesting that treatment of fucosterol 24,28-epoxide (136) with BF_3-etherate yielded desmosterol (137), possibly *via* a mechanism involving hydrogen migration from C-25 to C 24.[248] An analogous hydrogen migration occurs during dealkylation of 28-isofucosterol (139) to cholesterol

---

[243] M. Metzler, K. H. Dahm, D. Meyer, and H. Röller, *Z. Naturforsch.*, 1971, **26b**, 1270.
[244] M. Metzler, D. Meyer, K. H. Dahm, and H. Röller, *Z. Naturforsch.*, 1972, **27b**, 321.
[245] D. Whitmore, jun., E. Whitmore, and L. I. Gilbert, *Proc. Nat. Acad. Sci. U.S.A.*, 1972, **69**, 1592; J. B. Siddall, R. J. Anderson, and C. A. Henrick, 23rd International Congress of Pure and Applied Chemistry, 1971, vol. 3, p. 17.
[246] M. J. Thompson, S. A. Svoboda, J. N. Kaplanis, and W. E. Robbins, *Proc. Roy. Soc.*, 1972, **B180**, 203.
[247] M. Morisaki, H. Ohtaka, M. Okubayashi, N. Ikekawa, Y. Horie, and S. Nakasone, *J.C.S. Chem. Comm.*, 1972, 1275.
[248] N. Ikekawa, M. Morisaki, H. Ohtaka, and Y. Chiyoda, *Chem. Comm.*, 1971, 1498.

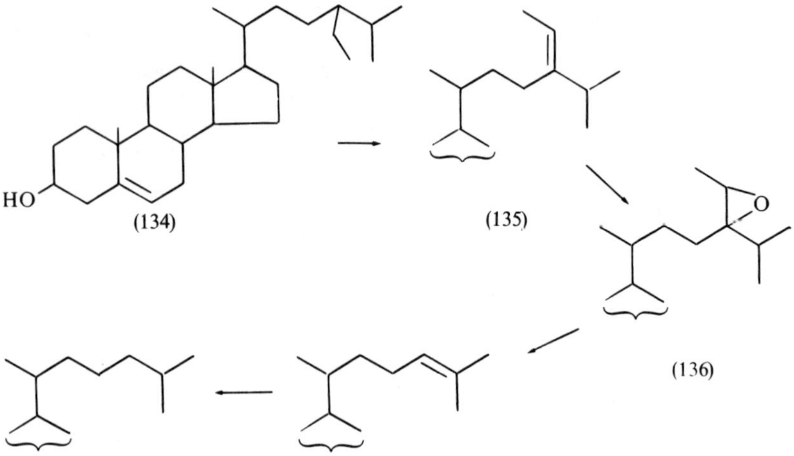

(134) (135) (136) (137) (138)

Scheme 10

(138) in the insect *Tenebrio molitor*, since the C-25 hydrogen of (139) is retained in (138), probably at C-24.[249] A possible alternative mechanism for this process, not involving the obligatory intermediacy of an epoxide, was considered. Evidence has also been furnished for the intermediacy of 24-methylenecholesterol (140) in the conversion of campesterol (141) into cholesterol (138) by the

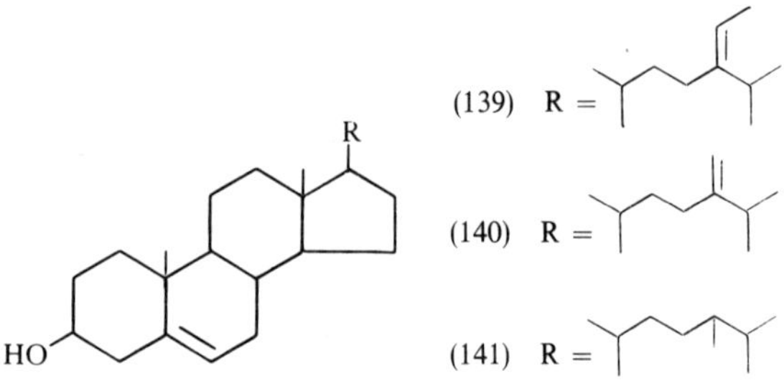

(139) R =

(140) R =

(141) R =

tobacco hornworm.[250] A homogenate of the insect *Corcyra cephalonica* catalyses conversion of ergosterol into a C_{27}-sterol.[251] Dealkylation in insects is

[249] P. J. Randall, J. G. Lloyd-Jones, I. F. Cook, H. H. Rees, and T. W. Goodwin, *J.C.S. Chem. Comm.*, 1972, 1296.
[250] J. A. Svoboda, M. J. Thompson, and W. E. Robbins, *Lipids*, 1972, **7**, 156.
[251] M. Jayaram and T. S. Raman, *Biochem. J.*, 1972, **128**, 46P.

Biosynthesis of Triterpenes, Steroids, and Carotenoids

inhibited by hypocholesterolemic agents[252] and various aza-sterols, the latter also possibly disrupting moulting-hormone biosynthesis.[253]

The biosynthetic pathway of ecdysones from cholesterol is incompletely worked out. Further evidence has been obtained suggesting that the prothoracic glands themselves do not produce ecdysterone (142).[1, 254-256] Such glands from *Manduca sexta* prepupae could not convert ecdysone (143) into ecdysterone (142).[254] Studies on the metabolism of 22,25-bisdeoxy[4-^{14}C]ecdysone (144) in *Manduca* indicated that conversion into ecdysone (143) and ecdysterone (142) was at most a minor pathway.[257] Significant differences were obtained in the metabolites from insects at different stages of development Neither 25-hydroxycholesterol nor (22R)-22-hydroxycholesterol are precursors of ecdysterone (142) in *Calliphora stygia*,[258] thus suggesting that some

	R^1	R^2	R^3	R^4
(142)	OH	OH	OH	β-OH, α-H
(143)	H	OH	OH	β-OH, α-H
(144)	H	H	H	β-OH, α-H
(145)	H	OH	OH	O

elaboration of the cholesterol nucleus occurs at an early stage. An earlier report of the occurrence of an ecdysone-like steroid in a snail has now been refuted.[259]

Further metabolism of ecdysones in insects can occur by conversion into both less and more polar metabolites as well as into glycoside and sulphate conjugates.[1, 257, 260] Ecdysone (143) is metabolized by a soluble enzyme from blowfly pupae into 2β,14α,22,25-tetrahydroxy-5β-cholest-7-ene-3,6-dione (145; 3-dehydroecdysone).[261] A non-specific enzyme system from larval gut tissues

---

[252] H. Hikino, Y. Ohizumi, T. Saito, E. Nakamura, and T. Takemoto, *Chem. and Pharm. Bull. (Japan)*, 1972, **20**, 851.
[253] J. A. Svoboda, M. J. Thompson, and W. E. Robbins, *Lipids*, 1972, **7**, 553.
[254] D. S. King, *Amer. Zoologist*, 1972, **12**, 343.
[255] P. E. Ellis, E. D. Morgan, and A. P. Woodbridge, *Nature*, 1972, **238**, 274.
[256] K. Nakanishi, H. Moriyama, T. Okauchi, S. Fujioka, and M. Koreeda, *Science*, 1972, **176**, 51.
[257] J. N. Kaplanis, M. J. Thompson, S. R. Dutky, W. E. Robbins, and E. L. Lindquist, *Steroids*, 1972, **20**, 105.
[258] M. N. Galbraith, D. H. S. Horn, E. J. Middleton, and J. A. Thomson, *J.C.S. Chem. Comm.*, 1972, 264.
[259] C-J. Bayne, *Parasitology*, 1972, **64**, 501.
[260] T. A. Gorell, L. I. Gilbert, and J. Tash, *Insect Biochem.*, 1972, **2**, 94.
[261] P. Karlson, H. Bugany, H. Döpp, and G.-A. Hoyer, *Z. physiol. Chem.*, 1972, **353**, 1610.

of an insect catalyses sulphation of several steroids, including cholesterol, sitosterol, and ecdysone.[262] Both ecdysone (143) and ecdysterone (142) have been identified in the frass of *Manduca sexta* larvae.[263]

Invertebrates other than Insects.—*Sterol Biosynthesis.* Some invertebrates can synthesize sterols whereas others cannot.[1] The position is still confusing, especially since some reports warrant re-investigation in view of the doubtful nature of the techniques employed. This short section attempts to bring the picture up to date by inclusion of recent *reported* results. Another Coelenterate, *Calliactis parasitica*, could not synthesize sterols from acetate.[264] In agreement with previous results,[1] another five members of the Gastropoda incorporated acetate into sterols.[265, 266] It is significant that biosynthesis of sterols in two of these animals (the neogastropods *Purpura lapillus* and *Murex brandaris*) proceeded at a much slower rate from acetate than from mevalonic acid,[266] and may possibly reflect a regulatory role of HMG-CoA reductase. Further support for sterol synthesis in Echinoidea has been furnished.[267]

Metabolism of Sterols. It is interesting that a heat-stable sterol carrier protein (SCP) has been detected in the protozoan *Tetrahymena pyriformis*.[268] This protozoan-SCP (P-SCP) was required, in addition to oxygen and pyridine nucleotides, for conversion of cholesterol into cholesta-5,7,22-trien-3β-ol by the protozoan microsomal enzymes (Δ^7- and Δ^{22}-dehydrogenase). It is interresting that both protozoan-SCP and liver-SCP are interchangeable in cholesterol biosynthesis by liver enzymes and the oxidation of cholesterol to the triene by protozoan enzymes. The effect of numerous hypocholesteraemic compounds on the cyclization of squalene to the pentacyclic triterpenoid tetrahymanol has also been studied in *Tetrahymena*.[269]

In common with many other invertebrates,[1] *Artemia salina* can dealkylate plant sterols, as demonstrated by the conversion of brassicasterol into cholesterol.[270] The transformation of cholest-5-en-3β-ol into 5α-cholestan-3β-ol by two starfish occurs *via* 3-oxo-steroids.[271] This is analogous to the saturation of the sterol Δ^5-bond in mammals.

The presence of numerous *in vitro* steroid-hormone-metabolizing enzyme activities has been demonstrated in invertebrate tissues.[272] However, such *in vitro* studies may not always reflect the importance of these enzymic activities

---

[262] R. S. H. Yang and C. F. Wilkinson, *Biochem. J.*, 1972, **130**, 487.
[263] J. N. Kaplanis, M. J. Thompson, W. E. Robbins, and E. L. Lindquist, *Steroids*, 1972, **20**, 621.
[264] J. P. Ferezou, M. Devys, and M. Barbier, *Experientia*, 1972, **28**, 407.
[265] P. A. Voogt and D. J. Van der Horst, *Arch. Internat. Physiol. Biochim.*, 1972, **80**, 293; D. J. Van der Horst, *Comp. Biochem. Physiol.*, 1972, **42B**, 1.
[266] P. A. Voogt, *Comp. Biochem. Physiol.*, 1972, **41B**, 831.
[267] P. A. Voogt, *Comp. Biochem. Physiol.*, 1972, **43B**, 457.
[268] T. Calimbas, *Fed. Proc.*, 1972, **31**, 430.
[269] J. D. Sipe and C. E. Holmlund, *Biochim. Biophys. Acta*, 1972, **280**, 145.
[270] S. Teshima and A. Kanazawa, *Bull. Jap. Soc. Sci. Fish.*, 1972, **38**, 1305.
[271] A. G. Smith, R. Goodfellow, and L. J. Goad, *Biochem. J.*, 1972, **128**, 1371.
[272] M. H. Briggs, *Biochim. Biophys. Acta*, 1972, **280**, 481; O. Bardon, P. Lubet, and M. A. Drosdowsky, *Steroidologia*, 1971, **2**, 366; M.-F. Blanchet, R. Ozon, and J.-J. Meusy, *Comp. Biochem. Physiol.*, 1972, **41B**, 251.

Biosynthesis of Triterpenes, Steroids, and Carotenoids 65

under normal *in vivo* physiological conditions, and may in some cases merely reflect a certain degree of low substrate specificity. It is interesting that cholesterol sulphate has been identified in a sea star, *Asterias rubens*.[273]

8 Carotenoids

Prephytoene.—The chemical synthesis of prephytoene pyrophosphate, the C_{40} analogue of presqualene pyrophosphate, has been reported in varying detail from two laboratories,[274,275] and it has been synthesized enzymically from geranylgeranyl pyrophosphate by a purified squalene synthase and by a preparation from tomato plastids.[276] Until recently it had been accepted that phytoene (146) was the first C_{40} hydrocarbon precursor of carotenoids and that lycopersene (15,15'-dihydrophytoene) was not involved[277] (see also the previous volume[1] in this series), and it has been reported that phytoene is formed from prephytoene pyrophosphate in an enzyme system from a *Mycobacterium* sp.[274] However, both squalene synthase and the tomato plastid system convert geranylgeranyl pyrophosphate into lycopersene in the presence of NADPH. In the absence of NADPH, prephytoene pyrophosphate accumulates[276] (the authors involved consider that prelycopersene is a more appropriate name than prephytoene). Previously, work from the same laboratory had shown that geranylgeranyl pyrophosphate was effectively converted by the plastid system into phytoene in the absence of NADPH.[277] The activity of purified squalene synthase on prelycopersene pyrophosphate suggests that the substrate specificity of this enzyme is not great and that the enzyme preponderates over the true phytoene synthase in the tomato plastids. However, in these latest experiments not only is geranylgeranyl pyrophosphate converted into lycopersene but this compound itself is effectively converted into phytoene and lycopene. A bacterial system, which would contain little, if any, squalene synthase, might help to resolve some of the questions raised by this work by simplifying the experimental system.

The Stereochemistry of Phytoene.—Naturally occurring phytoene has been reported as the 15-*cis*-isomer of (146);[278,279] this raises the interesting question as to where in the pathway from phytoene to lycopene does the central double bond revert to the *trans*-configuration. The recent demonstration that phytoene from *Flavobacterium dehydrogenans*[280-282] and *Mucor hiemalis*[283] was

---
[273] L. J. Björkmann, K.-A. Karlsson, and K. Nilsson, *Comp. Biochem. Physiol.*, 1972, **43B**, 409.
[274] L. J. Altman, L. Ash, R. C. Kowerski, W. W. Epstein, B. R. Larsen, H. C. Rilling, F. Muscio, and D. E. Gregonis, *J. Amer. Chem. Soc.*, 1972, **94**, 3257.
[275] L. Crombie, D. A. R. Findlay, and D. A. Whiting, *J.C.S. Chem. Comm.*, 1972, 1045.
[276] A. A. Qureshi, F. J. Barnes, and J. W. Porter, *J. Biol. Chem.*, 1972, **247**, 6730.
[277] J. W. Porter, *Pure Appl. Chem.*, 1969, **20**, 449.
[278] F. B. Jungalwala and J. W. Porter, *Arch. Biochem. Biophys.*, 1965, **110**, 291.
[279] J. B. Davis, L. M. Jackman, P. T. Siddons, and B. C. L. Weedon, *J. Chem. Soc. (C)*, 1966, 2154.
[280] O. B. Weeks, in 'Aspects of Terpenoid Chemistry and Biochemistry', ed. T. W. Goodwin, Academic Press, New York, 1971, p. 301.
[281] N. Khatoon, D. E. Loeber, T. P. Toube, and B. C. L. Weedon, *J.C.S. Chem. Comm.*, 1972, 996.
[282] O. B. Weeks, A. G. Andrewes, R. O. Brown, and B. C. L. Weedon, *Nature*, 1969, **224**, 879.
[283] R. Herber, B. Mandinas, and J. Villoutreix, *Compt. rend.*, 1972, **274**, *D*, 327.

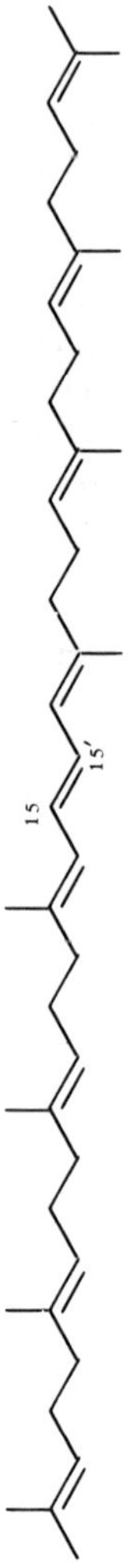

(146)

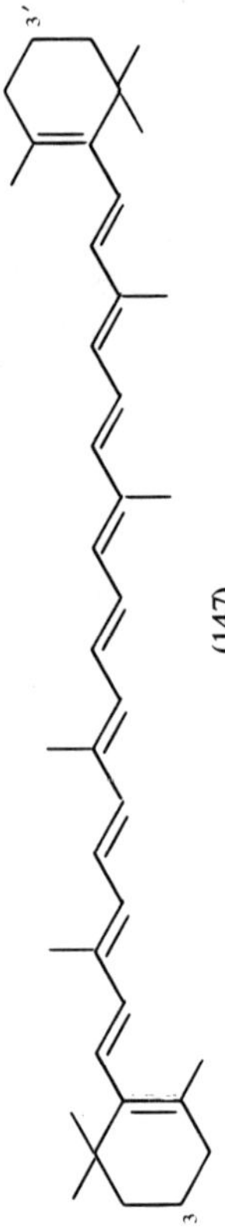

(147)

(148)

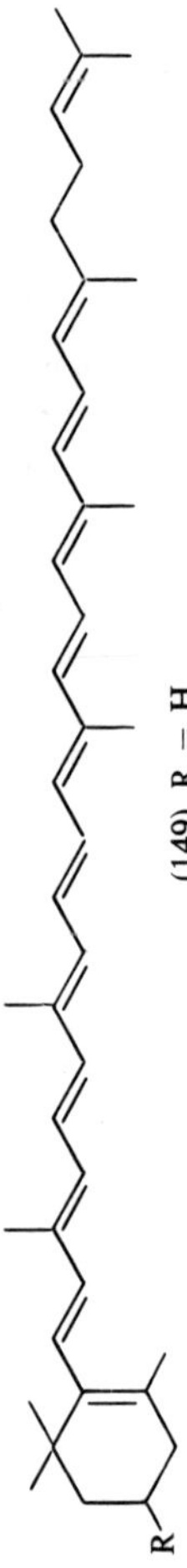

(149) R = H
(150) R = OH

essentially *trans* prompted a re-investigation of phytoene from other sources. Samples from *Neurospora crassa*, *Phycomyces blakesleeanus*, *Rhodospirillum rubrum*, tomato fruit, and carrot root were all predominantly 15-*cis* (85—99.8% of total)[284] as was a specimen from *Verticillium agaricinum* (98%).[285]

Cyclization of Acyclic Carotenes.—Nicotine[286,287] and the herbicide CPTA [2-(*p*-chlorophenylthio)triethylammonium chloride][288-290] inhibit the synthesis of cyclic compounds such as β-carotene (147) and cause accumulation of the acyclic lycopene (148). If cells of a *Mycobacterium* sp. grown in the presence of nicotine are washed free from the inhibitor and re-suspended in normal medium, then lycopene disappears as β-carotene accumulates.[286] The same phenomenon is observed in *Blakeslea trispora*, where there are indications that γ-carotene (149) is an intermediate between lycopene and β-carotene,[291] and in a *Flavobacterium* sp. grown in 7.5 mmol l^{-1} nicotine and re-suspended anaerobically.[287] If, in the last instance, oxygen is admitted, the reaction continues beyond β-carotene, and zeaxanthin (3,3'-dihydroxy-β-carotene) is formed almost exclusively; there is no indication that hydroxylycopenes are involved. Thus, in the *Flavobacterium*, cyclization is an anaerobic process and precedes the aerobic hydroxylation step, which only occurs after cyclization. These results suggested that the pathway was:

lycopene → β-carotene → zeaxanthin.

However, further details of this reaction were revealed by growing this organism in 1 mmol l^{-1} nicotine; in this case no lycopene was detected but the major pigment was rubixanthin (150) (3-hydroxy-γ-carotene),[287] which indicates that at this concentration cyclization is only inhibited at one end of the molecule and that hydroxylation can take place rapidly at the monocyclic level. If rubixanthin-containing cultures are freed from nicotine and re-suspended in normal medium, then rubixanthin (150) is rapidly converted into β-cryptoxanthin (3-hydroxy-β-carotene) (cyclization) under anaerobic conditions and into zeaxanthin (cyclization followed by hydroxylation).[287] From these experiments the overall pathway from lycopene to zeaxanthin in *Flavobacterium* sp. would appear to be:

lycopene → γ-carotene → rubixanthin → β-cryptoxanthin → zeaxanthin

The accumulation of lycopene in place of cyclic carotenoids has also been observed in a number of fruits, including citrus fruits, in roots of higher plants, and in fungi.[290,291]

---

[284] A. Than, P. M. Bramley, B. H. Davies, and A. F. Rees, *Phytochemistry*, 1972, **11**, 3187.
[285] L. R. G. Valadon, R. Herber, J. Villoutreix, and B. Mandinas, *Phytochemistry*, 1973, **12**, 161.
[286] C. D. Howes and P. P. Batra, *Biochim. Biophys. Acta*, 1970, **222**, 174.
[287] J. C. B. McDermott, A. Ben-Aziz, R. K. Singh, G. Britton, and T. W. Goodwin, *Pure Appl. Chem.*, in the press.
[288] C. W. Coggins, jun., G. L. Henning, and H. Yokoyama, *Science*, 1970, **168**, 1589.
[289] H. Yokoyama, C. W. Coggins, jun., and G. L. Henning, *Phytochemistry*, 1971, **10**, 1831.
[290] H. Yokoyama, C. W. Coggins, jun., G. L. Henning, and C. De. Benedict, *Phytochemistry*, 1972, **11**, 1721.
[291] W. J. Hsu, H. Yokoyama, and C. W. Coggins, jun., *Phytochemistry*, 1972, **11**, 2985.

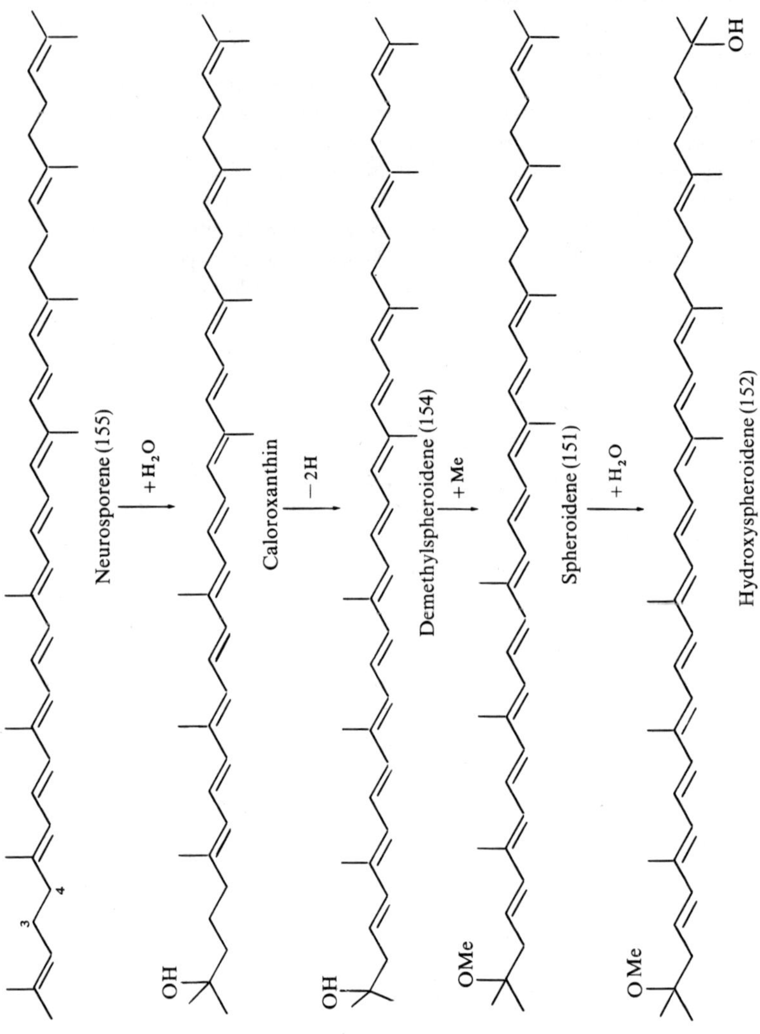

Scheme 11 *Postulated pathway for biosynthesis of spheroidene and hydroxyspheroidene in* **Rps**. *spheroides and* **Rps**. *gelatinosa*

Spheroidenone (153)

Synthesis in Photosynthetic Bacteria.—A characteristic of the carotenoids of purple photosynthetic bacteria is the hydration of the terminal 1 and 1′ double bonds.[292] This is frequently followed by methylation of the hydroxyls at C-1 and C-1′. Rarely does cyclization occur, but β-carotene is formed in *Rhodomicrobium vanneilii* alongside the main pigment 1-hydroxy-1,2-dihydrolycopene (rhodopin).[293–296] In the presence of 1 mmol l^{-1} nicotine, synthesis of both pigments is inhibited and that of lycopene stimulated by a concomitant amount;[297] so nicotine not only inhibits cyclization, which involves the terminal double bond, but also the hydration of this bond.[297] In *Rhodopseudomonas spheroides* grown under anaerobic conditions the main carotenoids are spheroidene (151) and 1-hydroxy-1,2-dihydrospheroidene (hydroxyspheroidene) (152), and it has been suggested that they are formed from neurosporene by the route shown in Scheme 11.

In the presence of nicotine (7.5 mmol^{-1}), *Rps. spheroides* accumulates neurosporene in place of the two main pigments.[290, 297] On re-suspending cells which had been washed free from nicotine, neurosporene disappeared and the normal pigments were synthesized.[297] Experiments with *Rps. gelatinosa* showed that the same pathway existed in this organism, and emphasized the precursor–product relationship between spheroidene (151) and spheroidenone (153).[297] These observations, combined with the characterization of the previously undetected demethylspheroidene (154)[298] from cultures of both species, provide strong direct support for the biosynthetic pathway just indicated. No 3,4-didehydro-derivative of neurosporene (155) was detected in nicotine-inhibited cells, which suggests that the insertion of the double bond at C-3 does not normally occur before the hydration of the double bond at C-1.

---

[292] T. W. Goodwin, in 'Carotenoids', ed. O. Isler, Burkhauser, Basel, 1971.
[293] W. A. Volk and D. Pennington, *J. Bacteriol.*, 1950, **59**, 169.
[294] S. F. Conti and C. R. Benedict, *J. Bacteriol.*, 1962, **83**, 929.
[295] L. Ryvarden and S. Liaaen-Jensen, *Acta Chem. Scand.*, 1964, **18**, 643.
[296] A. Ben-Aziz, G. Britton, and T. W. Goodwin, unpublished results,.
[297] R. K. Singh, A. Ben-Aziz, G. Britton, and T. W. Goodwin, *Biochem. J.*, 1973, **132**, 649.
[298] A. Ben-Aziz, H. C. Malhotra, R. K. Singh, G. Britton, and T. W. Goodwin, unpublished results.

3
Non-protein Amino-acids, Cyanogenic Glycosides, and Glucosinolates

BY A. KJAER AND P. OLESEN LARSEN

1 Introduction

This chapter is devoted to three groups of nitrogenous products, lacking, perhaps, obvious community of structure and distribution. Nonetheless, in a biosynthetic context they may profitably be held together, hopefully bringing to the fore a remarkable limitation in fundamental metabolic reaction types utilized in assembling a multitude of diverse and by simple inspection not obviously related structures.

Protein amino-acids, with their by and large well understood *in vivo* origin, are not part of the chapter. Strong emphasis is placed on products from higher plants, almost to the extent of exclusion of constituents from other natural sources. The delineation from alkaloids is obscure. Hence, occasional overlaps may occur.

The cyanogenic glycosides, naturally not limited to higher plants, are discussed here in the same restricted sense. They share with the glucosinolates a strikingly parallel anabolic pathway, limited, as far as we know, to higher plants, with α-amino-acids as points of departure in both series. Hence, the two groups of glycosides are discussed in adjacent sections.

Generally, the discussion will be focused on knowledge acquired within the past few years. Where desirable, however, lines have been drawn to slightly less recent results, the purpose being to span the gap to the most recent comprehensive reviews in the field.

2 Non-protein Amino-acids

General.—Non-protein amino-acids have continued to attract interest during recent years not only because of their potential significance as metabolic inhibitors, toxins, intermediates in biosynthetic sequences, and taxonomic markers, but also because their interesting structural features present numerous challenges to biosynthetic exploration. Several reviews are available dealing with the structure, distribution, and biosynthesis of free plant amino-acids.

Hence, in the present context, the main emphasis will be placed on work published subsequent to the appearance of the most recent reviews.[1–3]

The trend in this field, as in other biosynthetic areas, is clearly towards investigations on the enzymic level, though traditional incorporation studies on intact plants still play an important role. It is apparent from recent results that the types of enzymic process involved in the biosynthesis of the ca. 200 known free plant amino-acids are rather small in number, most of them representing minor deviations from the biosynthesis of protein amino-acids. The operation of specific enzymes has been established in a number of cases, indicating that non-protein amino-acids are not generally the results of low specificity in the enzymic syntheses of protein amino-acids.

Serine and Cysteine Derivatives.—L-Cysteine (1) is derived from L-serine (2) via O-acetyl-L-serine (3) in bacteria.[4,5] Recent data indicate that the same pathway is utilized for the production of cysteine in plants. Thus, though never encountered in natural material, O-acetylserine is synthesized from serine and acetyl-CoA by enzyme preparations from kidney beans, lupin, barley, pea seedlings, leaves of swiss chard, turnip, and radishes.[6–8] The transformation of O-acetylserine and sulphide into cysteine has been accomplished with enzyme systems from spinach,[9] turnip,[10] and kidney bean.[7,8] Tracer experiments in turnip with [^{35}S]sulphide and [^{14}C]homoserine indicate that the reaction between O-acetylserine and sulphide constitutes the main entrance for sulphur into organic compounds in plants.[11]

Spinach contains two enzymes catalysing the sulphuration of O-acetylserine. One enzyme possesses the additional ability to produce S-methyl-L-cysteine (4) or S-ethyl-L-cysteine from O-acetyl-L-serine and methanethiol or ethanethiol, respectively. The second enzyme also catalyses these processes but, in addition, catalyses the formation of L-methionine, L-ethionine, and L-homocysteine from O-acetyl-L-homoserine and methanethiol, ethanethiol, and sulphide, respectively.[12] Enzymes from turnip and kidney beans are able to catalyse the formation of S-methylcysteine from O-acetylserine and methanethiol;[7,10] ^{14}C-labelled serine, when fed to *Allium cepa* and *A. sativum* along with thiols, is converted into S-substituted cysteines.[13]

---

[1] L. Fowden, in 'Progress in Phytochemistry', ed. L. Reinhold and Y. Liwschitz, Interscience, New York, 1970, vol. 2, p. 203.
[2] L. Fowden, in 'Biosynthetic Pathways in Higher Plants', ed. J. B. Pridham and T. Swain, Academic Press, New York, 1971, p. 73.
[3] J. F. Thompson, C. J. Morris, and I. K. Smith, *Ann. Rev. Biochem.*, 1969, **38**, 137.
[4] N. M. Kredich and G. M. Tomkins, *J. Biol. Chem.*, 1966, **241**, 4955.
[5] N. M. Kredich and M. A. Mecker, in 'Methods in Enzymology', ed. H. Tabor and C. W. Taylor, Academic Press, New York and London, 1971, vol. XVIIB, p. 459.
[6] I. K. Smith and J. F. Thompson, *Biochem. Biophys. Res. Comm.*, 1969, **35**, 939.
[7] I. K. Smith and J. F. Thompson, *Biochim. Biophys. Acta*, 1971, **227**, 288.
[8] I. K. Smith, *Plant Physiol.*, 1972, **50**, 477.
[9] J. Giovanelli and S. H. Mudd, *Biochem. Biophys. Res. Comm.*, 1967, **27**, 150.
[10] J. F. Thompson and D. P. Moore, *Biochem. Biophys. Res. Comm.*, 1968, **31**, 281.
[11] T. T. Ngo and P. D. Shargool, *Biochem. J.*, 1972, **126**, 985.
[12] J. Giovanelli and S. H. Mudd, *Biochem. Biophys. Res. Comm.*, 1968, **31**, 275.
[13] B. Granroth, *Ann. Acad. Sci. Fennicae. Ser. A2*, 1970, no. 154.

O-Acetyl-L-serine presumably represents a branching point in plant biochemistry, with paths leading not only to cysteine or S-methylcysteine by displacement at C-3 with a sulphur atom, but also to several other amino-acids by alternative displacements with a nitrogen or carbon atom. Thus, extracts from *Leucaena leucocephala* (Leguminosae) seedlings catalyse the production of L-mimosine (5), a constituent of this species, from 3,4-dihydroxypyridine (6) and O-acetyl-L-serine. Similarly, extracts from watermelon (*Citrullus vulgaris*) (Cucurbitaceae) seedlings promote the synthesis of L-pyrazolylalanine (7), present in this species, from pyrazole and O-acetyl-L-serine. A number of other amino-acids, including 2,3-diaminopropionic acid (8) and albizziine (9) may result from analogous reactions, although the appropriate enzyme systems have not yet been identified.[14] Willardiine (10) and isowillardiine [3-(2,4-dihydroxypyrimidin-3-yl)alanine] have been shown to derive from uracil and serine, probably via O-acetylserine, in pea seedlings.[14a] A possible example of C-substitution may be L-orcylalanine (11), present in *Agrostemma githago* (Caryophyllaceae), where the methylphenyl part is known to derive from acetate units via orsellinic acid (12), whereas the side-chain arises from serine.[15a] A second pertinent example of C-substitution is the production of β-cyanoalanine (13) from cyanide and O-acetylserine, though cysteine seems to be the preferred substrate for the formation of β-cyanoalanine (*vide infra*). The derivation of fluoroacetic acid by nucleophilic attack of fluoride ions on enzyme-bound dehydroalanine and subsequent oxidative decarboxylation of the keto-acid corresponding to 3-fluoroalanine has been proposed. Fluoroacetic acid occurs in *Acacia georginae* together with albizziine, among other non-protein amino-acids.[15b]

The degradation of O-acetylserine has been studied with extracts from various *Brassica* species (broccoli buds, turnip roots, cabbage leaves, and cauliflower buds) and from *Leucaena leucocephala* seedlings.[16] Observed products were pyruvic acid, ammonia, and acetate formed through dehydroalanine (14) as an intermediate. Mimosine and methanethiol were transformed, by an extract from *L. leucocephala*, into S-methylcysteine and 3,4-dihydroxypyridine, supposedly by an elimination process to give an enzymically bound dehydroalanine residue which ultimately reacts with methanethiol.[17] In fact, this reaction may signify reversibility in the biosynthesis of mimosine described above, suggesting a general intermediary role of dehydroalanine.

A similar explanation may obtain for the degradation of dichrostachinic acid (15) by an enzyme preparation from seedlings of *Leucaena leucocephala*,

---

[14] I. Murakoshi, H. Kuramoto, J. Haginiwa, and L. Fowden, *Phytochemistry*, 1972, **11**, 177.
[14a] T. S. Asworth, E. G. Brown, and F. M. Roberts, *Biochem. J.*, 1972, **129**, 897.
[15a] H. R. Schütte and P. Müller, *Biochem. Physiol. Pflanz.*, 1972, **163**, 518.
[15b] R. J. Mead and W. Segal, *Austral. J. Biol. Sci.*, 1972, **25**, 327.
[16] M. Mazelis and L. Fowden, *Phytochemistry*, 1972, **11**, 619.
[17] I. Murakoshi, H. Kuramoto, J. Haginiwa, and L. Fowden, *Biochem. Biophys. Res. Comm.*, 1970, **41**, 1009.

AcCoA
MeSH
S^{2-}
Oxalyl-CoA
Oxalate + ATP + CoA
Pyruvate + Acetate + NH_3
(1)
(2)
(3)
(4)
(5)
(6)
(8)
(9)
(10)
(14)
(15)
(16)

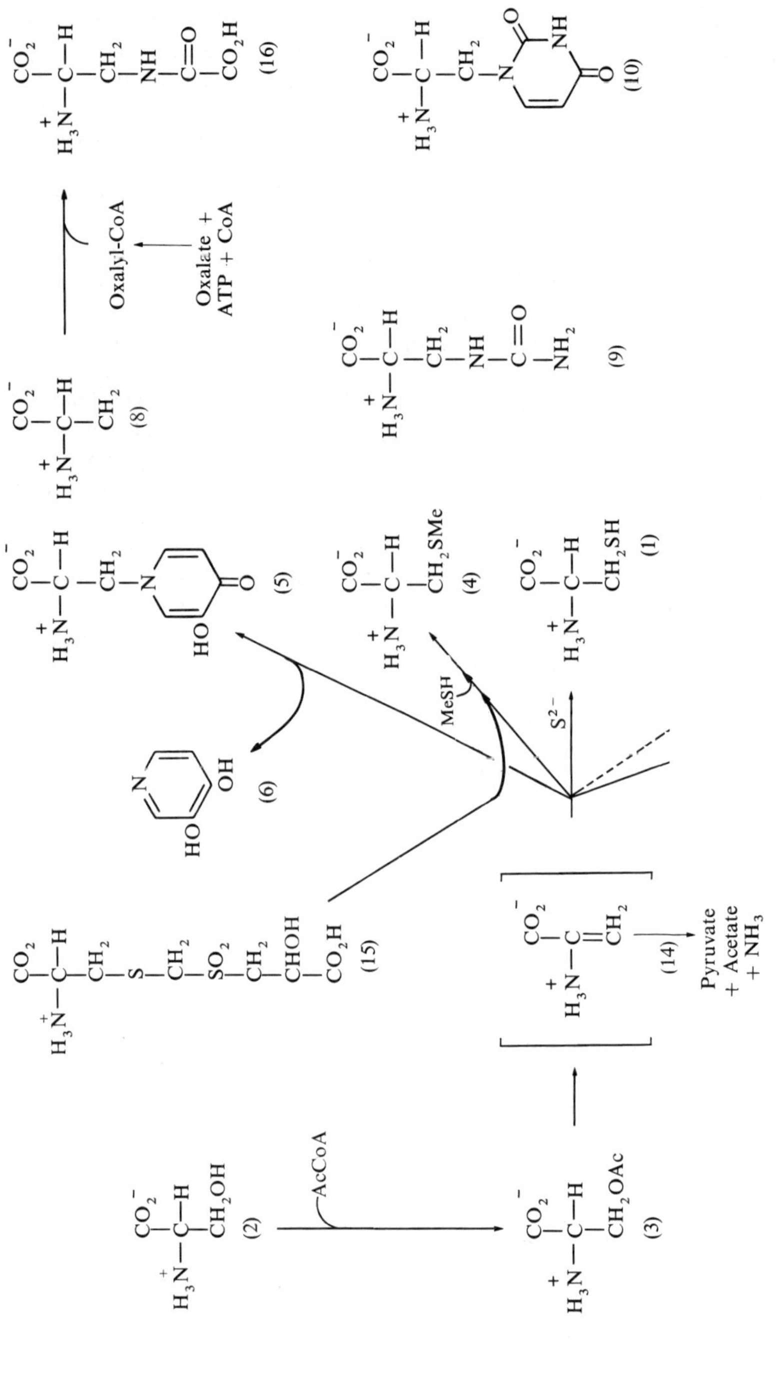

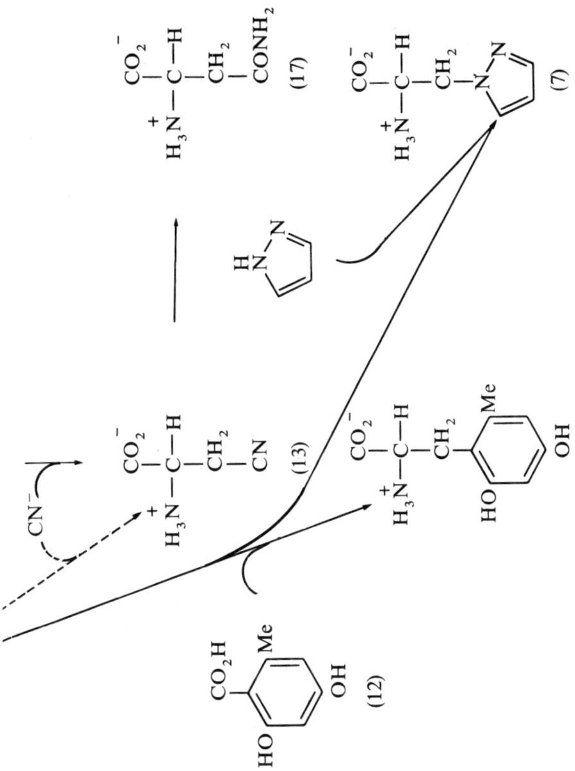

Scheme 1 *Serine and cysteine derivatives*

yielding S-alkylcysteines in the presence of the appropriate thiols (methanethiol, 2-hydroxyethanethiol, carboxymethanethiol).[17a] On combined administration of S-ethyl- or S-allyl-cysteine and [^{14}C]serine to *Allium cepa*, radioactivity is introduced into the cysteine derivative.[13] Reversibility of the reaction between O-acetylserine and thiol may account for this observation.

Though diaminopropionic acid could conceivably arise from serine, ^{14}C-labelled serine was not incorporated into N^3-oxalyldiaminopropionic acid (16) in *Lathyrus sativus*.[18] The biosynthesis of the latter from diaminopropionic acid has been studied with enzyme preparations from *L. sativus*. One enzyme catalyses the production of oxalyl-CoA from oxalate, ATP, and CoA, whereas a second enzyme catalyses the conversion of diaminopropionic acid and oxalyl-CoA into N^3-oxalyldiaminopropionic acid and CoA.[19]

L-β-Cyanoalanine and its γ-glutamyl derivative are well-known constituents of species of Leguminosae.[20] Their biosynthesis takes place from L-cysteine and cyanide, catalysed by β-cyanoalanine synthetase, an enzyme partly purified from seedlings of blue lupin. O-Acetyl-L-serine may serve as an alternative substrate but is utilized at a much lower rate. Methanethiol may take over the role of cyanide, resulting in the production of S-methylcysteine. O-Acetylserine sulphhydrolase has been isolated from the same material but is a distinctly different enzyme, which can also catalyse a slow transformation of O-acetylserine and cyanide into β-cyanoalanine.[21,22] In *Bacillus megaterium*, tracer experiments indicate that cyanide and serine are transformed into β-cyanoalanine,[23] but here one and the same enzyme catalyses the transformation of O-acetylserine and sulphide into cysteine, of O-acetylserine and cyanide into β-cyanoalanine, and of cysteine and cyanide into β-cyanoalanine.[24] β-Cyanoalanine undergoes enzymatic hydrolysis to give asparagine (17) by β-cyanoalanine hydrolase, an enzyme presumably absent from plants accumulating β-cyanoalanine and γ-glutamyl-β-cyanoalanine. The enzyme, which has been partially purified from etiolated lupin seedlings, is highly specific and different from other plant enzymes, including asparaginases.[25]

The assumption that cysteine and not O-acetylserine is the immediate precursor for β-cyanoalanine is supported by tracer experiments with incorporation of [^{14}C]cyanide, [^{14}C]serine, and [^{14}C]cysteine into asparagine

---

[17a] I. Murakoshi, H. Kuramoto, J. Haginiwa, and L. Fowden, *Chem. and Pharm. Bull. (Japan)*, 1971, **19**, 209.
[18] D. N. Roy, *Indian J. Biochem.*, 1969, **6**, 147.
[19] K. Malathi, G. Padmanaban, and P. S. Sarma, *Phytochemistry*, 1970, **9**, 1603.
[20] E. A. Bell, in 'Chemotaxonomy of the Leguminosae', ed. J. B. Harborne, D. Boulter, and B. L. Turner, Academic Press, London and New York, 1971, p. 179.
[21] H. R. Hendrickson and E. E. Conn, *J. Biol. Chem.*, 1969, **244**, 2632.
[22] H. R. Hendrickson and E. E. Conn, in 'Methods in Enzymology' ed. H. Tabor and C. W. Taylor, Academic Press, New York and London, 1971, vol. XVIIB, p. 233.
[23] P. A. Castric and G. A. Strobel, *J. Biol. Chem.*, 1969, **244**, 4089.
[24] P. A. Castric and E. E. Conn, *J. Bacteriol.*, 1971, **108**, 132.
[25] P. A. Castric, K. J. F. Farnden, and E. E. Conn, *Arch. Biochem. Biophys.*, 1972, **152**, 62.

in corn (*Zea mays*) roots,[26] and of [^{14}C]cysteine and [^{14}C]serine into β-cyanoalanine in *Vicia sativa*.[27]

Tracer experiments in etiolated shoots of lupin[28] and investigations of the level of β-cyanoalanine synthetase in seedlings of the same plant[29] indicate, however, that the major part of asparagine in this plant is derived from aspartic acid.

The various transformations of serine derivatives are summarized in Scheme 1.

Apart from the biosynthesis of S-methylcysteine from O-acetylserine and methanethiol discussed above, direct methylation of cysteine appears to be an alternative possibility, since a methyl group is carried from methionine into S-methylcysteine in radish leaves.[30] S-Methylcysteine and S-methylcysteine

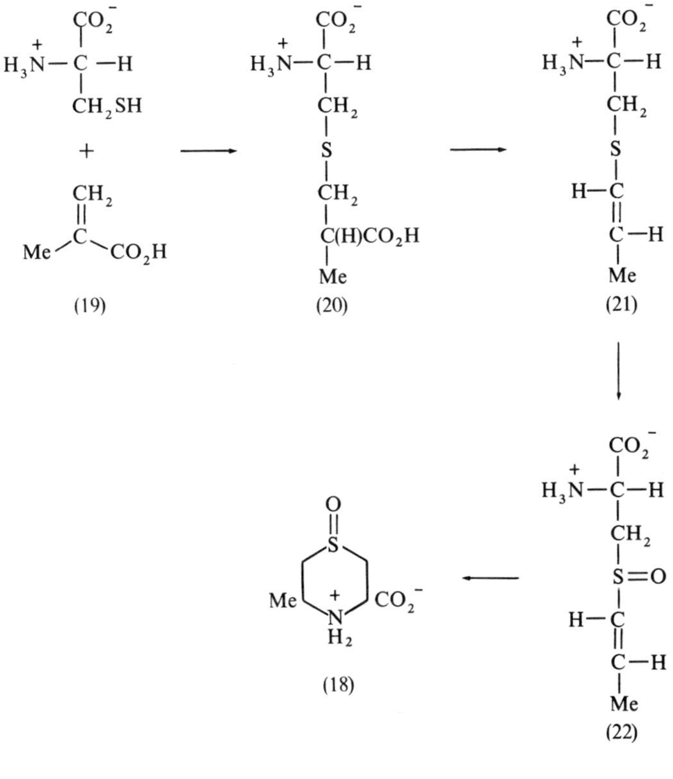

[26] A. Dalis and F. J. Johnson, *Phytochemistry*, 1972, **11**, 3465.
[27] S. N. Nigam and W. B. McConnell, *Canad. J. Biochem.*, 1971, **49**, 799.
[28] M. Lever and G. W. Butler, *J. Exp. Bot.*, 1971, **22**, 279.
[29] M. Lever and G. W. Butler, *J. Exp. Bot.*, 1971, **22**, 285.
[30] J. F. Thompson and R. K. Gering, *Plant Physiol.*, 1966, **41**, 1301.

sulphoxide occur in many higher plants. Experiments demonstrate that, in *Allium cepa* and *A. sativum*, S-methylcysteine arises partly by methyl-group transfer from methionine to cysteine and partly by methylthio-group transfer from methionine to serine. Administration of ethionine to the same plants results in the production of S-ethylcysteine, not normally a natural constituent.[13] The production of the sulphoxide from S-methylcysteine has been established in crucifer species,[31] but the reverse, reduction of the sulphoxide to methylcysteine, has also been recognized as a natural process in turnip and bean leaves.[32]

Stereospecific S-oxidation to the sulphoxide has been observed in *Allium cepa* with S-methylcysteine, S-ethylcysteine, S-propylcysteine, S-*trans*-propenylcysteine, S-*cis*-propenylcysteine, and deoxycycloallinin as the sulphidic precursors.[13]

Tracer experiments in *Brassica pekinensis* have shown that S-methylcysteine sulphoxide can function as a reservoir, from which sulphur can easily be transported into sulphur-containing protein amino-acids.[33] Thus S-methylcysteine sulphoxide, doubly labelled with ^{35}S and with ^{3}H in the C_4-chain, undergoes demethylation and reduction to cysteine.[34] The methyl group in S-methylcysteine and its sulphoxide is utilized for methylating pectic substances in *B. pekinensis*.[33,35] In kidney beans, on the other hand, S-methylcysteine, doubly labelled with ^{14}C and ^{3}H in the methyl group, donates its methyl group to methionine, though not in intact form.[36] It is postulated that degradation starts with production of methanethiol, pyruvic acid, and ammonia.

Enzymes, *e.g.* alliinase, catalysing elimination reactions from S-alkylcysteine sulphoxides are well-known in higher plants, including onions.[37—40] An enzyme which degrades S-methylcysteine sulphoxide to pyruvic acid, ammonia, and methanethiol sulphinate has been studied in crucifers.[41] Again, an enzyme-catalysed degradation of cysteine to give pyruvic acid and thiocysteine has been described from *Brassica napobrassica*, promoting also the above described degradation of S-methylcysteine sulphoxide.[42]

The biosynthesis of S-(*trans*-propen-1-yl)cysteine sulphoxide (22) has been studied in *Allium cepa*. A pathway involving addition of methacrylic acid (19), deriving from valine, to cysteine to give S-(2-carboxypropyl)cysteine (20), decarboxylation to S-(*trans*-propen-1-yl)cysteine (21), and oxidation to (22),

---

[31] W. N. Arnold and J. F. Thompson, *Biochim. Biophys. Acta*, 1962, **57**, 604.
[32] R. C. Doney and J. F. Thompson, *Biochim. Biophys. Acta*, 1966, **124**, 39.
[33] T. Mae, K. Ohira, and A. Fujiwara, *Plant Cell. Physiol.*, 1971, **12**, 1.
[34] T. Mae, K. Ohira, and A. Fujiwara, *Plant Cell. Physiol.*, 1971, **12**, 881.
[35] T. Mae, K. Ohira, and A. Fujiwara, *Plant Cell Physiol.*, 1972, **13**, 407.
[36] R. C. Doney and J. F. Thompson, *Phytochemistry*, 1971, **10**, 1745.
[37] A. I. Virtanen, *Phytochemistry*, 1965, **4**, 207.
[38] M. Mazelis and L. Crews, *Biochem. J.*, 1968, **108**, 725.
[39] J. V. Jacobsen, Y. Yamaguchi, L. K. Mann, F. D. Howard, and R. A. Bernhard, *Phytochemistry*, 1968, **7**, 1099.
[40] S. Schwimmer, *Phytochemistry*, 1968, **7**, 401.
[41] M. Mazelis, *Phytochemistry*, 1963, **2**, 15.
[42] M. Mazelis, N. Beimer, and R. K. Creveling, *Arch. Biochem. Biophys.*, 1967, **120**, 371.

has been proposed. Further transformation, supposedly non-enzymic, to cycloalliin (18) takes place in wilting leaves.[13]

Selenium-containing Amino-acids.—Selenium metabolism in higher plants has recently been reviewed.[43] Experiments in *Astragalus bisulcatus* suggest that Se-methylselenocysteine is formed from serine *via* selenocysteine and subsequent methylation, in parallel to the formation of S-methylcysteine.[44, 45] Experiments with addition of either selenate or sulphate to the growth medium of *A. bisulcatus* indicate that a common enzyme system is involved in the synthesis of both S-methylcysteine and Se-methylselenocysteine and that the syntheses are competing.[46] The methyl group from selenomethionine can be transferred to selenocysteine in the same plant to give Se-methylselenocysteine. Selenocystathionine presumably is an intermediate in the transfer of selenium from selenomethionine to Se-methylselenocysteine.[45] Seeds of *Phaseolus lunatus* contain S-methylcysteine. Plants grown with the addition of selenate, however, produce Se-methylselenocysteine in the ripening seeds at the expense of the amount of S-methylcysteine, obviously in a competitive reaction. In the leaves, where no S-methylcysteine is present, selenium is incorporated into selenomethionine.[47] Both dimethyl selenide and Se-methylselenomethionine are found in *Brassica oleraceae* grown with the addition of selenite or selenate. An enzyme preparation from this plant catalyses the splitting of Se-methylselenomethionine into dimethyl selenide and homoserine.[48]

Homoserine, Homocysteine, Methionine, and Derivatives.—Homoserine is widely distributed in higher plants and occupies a key position as a precursor for threonine, methionine, and isoleucine in micro-organisms and plants. The isolation of L-homoserine and O-acetyl-L-homoserine from pea shoots has been described.[49] The *in vivo* synthesis of homoserine from aspartic acid has been established in several higher plants. Recently, formation of homoserine in pea roots from aspartic acid was shown to proceed *via* extramitochondrial fumarate.[50] Other studies are concerned with the same conversion and its biological significance in pea seedlings[51] and in *Lathyrus sativus*,[52] and still others[53] with the influence of temperature and light on the accumulation of homoserine in pea seedlings.

Sulphuration of O-acetylhomoserine to homocysteine is catalysed by enzymes in spinach.[9, 12] The product is further transformed into S-adenosylhomocysteine and methionine.[9] The main pathway of sulphur to methionine,

---

[43] A. Schrift, *Ann. Rev. Plant Physiol.*, 1969, **20**, 475.
[44] D. M. Chen, S. N. Nigam, and W. B. McConnell, *Canad. J. Biochem.*, 1970, **48**, 1278.
[45] C. M. Chow, S. N. Nigam, and W. B. McConnell, *Biochim. Biophys. Acta*, 1972, **273**, 91.
[46] C. M. Chow, S. N. Nigam, and W. B. McConnell, *Phytochemistry*, 1971, 1971, **10**, 2693.
[47] S. N. Nigam and W. B. McConnell, *Phytochemistry*, 1973, **12**, 359.
[48] B. G. Lewis, C. M. Johnson, and T. C. Broyer, *Biochim. Biophys. Acta*, 1971, **237**, 603.
[49] N. Grobbelaar and F. C. Steward, *Phytochemistry*, 1969, **8**, 553.
[50] D. J. Mitchell and R. G. S. Bidwell, *Canad. J. Bot.*, 1970, **48**, 2001.
[51] D. J. Mitchell and R. G. S. Bidwell, *Canad. J. Bot.*, 1970, **48**, 2037.
[52] J. Przybylska, *Acta Soc. Bot. Polon.*, 1972, **41**, 71.
[53] D. R. Grant and E. Voelkert, *Phytochemistry*, 1970, **9**, 985.

however, supposedly includes cystathionine,[11] the enzymic synthesis of which from O-acetylhomoserine or O-succinylhomoserine has been demonstrated in spinach.[54] In bacteria, cystathionine is formed from O-succinylhomoserine and cysteine,[55,56] whereas in *Saccharomyces cerevisiae*, O-acetylhomoserine seems to be involved.[57,58] Cystathionine has never been recognized as a constituent of higher plants but has recently been identified in the fungus *Boletus erythropus*.[59] The enzyme β-cystathionase, which catalyses the degradation of cystathionine into homocysteine, pyruvic acid, and ammonia, has been identified in broccoli, barley, spinach, and bush beans and has been purified from spinach leaves.[60]

S-Methylmethionine, a common constituent in various higher plants, is formed from S-adenosylmethionine and methionine as demonstrated with cell-free extracts from jack bean roots,[61] and with an enzyme preparation from wheat germ.[62] [^{14}C]methyl-labelled methionine affords ^{14}C-labelled methionine sulphoxide, S-methylmethionine, and serine in seedlings of peas, of *Cucurbita pepo*, of *Brassica caulocarpa*, and of *Sesbania macrocarpa* (Leguminosae).[63] Methionine sulphoxide can be reduced to methionine in *S. macrocarpa*.[63] On the basis of tracer experiments with doubly labelled [^{3}H,^{14}C]-methylmethionine and [^{35}S]methionine in germinating pea seeds it was concluded that methionine is transformed into S-adenosylmethionine, which in its turn is demethylated to S-adenosylhomocysteine, and the latter split into adenine and homocysteine, undergoing remethylation with 5-methyltetrahydrofolate to methionine,[64] a cycle similar to that prevailing in microorganisms. The transfer of label from the main carbon chain in methionine to homoserine, O-acetylhomoserine, and cystathionine has been observed in germinating peas.[64a] Labelled ethionine has been identified in tissues of the apple *Malus pumila*, fed ^{14}C- or ^{35}S-labelled homocysteine, indicating a rather unusual ethylation reaction. Ethylation of cysteine did not take place. Methionine was transformed into S-methylmethionine in the same tissue.[65]

Canavanine (23) is a well-known constituent of Leguminosae.[20] Its degradation product, canaline (24), has also been identified in several leguminous species.[66] Fission of canavanine in *Caragana spinosa* to canaline and urea has been studied both *in vivo* and *in vitro*.[67] Part of the canaline is further degraded

---

[54] J. Giovanelli and S. H. Mudd, *Biochem. Biophys. Res. Comm.*, 1966, **25**, 366.
[55] M. M. Kaplan and M. Flavin, *J. Biol. Chem.*, 1966, **241**, 4463.
[56] B. I. Posner, *Biochim. Biophys. Acta*, 1972, **276**, 277.
[57] H. de Robichon-Szulmaister and H. Cherest, *Biochem. Biophys. Res. Comm.*, 1967, **28**, 256.
[58] M. A. Savin and M. Flavin, *J. Bacteriol.*, 1972, **112**, 299.
[59] J. Jadot, J. Casimir, and G. Maghuin, *Bull. Soc. roy. Sci. Liège*, 1971, **40**, 355.
[60] J. Giovanelli and S. H. Mudd, *Biochim. Biophys. Acta*, 1971, **227**, 654.
[61] R. C. Greene and N. B. Davis, *Biochim. Biophys. Acta*, 1960, **43**, 360.
[62] D. Karr, J. Tweto, and P. Albersheim, *Arch. Biochem. Biophys.*, 1967, **121**, 732.
[63] W. E. Splittstoesser and M. Mazelis, *Phytochemistry*, 1967, **6**, 39.
[64] W. A. Dodd and E. A. Cossins, *Arch. Biochem. Biophys.*, 1969, **133**, 216.
[64a] D. R. Grant and E. Voelkert, *Canad. J. Biochem.*, 1971, **49**, 795.
[65] A. H. Baur and S. F. Yang, *Phytochemistry*, 1972, **11**, 2503.
[66] J. Miersch, *Naturwiss.*, 1967, **54**, 169.
[67] R. Toepfer, J. Miersch, and H. Reinbothe, *Biochem. Physiol. Pflanz.*, 1970, **161**, 231.

to homoserine. An enzyme preparation from seedlings of the jack bean, *Canavalia ensiformis*, also catalyses the hydrolysis of canavanine to canaline and urea.[68] Again, experiments in jack bean cotyledons with canavanine, labelled with ^{14}C in the guanidino carbon atom, demonstrate conversion into canaline and urea, with subsequent urease-catalysed hydrolysis of the urea.[69] In pericarp dishes from ripening seeds of jack bean, the incorporation of $^{14}CO_2$ into the guanidino-group as well as the C_4-chain of canavanine has been

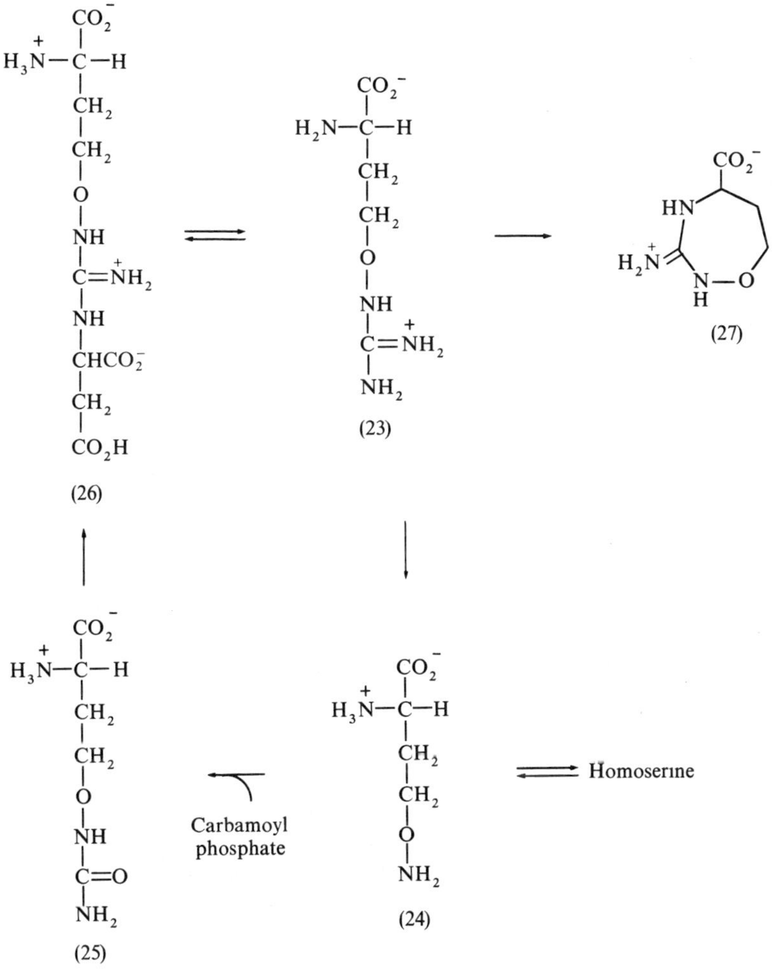

[68] G. A. Rosenthal, *Plant Physiol.*, 1970, **46**, 273.
[69] J. A. Whiteside and D. A. Thurman, *Planta*, 1971, **98**, 279.

demonstrated.[70] Applying tracer techniques to developing leaves of the jack bean and enzyme preparations from the same sources, the formation of O-ureidohomoserine (25) from canaline and carbamoyl phosphate and of canavaninosuccinate (26) from canavanine and fumarate has been established, pointing to the operation of a canavanine cycle parallel to that of arginine; the same enzymes may, in fact, catalyse both cycles.[71] The spontaneous conversion of canavanine into deaminocanavanine (27) is probably a reaction of no biological consequence.[72]

Chain-lengthening of Amino-acids.—Lengthening of amino-acids is a well-established biosynthetic feature in plants, operating in the biosynthesis of leucine from valine, and in the assemblage of several glucosinolates (*vide infra*). Thus, L-homomethionine (2-amino-5-methylmercaptovaleric acid), present in cabbage, is derived from methionine.[73] Again, chain-lengthening might conceivably be involved in the biosynthesis of several C_7 amino-acids from leucine or isoleucine in species of Sapindaceae, although recent results suggest another pathway (*vide infra*). The occurrence of 2-aminoisoheptanoic acid, 2-aminoiso-octanoic acid, and 2-aminoisononanoic acid in hydrolysates of antibiotics occurring in a *Streptomyces* species may reflect the operation of a chain-lengthening process in micro-organisms.[74]

Branched C_6 and C_7 Amino-acids.—4-Methylglutamic acid, 4-methyleneglutamic acid, and 4-hydroxy-4-methylglutamic acid occur in various, systematically non-related groups of higher plants. Since they often co-occur, a common biosynthetic origin has long been suspected. Formerly, it was assumed that 4-hydroxy-4-methylglutamic acid derives from the condensation of two molecules of pyruvic acid, followed by transamination of the resulting 4-hydroxy-4-methyl-2-oxoglutaric acid. However, the observed incorporation of labelled pyruvate into amino-acid may be explained by an exchange reaction and hardly reflects a true precursor role.[1] A renewed study of the biosynthesis of 4-substituted glutamic acids has been performed using shoots of *Gleditsia triacanthos* (Leguminosae).[75] No radioactivity is introduced from [*methyl*-^{14}C] methionine or from [^{14}C]glutamic acid, excluding the possibility of addition of a C_1 unit to glutamic acid. Incorporation of [1-^{14}C]pyruvic acid into (2*S*,4*S*)-4-hydroxy-4-methylglutamic acid is interpreted as resulting from the above-mentioned exchange reaction. [1-^{14}C]Leucine (28) is incorporated into (2*S*,4*R*)-4-methylglutamic acid (29) with all radioactivity residing in C-1, suggesting a direct route. Leucine is also incorporated into (2*S*,4*S*)-4-hydroxy-4-methylglutamic acid (30) obviously *via* the methylglutamic acid, since (2*S*,4*R*)-4-methylglutamic acid is converted into (2*S*,4*S*)-4-hydroxy-4-methylglutamic acid by direct hydroxylation, as apparent from double-labelling

[70] R. P. Warren and G. E. Hunt, *Planta*, 1971, **100**, 258.
[71] G. A. Rosenthal, *Plant Physiol.*, 1972, **50**, 328.
[72] G. A. Rosenthal, *Phytochemistry*, 1972, **11**, 2827.
[73] Y. Suketa, M. Sugii, and T. Suzuki, *Chem. and Pharm. Bull. (Japan)*, 1970, **18**, 249.
[74] J. Shoji and R. Sakazaki, *J. Antibiot.*, 1970, **23**, 519.
[75] P. J. Peterson and L. Fowden, *Phytochemistry*, 1972, **11**, 663.

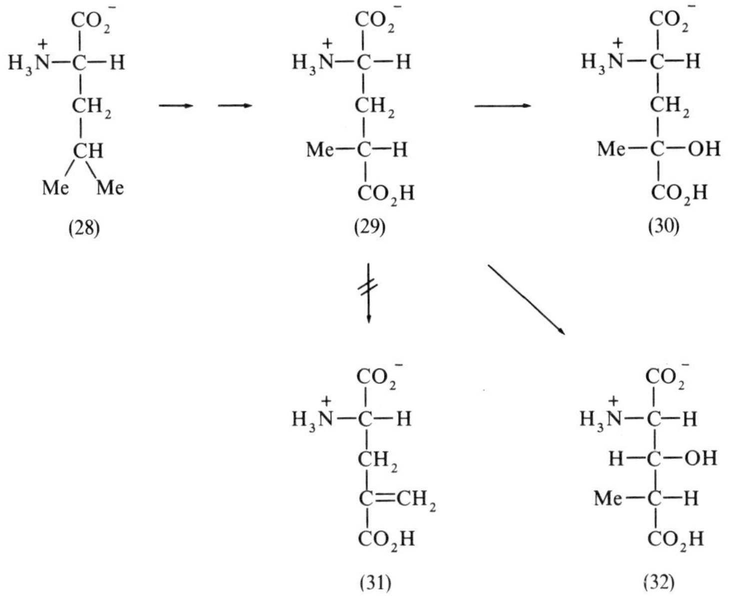

Structures

(28): H$_3$N$^+$—C(H)(CO$_2^-$)—CH$_2$—CH(Me)$_2$

(29): H$_3$N$^+$—C(H)(CO$_2^-$)—CH$_2$—C(Me)(H)—CO$_2$H

(30): H$_3$N$^+$—C(H)(CO$_2^-$)—CH$_2$—C(Me)(OH)—CO$_2$H

(31): H$_3$N$^+$—C(H)(CO$_2^-$)—CH$_2$—C(=CH$_2$)—CO$_2$H

(32): H$_3$N$^+$—C(H)(CO$_2^-$)—C(H)(OH)—C(Me)(H)—CO$_2$H

(28) → → (29) → (30); (30) ↛ (31); (30) → (32)

experiments. The reverse process, reduction of 4-hydroxy-4-methylglutamic acid, does not occur. The 4-methylglutamic acid is not present in the seedlings but, when added, is transformed into 4-hydroxy-4-methylglutamic acid. No radioactivity was observed in either 4-methyleneglutamic acid (31) or 4-methyleneglutamine in any of the experiments, a result suggesting the operation of a separate route to the methylene derivatives. Radioactivity from (2S,4R)-4-methylglutamic acid is also incorporated into (2S,3S,4R)-3-hydroxy-4-methylglutamic acid (32), suggesting that this amino-acid also is formed by hydroxylation of methylglutamic acid. Both (2S,3S,4R)- and (2S,3R,4R)-3-hydroxy-4-methylglutamic acid have been isolated from *Gymnocladus dioica* (Leguminosae).[76,77]

4-Methyleneglutamic acid and 4-amino-2-methylenebutyric acid occur in tulips.[78,79] The lactone of 4-hydroxy-2-methylenebutyric acid, isolated from tulip bulbs,[80] supposedly derives from one of these amino-acids. Incorporation of labelled acetate and pyruvic acid into the lactone has been taken as evidence of a condensation reaction,[81] although the results are also in keeping with the operation of the above-mentioned exchange reaction.

[76] G. A. Dardenne, J. Casimir, E. A. Bell, and J. R. Nulu, *Phytochemistry*, 1972, **11**, 787.
[77] G. A. Dardenne, E. A. Bell, J. R. Nulu, and C. Cone, *Phytochemistry*, 1972, **11**, 791.
[78] R. M. Zacharius, J. K. Pollard, and F. C. Steward, *J. Amer. Chem. Soc.*, 1954, **76**, 1961.
[79] L. Fowden, *J. Exp. Bot.*, 1954, **5**, 28.
[80] B. H. H. Bergman, J. C. M. Beijersbergen, J. C. Overeem, and A. K. Sijpensteijn, *Rec. Trav. chim.*, 1967, **86**, 709.
[81] C. R. Hutchinson and E. Leete, *Chem. Comm.*, 1970, 1189.

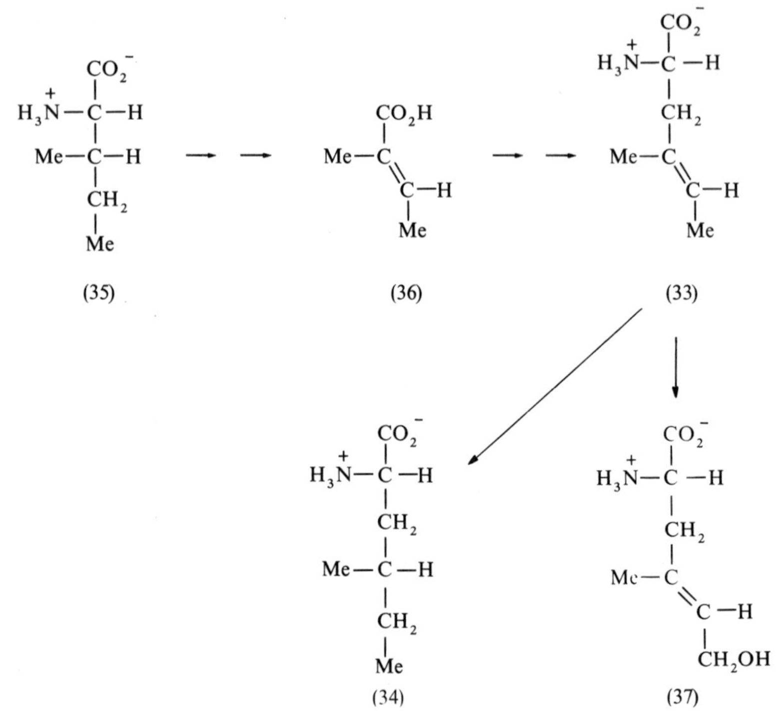

$$\underset{(35)}{\overset{+}{H_3N}-\overset{CO_2^-}{\underset{|}{C}}-H \atop Me-\overset{|}{\underset{|}{C}}-H \atop \underset{|}{CH_2} \atop Me} \longrightarrow \longrightarrow \underset{(36)}{Me-\overset{CO_2H}{\underset{\|}{C}}\diagdown_{C-H}^{} \atop Me} \longrightarrow \longrightarrow \underset{(33)}{\overset{+}{H_3N}-\overset{CO_2^-}{\underset{|}{C}}-H \atop \underset{|}{CH_2} \atop Me-\overset{|}{\underset{\|}{C}}\diagdown_{C-H}^{} \atop Me}$$

(35) → → (36) → → (33)

(34)   (37)

where (34) is the saturated homoisoleucine structure and (37) has a CH$_2$OH terminus.

The biosynthesis of 2-amino-4-methylhex-4-enoic acid (33) and homoisoleucine (34) has been studied in *Aesculus californica* (Sapindaceae).[82] Whereas radioactivity is not incorporated from [*methyl*-^{14}C]methionine, [U-^{14}C]-leucine, or [1-^{14}C]isoleucine, [U-^{14}C]isoleucine (35) serves as a good precursor, indicating that chain-elongation from isoleucine takes place, with loss of the carbonyl group and addition of an acetate group, either by the normal pathway mentioned above or *via* tiglic acid (36). The latter possibility seems the more attractive inasmuch as the specific activity of homoisoleucine was identical to that of 2-amino-4-methylhex-4-enoic acid. This conclusion is supported by subsequent experiments performed with the same plant, demonstrating that tiglic acid is a good precursor for both amino-acids and that radioactivity can be transferred from 2-amino-4-methylhex-4-enoic acid to homoisoleucine whereas the reverse process is not observed. 2-Amino-6-hydroxy-4-methylhex-4-enoic acid (37), present in the same plant, seems to be derived from 2-amino-4-methylhex-4-enoic acid (33). A study of the transaminases involved in the proposed transformations suggest that a single

---

[82] L. Fowden and M. Mazelis, *Phytochemistry*, 1971, **10**, 359.

enzyme catalyses the transamination of leucine, isoleucine, and 2-amino-4-methylhex-4-enoic acid.[83]

The biosynthesis of N,3,4-trimethylpentanoic acid in a *Streptomyces* species has been shown to proceed from leucine with transfer of two methyl groups from methionine.[84,85]

Numerous additional branched C_6 and C_7 amino-acids have been isolated within recent years, reflecting great variability in the metabolism of valine, leucine, and isoleucine in different plant groups. Thus, 2-amino-4-methylhex-5-ynoic acid and 2-amino-4-hydroxymethylhex-5-ynoic acid have been isolated from *Euphoria longan*,[86] 2-amino-6-hydroxy-5-methylhex-4-enoic acid and α-(2-carboxymethylcyclopropyl)glycine from *Blighia unijugata*,[87] L-2-amino-4-methylhex-5-enoic acid from a *Boletus*[88] and a *Streptomyces* species,[89] L-2-amino-5-methylhex-4-enoic acid from *Leucocortinarius bulbiger*,[90] L-2-amino-3-methylenepentanoic acid from *Lactarius helveticus*,[91] L-2-amino-3-methylenehexanoic acid from *Amanita vaginata var. fulva*,[92] and, finally, L-2-amino-3-hydroxymethylpent-3-enoic acid and L-2-amino-3-formylpent-3-enoic acid from *Bankera fuligineoalba*.[93]

A number of unbranched C_6 and C_7 amino-acids have also been encountered, though no information is available on their biosynthesis. The occurrence of 2-amino-4-hydroxyhept-6-ynoic acid in *Euphoria longan*,[86] along with the branched-chain C_7 amino-acids (*vide supra*) may be indicative of a rearrangement reaction. 2-Aminohex-4-ynoic acid has been isolated from *Tricholomopsis rutilans*[94] and 2-aminohexa-4,5-dienoic acid from *Amanita solitaria*,[95] where it occurs together with 2-amino-5-chlorohex-4-enoic acid.[96]

Imino-acids.—The transformation of DL-[6-^{3}H,6-^{14}C]lysine into pipecolic acid with retention of tritium has been demonstrated in rats, in *Neurospora crassa*, in kidney beans, and in *Sedum acre*, and into the piperidine alkaloid sedamine with retention of tritium in *S. acre*.[97] Correspondingly, [2-^{3}H,6-^{14}C]lysine is incorporated into pipecolic acid in kidney beans and *S. acre* with loss of tritium, indicating that the transformation takes place *via* 6-amino-2-ketohexanoic acid and Δ^1-piperideine-2-carboxylic acid.[98] In *S. acre*, however,

---

[83] J. E. Boyle and L. Fowden, *Phytochemistry*, 1971, **10**, 2671.
[84] J. E. Walker and D. Perlman, *Biotechnol. and Bioeng.*, 1971, **13**, 371.
[85] J. E. Walker, M. Bodanszky, and D. Perlman, *J. Antibiot.*, 1970, **23**, 255.
[86] M.-L. Sung, L. Fowden, D. S. Millington, and R. C. Sheppard, *Phytochemistry*, 1969, **8**, 1227.
[87] L. Fowden, C. M. MacGibbon, F. A. Mellon, and R. C. Sheppard, *Phytochemistry*, 1972, **11**, 1105.
[88] R. Rudzats, E. Gellert, and B. Halpern, *Biochem. Biophys. Res. Comm.*, 1972, **47**, 290.
[89] R. B. Kelly, D. G. Martin, and L. J. Hanka, *Canad. J. Chem.*, 1969, **47**, 2504.
[90] G. Dardenne and J. Casimir, *Phytochemistry*, 1968, **7**, 1401.
[91] B. Levenberg, *J. Biol. Chem.*, 1968, **243**, 6009.
[92] R. Vervier and J. Casimir, *Phytochemistry*, 1970, **9**, 2059.
[93] R. R. Doyle and B. Levenberg, *Biochemistry*, 1968, **7**, 2457.
[94] S. Hatanaka, Y. Niimura, and K. Taniguchi, *Phytochemistry*, 1972, **11**, 3327.
[95] W. S. Chilton, G. Tsou, and L. Kirk, *Tetrahedron Letters*, 1968, 6283.
[96] W. S. Chilton and G. Tsou, *Phytochemistry*, 1972, **11**, 2953.
[97] R. N. Gupta and I. D. Spenser, *J. Biol. Chem.*, 1969, **244**, 88.
[98] R. N. Gupta and I. D. Spenser, *Phytochemistry*, 1970, **9**, 2329.

DL-[2-^{3}H,6-^{14}C]lysine was incorporated into sedamine without loss of tritium, indicating an alternative pathway in plants from lysine to sedamine and other alkaloids. In this connection it is of interest that D-lysine is transformed into pipecolic acid whereas L-lysine is converted into the alkaloid anabasin in *Nicotiana glauca*.[99] Analogously, in *Pseudomonas putida*, L-lysine is transformed into 5-aminopentanoic acid but D-lysine into pipecolic acid.[100] The latter transformation has also been observed in corn and ryegrass seedlings.[101]

Extracts of crown galls from *Nicotiana tabacum*, *Parthenocissus tricuspidata*, and *Scorzonera hispanica* catalyse the formation of lysopine [N^2-(1-carboxyethyl)lysine] from lysine, pyruvic acid, and NADH, and of octopinic acid [N^2-(1-carboxyethyl)ornithine] from ornithine, pyruvic acid, and NADH. Extracts from normal tissues of the same plants are unable to catalyse the reactions.[102] [^{14}C]Lysine and diamino[^{14}C]pimelic acid are incorporated into lysopine and into pipecolic acid in crown gall tissue,[103] and it appears likely that lysine in crown gall tissue is not catabolized in the normal way, through pipecolic acid or α-aminoadipic acid, but rather transferred into lysopine.

The biosynthesis of lycomarasmine (38) has been studied in *Fusarium oxysporium f. lycopersici*. Aspartic acid is a good precursor for the C_4 unit, and

$$\begin{array}{c} NH_2 \\ | \\ C=O \\ | \quad + \\ H_2C-NH_2-CH \\ | \\ CO_2^- \end{array} \quad \begin{array}{c} \quad \quad CO_2^- \\ \quad \quad + \quad | \\ CH_2-NH_2-CH \\ | \\ CH_2 \\ | \\ CO_2H \end{array}$$

(38)

glycine and serine for the C_2 unit, whereas glycine, serine, and alanine all serve as precursors for the C_3 unit.[104]

DL- and DL-*allo*-5-hydroxy[6-^{14}C]lysine is transformed into *cis*- and *trans*-5-hydroxypipecolic acid and into N^6-acetyl-5-hydroxylysine in *Gleditsia triacanthos*.[105] 5-Hydroxypipecolic acid is not further metabolized and 90% of the radioactivity was retained in 5-hydroxypipecolic acid, indicating that all four diastereomers of the added hydroxylysine were utilized.

2,4-Diaminobutyric acid is a better precursor for azetidine-2-carboxylic

---

[99] T. T. Gilbertson, *Phytochemistry*, 1972, **11**, 1737.
[100] Y.-F. Chang and E. Adams, *Biochem. Biophys. Res. Comm.*, 1971, **45**, 570.
[101] R. Aldag and J. L. Young, *Planta*, 1970, **95**, 187.
[102] B. Lejeune, *Compt. rend.*, 1967, **265**, D, 1753.
[103] B. Lejeune and M. F. Jubier, *Compt. rend.*, 1968, **266**, D, 1189.
[104] C. R. Poppelstone and A. M. Unrau, *Phytochemistry*, 1971, **10**, 2723.
[105] J. F. Thompson and C. J. Morris, *Arch. Biochem. Biophys.*, 1968, **125**, 362.

acid in *Delonix regia* than methionine, aspartic acid, or homoserine, supporting the view that azetidine-2-carboxylic acid is not produced *via* S-adenosylmethionine as previously proposed.[106] Proline is incorporated into 3-hydroxyproline in the same plants, suggesting a direct hydroxylation.[106]

Aromatic Amino-acids.—Hydroxylation of 3-(3-carboxyphenyl)alanine (39) to give 3-(3-carboxy-4-hydroxyphenyl)alanine (40) has been demonstrated in *Reseda lutea* and *R. odorata*.[107] The two amino-acids have been previously shown to be shikimic acid-derived.[108] On the basis of experiments in *R. lutea* with stereospecifically labelled shikimic acids, it has been proposed that 3-(3-carboxyphenyl)alanine is formed *via* chorismic acid (41), isochorismic acid (42), isoprephenic acid (43), and 3-(3-carboxyphenyl)pyruvic acid (44).[109] Isochorismic acid has previously been identified in *Aerobacter aerogenes*, and chemical transformation into the keto-acid (44) has been observed.[110] The *pro-6S*-hydrogen atom in shikimic acid is retained and the *pro-6R*-hydrogen lost in the transformation of shikimic acid into 3-(3-carboxyphenyl)alanine, phenylalanine, and tyrosine in *R. lutea*,[109] *i.e.* the steric course of the elimina-

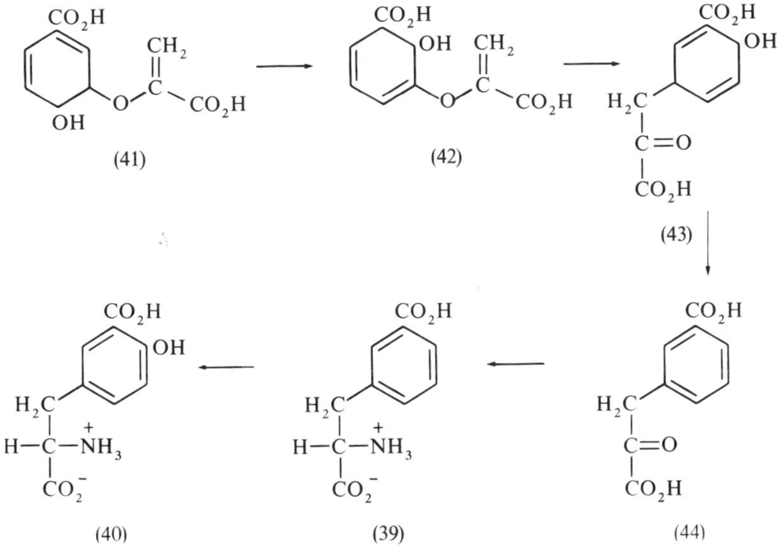

---

[106] M. Sung and L. Fowden, *Phytochemistry*, 1971, **10**, 1523.
[107] P. O. Larsen and H. Sørensen, *Biochim. Biophys. Acta*, 1968, **156**, 190.
[108] P. O. Larsen, *Biochim. Biophys. Acta*, 1967, **141**, 27.
[109] P. O. Larsen, D. K. Onderka, and H. G. Floss, *J.C.S. Chem. Comm.*, 1972, 842.
[110] I. G. Young, T. J. Batterham, and F. Gibson, *Biochim. Biophys. Acta*, 1969, **177**, 389.

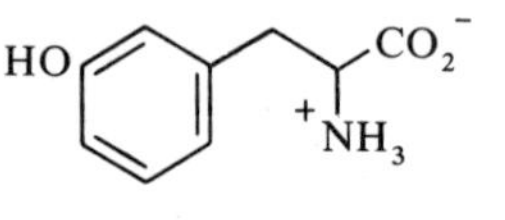

(45) 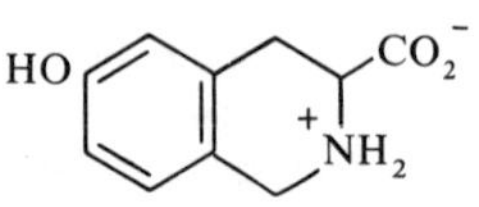 (46) 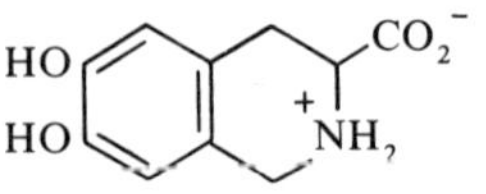 (47) 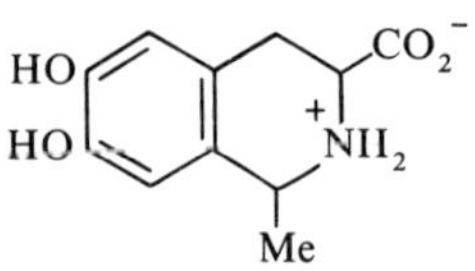 (48) 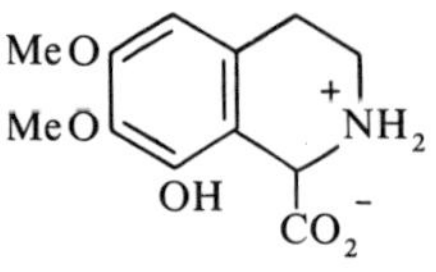 (49)

tion reaction leading to chorismic acid is the same in higher plants and in micro-organisms.[111−113]

m-Tyrosine (45), present in the latex of *Euphorbia myrsinitis*, is known to derive from shikimic acid, yet supposedly not with phenylalanine as an intermediate.[114] If administered to shoots of *E. myrsinitis*, ^{14}C-labelled m-tyrosine is mostly excreted into the latex. A smaller part, however, is transformed into 1,2,3,4-tetrahydro-6-hydroxy-1-methylisoquinoline-3-carboxylic acid (46),[115] found in labelled form both in the latex and in the plant tissues.[116] When m-tyrosine is introduced directly into either fresh or boiled latex, the same conversion into the isoquinoline derivative takes place, presumably through a non-enzymic Mannich condensation with formaldehyde or its equivalent.[116] Most likely, a similar condensation between formaldehyde and 3′,4′-dihydroxyphenylalanine is responsible for the formation of 1,2,3,4-tetrahydro-6,7-dihydroxyisoquinoline-3-carboxylic acid (47), isolated from *Mucuna mutsiana* (Leguminosae) together with large quantities of 3′,4′-dihydroxyphenylalanine.[117] A third example of this new group of tetrahydroisoquinoline acids is 1,2,3,4-tetrahydro-6,7-dihydroxy-1-methylisoquinoline-3-carboxylic acid (48), isolated from velvet beans.[118]

Another type of naturally occurring tetrahydroisoquinoline acid, represented by 1,2,3,4-tetrahydro-8-hydroxy-6,7-dimethoxyisoquinoline-1-carboxylic acid (49) and its 1-methyl derivative, both constituents of peyote cactus, is produced by condensation of 3′-hydroxy-4′,5′-dimethoxyphenylethylamine with glyoxalic acid or pyruvic acid, with subsequent decarboxylation to give the alkaloids

[111] D. K. Onderka and J. G. Floss, *J. Amer. Chem. Soc.*, 1969, **91**, 5894.
[112] H. G. Floss, D. K. Onderka, and M. Caroll, *J. Biol. Chem.*, 1972, **247**, 736.
[113] R. K. Hill and G. R. Newkome, *J. Amer. Chem. Soc.*, 1969, **91**, 5893.
[114] P. Müller and H. R. Schütte, *Flora (A)*, 1967, **158**, 421.
[115] P. Müller and H. R. Schütte, *Z. Naturforsch.*, 1968, **23b**, 491.
[116] P. Müller and H. R. Schütte, *Biochem. Physiol. Pflanz.*, 1971, **162**, 234.
[117] E. A. Bell, J. R. Nulu, and C. Cone, *Phytochemistry*, 1971, **10**, 2191.
[118] M. E. Daxenbichler, R. Kleiman, D. Weisleder, C. H. VanEtten, and K. D. Carlson, *Tetrahedron Letters*, 1972, 1801.

anhalamine or anhalonidine. Supposedly, other structurally related amino-acids are intermediates in similar alkaloid biosyntheses.[119]

m-Hydroxyphenylglycine and 3′,5′-dihydroxyphenylglycine are present in the latex of *Euphorbia helioscopia*.[120] The monophenol can be produced from m-tyrosine and from phenylalanine in shoots of the plant,[116] thus providing a new example of chain-shortening in the amino-acid series. Non-observed incorporation into the diphenol is supposedly caused by technical difficulties in the feeding experiments. No incorporation into the glycine derivatives was observed when labelled m-tyrosine was introduced directly into the latex. A similar chain-shortening has previously been observed in the transformation of 3-(3-carboxyphenyl)alanine into (3-carboxyphenyl)glycine in *Iris*,[121] and of 3-(3-carboxy-4-hydroxyphenyl)alanine into (3-carboxy-4-hydroxyphenyl)-glycine.[108, 122]

5-Hydroxy-L-tryptophan and 5-hydroxytryptamine occur in *Griffonia simplicifolia* (Leguminosae),[123] from which an enzyme has been partly purified which catalyses the hydroxylation of tryptophan to 5-hydroxytryptophan.[124]

An O-β-D-glucoside of L-mimosine (5), given the trivial name mimoside, has been found in seeds and shoots of *Mimosa pudica* and *Leucaena leucocephala*, and has been shown to derive from mimosine in shoots of both species.[125] An enzyme catalysing the formation of the glucoside from mimosine and UDP-glucose has been partly purified from seedlings of *L. leucocephala*. The same enzyme preparation catalyses – although to a minor degree – the formation from tyrosine and UDP-glucose of the O-β-D-glucoside of L-tyrosine,[126] found, together with a mono-O-β-D-glucoside of dopamine, in seeds of *Entada scandens*.[127] Conversely, extracts of seedlings from *L. leucocephala* catalyse the hydrolysis of the mimosine glucoside.[126]

D-Amino-acids.—In recent years it has become obvious that, in higher plants, various D-amino-acids are metabolized by routes differing from those of the L-amino-acids. The deviating metabolism of D-lysine has been discussed above. The principal pathway for transformation of D-amino-acids seems to be malonylation. This was first demonstrated for D-tryptophan, undergoing acylation to N^{α}-malonyl-D-tryptophan, a compound widely occurring in

---

[119] G. J. Kapadia, G. S. Rao, E. Leete, M. B. E. Fayez, Y. N. Vaishnav, and H. M. Fales, *J. Amer. Chem. Soc.*, 1970, **92**, 6943.
[120] P. Müller and H. Schütte, *Z. Naturforsch*, 1968, **23b**, 659.
[121] C. J. Morris and J. F. Thompson, *Arch. Biochem. Biophys.*, 1965, **110**, 506.
[122] C. J. Morris and J. F. Thompson, *Arch. Biochem. Biophys.*, 1967, **119**, 269.
[123] E. A. Bell and L. E. Fellows, *Nature*, 1966, **210**, 529.
[124] L. E. Fellows and E. A. Bell, *Phytochemistry*, 1970, **9**, 2389.
[125] I. Murakoshi, S. Ohmiya, and J. Haginiwa, *Chem. and Pharm. Bull. (Japan)*, 1971, **19**, 2655.
[126] I. Murakoshi, H. Kuramoto, S. Ohmiya, and J. Haginiwa, *Chem. and Pharm. Bull. (Japan)*, 1972, **20**, 855.
[127] P. O. Larsen, H. Sørensen, and P. Sørup, presented at the 'IV International Symp. Biochem. u. Physiol. d. Alkaloide,' Halle, 1969, Abh. Dtsch. Akad. Wiss. Symposiumsbericht Band b. ed. K. Mothes, K. Schreiber, and H. R. Schütte, Akademieverlag, Berlin, 1972, p. 113.

higher plants.[128–130] Transformation of D-methionine into N-malonyl-D-methionine has been demonstrated in *Nicotiana rustica*[131, 132] and various higher plants,[133] whereas there is no difference in the metabolism of D- and L-methionine in bacteria, algae, liverworts, and lichens. In fungi, lichens, and mosses, both acylation and deamination of D-methionine can occur.[133] In the same biological systems, L- and D-serine behave similarly.[134] Malonylation of D-phenylalanine, D-valine, D-isoleucine, D-tyrosine, D-tryptophan, D-alanine, and D-glutamic acid has been demonstrated in barley, and the malonylation of D-phenylalanine and D-leucine in seven other plant species. It is not known, however, if the malonyl derivatives contain the amino-acids with D-configuration or whether inversion or racemization has taken place.[135] The malonylation of D-phenylalanine to N-malonyl-D-phenylalanine has been demonstrated in *Vicia faba*.[136] Stereomutation of D-tryptophan to L-tryptophan has been established in tissue cultures and cell-free extracts from tobacco. Hence, it seems conceivable that L-tryptophan normally is transformed into D-tryptophan, and thence into N^α-malonyl-D-tryptophan, by a reversion of this process.[137] In wheat, both L- and D-tryptophan are precursors for indoleacetic acid. Malonylation of D-, but not L-tryptophan, takes place.[138] In cabbage, corn, and peas, D-tryptophan is malonylated whereas L-tryptophan yields indoleacetic acid and other indole derivatives, including glucobrassicin.[139] In this connection, the identification in peas of N^α-malonyl-4-chloro-D-tryptophan, esterified at the free malonyl carboxy-group with either methanol or ethanol, is of interest.[140] 4-Chloroindole-3-acetic acid has previously been identified in peas.[141] D-Tyrosine seems in some cases to have special metabolic pathways. Thus, D-tyrosine was a more efficient precursor than L-tyrosine for plastoquinone and related substances in corn.[142] In *Sorghum vulgare*, D-tyrosine is not incorporated into *p*-hydroxymandelonitrile-β-glucoside, but into two non-identified products, not obtained from L-tyrosine.[143] Tracer experiments with stereospecifically labelled D-tyrosine indicate that only one of the products has lost the carboxy-group. None of the products seems to be a malonyl derivative.

[128] M. H. Zenk and H. Scherf, *Biochim. Biophys. Acta*, 1963, **71**, 737.
[129] M. H. Zenk and H. Scherf, *Planta*, 1964, **62**, 350.
[130] M. H. Zenk and J. H. Schmitt, *Biochem. Z.*, 1965, **342**, 54.
[131] D. Keglević, B. Ladešić, and M. Pokorny, *Arch. Biochem. Biophys.*, 1968, **124**, 443.
[132] B. Ladešić, M. Pokorny, and D. Keglević, *Phytochemistry*, 1970, **9**, 2105.
[133] M. Pokorny, E. Marcenko, and D. Keglević, *Phytochemistry*, 1970, **9**, 2175.
[134] B. Ladešić, M. Pokorny, and D. Keglević, *Phytochemistry*, 1971, **10**, 3085.
[135] N. Rosa and A. C. Neish, *Canad. J. Biochem.*, 1968, **46**, 797.
[136] W. Escherich and T. Hartmann, *Planta*, 1969, **85**, 213.
[137] G. A. Miura and S. E. Mills, *Plant Physiol.*, 1971, **47**, 483.
[138] W. K. Kim and R. Rohringer, *Canad. J. Bot.*, 1969, **47**, 1425.
[139] M. Kutáček and V. I. Kefeli, *Biol. Plant.*, 1970, **12**, 145.
[140] S. Marumo and H. Hattori, *Planta*, 1970, **90**, 208.
[141] S. Marumo, H. Hattori, H. Abe, and K. Munakata, *Nature*, 1968, **219**, 959.
[142] G. R. Whistance and D. R. Threlfall, *Biochem. J.*, 1968, **109**, 577.
[143] L. V. Balce and J. E. Gander, *Nature*, 1965, **207**, 759.

The deviating metabolism of D-amino-acids in plants clearly calls for caution in the indiscriminate use of racemic amino-acid precursors in biosynthetic experiments.

3 Cyanogenic Glycosides

General.—The discontinuous distribution in living matter of principles releasing hydrocyanic acid, enzymically or otherwise, was recognized a long time ago. The chemical nature of the progenitors and enzymes involved, as well as their distribution within the plant and animal kingdom, continues to attract interest. Though not unique among natural products in their ability to release hydrocyanic acid, the cyanohydrin glycosides (50) constitute the largest

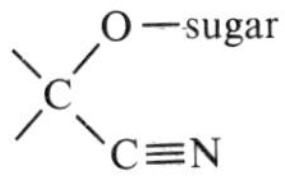

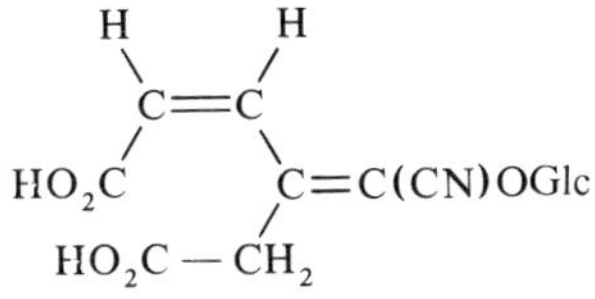

and by far the best known group of natural precursors. Structural variations within about a score of such glycosides known today comprise: the sugar moiety; the substitution pattern at the carbinol carbon atom; or, in three cases, solely the chirality of the latter. Recent reviews[144, 145] on the structure and distribution of such glycosides are available. Chemical contributions, published subsequently, comprise: a corrected structure for triglochinin (51);[146] the finding of a similar glucoside, formulated as (52), in *Thalictrum aquilegifolium*;[147] the description of a new β-glucoside, holocalin, derived from L-*m*-hydroxymandelonitrile, the D-enantiomer of which constitutes the aglucone of the long-known glucoside zierin;[148] establishment of the configuration of lotaustralin (53);[149] and final structure assignment to gynocardin (54)[150] and,

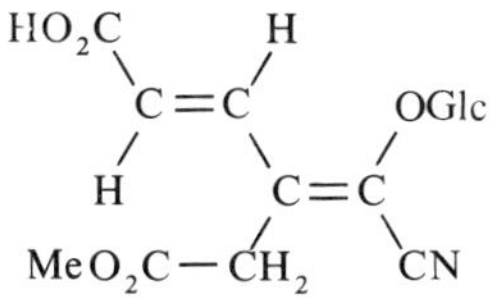

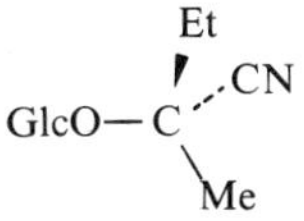

[144] R. Eyjólfsson, *Fortschr. Chem. org. Naturstoffe*, 1970, **28**, 74.
[145] R. Hegnauer, *Pharm. Acta Helv.*, 1971, **46**, 585.
[146] M. Ettlinger and R. Eyjólfsson, *J.C.S. Chem. Comm.*, 1972, 572.
[147] D. Sharples, M. S. Spring, and J. R. Stoker, *Phytochemistry*, 1972, **11**, 3069.
[148] R. Gmelin, M. Schüler, and E. Bordas, *Phytochemistry*, 1973, **12**, 457.
[149] F. H. Bissett, R. C. Clapp, R. A. Coburn, M. G. Ettlinger, and L. Long, jun., *Phytochemistry*, 1969, **8**, 2235.
[150] H. S. Kim, G. A. Jeffrey, D. Panke, R. C. Clapp, R. A. Coburn, and L. Long, jun., *Chem. Comm.*, 1970, 381.

(54)

(55) R = OH
(56) R = H

save for the configuration at the tertiary ring carbon atom, barterin (55)[151, 152] and deidaclin (56),[152] the last two compounds presumably stereoisomeric and identical with tetraphyllin B and A, respectively.[153]

Another uniform group of naturally occurring cyanogenic compounds, the cyanolipids, which are limited in distribution, with one exception,[154] to seed oils of members of the family Sapindaceae, has only recently been structurally clarified. Collectively, they are esters, divisible, according to their alcohol moieties, into four obviously related groups of undecided steric compositions, (57)—(60).[154-158] Striking, and repeatedly commented upon,[157, 158] is the

(57) R^1 = H; R^2 = Acyl[155]
(58) R^1 = Acyl—O; R^2 = Acyl[154, 158]
(59) R^1 = H; R^2 = Acyl[156, 157]
(60) R^1 = Acyl—O; R^2 = Acyl[157, 158]

aptitude for C_{20} acids, predominantly monoenoic, to enter into the ester linkages of cyanolipids, usually to the extent of dominance.

Biosynthesis.—An unabated interest during the past 15 years in the biosynthetic pathways leading to cyanogenic glycosides (50) has provided results which, though far from complete, compare favourably regarding detail with those of other sequences in higher plants. Studies carried out to the early part of 1968 have been comprehensively surveyed by pioneers in the field.[159, 160] In brief, the conclusions reached at that time were as follows. Protein amino-acids (61) (L-valine, L-isoleucine, L-phenylalanine, and L-tyrosine) are effectively converted into cyanogenic glycosides (62; R^1 = R^2 = Me; R^1 = Me, R^2 = Et;

---

[151] M. Paris, A. Bouquet, and R.-R. Paris, *Compt. rend.*, 1969, **268**, D, 2804.
[152] R. C. Clapp, M. G. Ettlinger, and L. Long, jun., *J. Amer. Chem. Soc.*, 1970, **92**, 6378.
[153] G. B. Russell and P. F. Reay, *Phytochemistry*, 1971, **10**, 1373.
[154] D. S. Seigler, K. L. Mikolajczak, C. R. Smith, jun., I. A. Wolff, and R. B. Bates, *Chem. and Phys. Lipids*, 1970, **4**, 147.
[155] D. Seigler, F. Seaman, and T. J. Mabry, *Phytochemistry*, 1971, **10**, 485.
[156] K. L. Mikolajczak, C. R. Smith, jun., and L. W. Tjarks, *Biochim. Biophys. Acta*, 1970, **210**, 306.
[157] K. L. Mikolajczak, C. R. Smith, jun., and L. W. Tjarks, *Lipids*, 1970, **5**, 672.
[158] K. L. Mikolajczak, C. R. Smith, jun., and L. W. Tjarks, *Lipids*, 1970, **5**, 812.
[159] E. E. Conn and G. W. Butler, in 'Perspectives in Phytochemistry', ed. J. B. Harborne and T. Swain, Academic Press, London and New York, 1969, p. 47.
[160] E. E. Conn, *J. Agric. Food Chem.*, 1969, **17**, 519.

$R^1 = H$, $R^2 = Ph$; and $R^1 = H$, $R^2 = p\text{-OHC}_6H_4$, respectively) when fed to appropriate, intact plants. Labelling experiments indicate the C-1—C-2 bond of the precursor α-amino-acids to be the sole linkage severed during the transformations. Even when attempts to isolate intermediates, of necessity nitrogenous, proved virtually futile, an observed *in vivo* conversion of aldoximes (63)

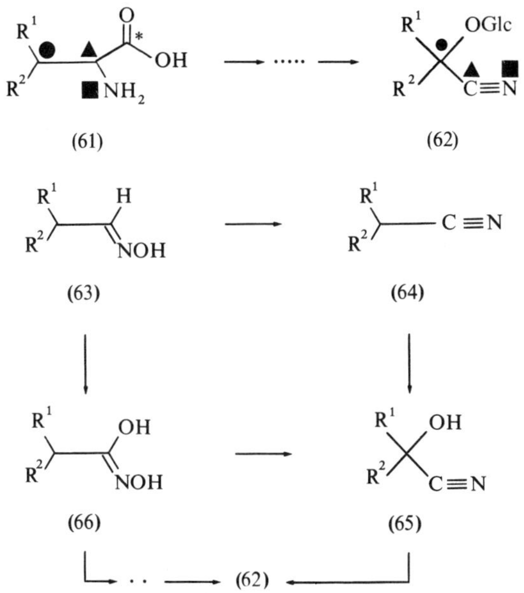

into glucosides was suggestive of a pathway involving dehydration to nitriles (64), oxidation to cyanohydrins (65) and, lastly, glucosylation to give the glucosides (62). However, a partly reversed sequence of these steps is not excluded by experiment. To what extent the established precursors are identical with the true intermediates on the natural pathways has been a central question in more recent studies.

In a full account[161] of preliminarily published results, detailed evidence is presented for the above conclusions, reached through experiments with linen flax seedlings, producing linamarin (62; $R^1 = R^2 = Me$), and cherry laurel or peach shoots, synthesizing prunasin (62; $R^1 = H$, $R^2 = Ph$). A notable incorporation of 2-oximinoisovaleric acid and 2-oximino-3-phenylpropionic acid into linamarin and prunasin, respectively, may conceivably reflect the role of the oximino-acids as sources of nitriles, formed by a long-known, non-enzymic conversion.[162] Experiments aimed at inducing accumulation of intermediates

---

[161] B. A. Tapper and G. W. Butler, *Biochem. J.*, 1971, **124**, 935.
[162] A. Ahmad and I. D. Spenser, *Canad. J. Chem.*, 1961, **39**, 1340.

in flax by administration of structural analogues of valine, or by trapping techniques, provided evidence for *in vivo* synthesis of isobutyraldoxime (63; $R^1 = R^2 = Me$), partly in *O*-glucosylated form. Likewise, isobutyronitrile was induced to accumulate, though to a smaller degree. Taken together, the results support the biosynthetic sequence proposed on the basis of precursor studies,[163] yet without excluding the operation of an additional pathway involving oxidation of the aldoxime (63) to a 2-hydroxyaldoxime (66), with subsequent transformation into the glucoside (62) *via* the cyanohydrin (65), or otherwise. Experiments undertaken to test the possibility in linen flax and cherry laurel afforded results which, though inconclusive, were not at variance with 2-hydroxyaldoximes (66) playing a role as biosynthetic intermediates.[164] In cases such as these, where anabolic intermediates cannot be isolated under normal circumstances, studies at the enzymic level become mandatory. Thus far, an enzyme catalysing the synthesis of linamarin (62; $R^1 = R^2 = Me$) by transfer of glucose from UDP-glucose to acetone cyanohydrin has been isolated from flax seedlings and partly purified. The transferase exhibited a high degree of specificity for the aliphatic side-chains of acetone and butanone cyanohydrins, although it remains undecided whether the same or two different, non-separated, enzymes are involved in the biosynthesis of linamarin (62; $R^1 = R^2 = Me$) and a methyl-linamarin (62; $R^1 = Me, R^2 = Et$).[165]

Assuming a common *in vivo* pathway for all known cyanogenic glycosides in higher plants, the one outlined above, mere structural inspection points to valine, leucine, isoleucine, phenylalanine, and tyrosine as prime parent candidates for the majority of these. Less obvious, however, is the biochemical origin of triglochinin (51) and the closely related natural glycoside (52). The interesting suggestion has been made, and supported by experiment, that both derive from a 3,4-dihydroxymandelonitrile glucoside (67) by oxidative cleavage between the aromatic hydroxy-groups.[146, 166]

Tyrosine,[146, 166] but not phenylalanine or Dopa,[146] serves as the amino-acid precursor in plant-feeding experiments. The possibility of the oxidative ring cleavage being associated with the release of hydrocyanic acid in the catabolism of tyrosine-derived glucosides in higher plants has been discussed.[166]

Even less certain is the biosynthetic derivation of deidaclin (56) and its hydroxylated counterparts, barterin (tetraphyllin B) (55) and gynocardin (54). L-2-Cyclopentene-1σ-glycine (68) has been suggested as a logical biosynthetic precursor, conceivably carried through to the cyclopentenoid glucosides by enzymes appreciating the structural similarity of (68) to valine and isoleucine.[152] Experimental verification appears inviting. Conversion of (68) into gynocardin (54), and probably also the various hydroxylated counterparts, demands

---

[163] B. A. Tapper and G. W. Butler, *Phytochemistry*, 1972, **11**, 1041.
[164] B. A. Tapper, H. Zilg, and E. E. Conn, *Phytochemistry*, 1972, **11**, 1047.
[165] K. Hahlbrock and E. E. Conn, *J. Biol. Chem.*, 1970, **245**, 917.
[166] D. Sharples, M. S. Spring, and J. R. Stoker, *Phytochemistry*, 1972, **11**, 2999.

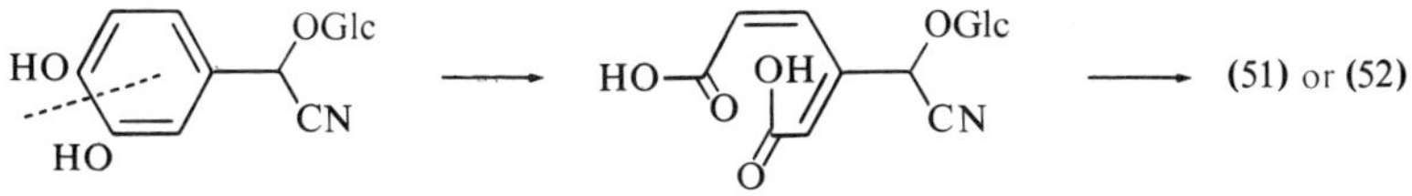

retention of configuration in the oxidation of the methine group, a reasonable steric course,[152] prevailing also in the biological conversion in *Trifolium repens* of L-isoleucine, with (S)-configuration at C-3 (69), into the aglucone of lotaustralin (70).[149]

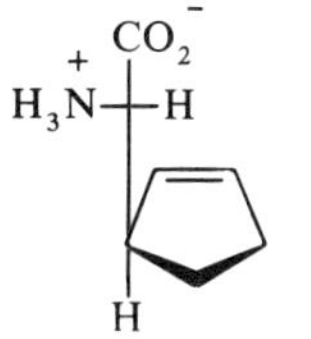

The cyanolipids, formally isoprenoid in their carbinol moieties, are of unknown biogenetic extraction. Assuming anabolic parallelism with the cyanogenic glycosides, then leucine, or, possibly, oxidized leucines, deserve attention as possible precursors.[158]

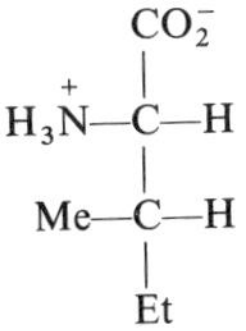

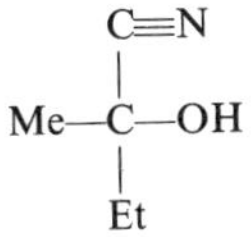

The established role of aldoximes as obligatory intermediates in the biosynthesis of cyanogenic glycosides is curiously duplicated in the *in vivo* synthesis of glucosinolates, another uniform group of constituents of higher plants discussed in the following section.

4 Glucosinolates

General.—Glucosinolates, a collective term for anions possessing the structure (71), occur in Nature, limited though, as far as we know, to certain higher plant families. For knowledge about glucosinolates extant prior to 1968, a comprehensive review published in that year[167] should be consulted. It contains a list of all known, naturally occurring, structurally identified glucosinolates, encompassing 50 compounds, divisible, according to their putative amino-acid derivation, into four major categories. Today, five years later, 21 additional glucosinolates have found their natural place within the same four main groups.

Additions to group (I) ('Aliphatic Glucosinolates containing a *C*-Methyl

---

[167] M. G. Ettlinger and A. Kjær, in 'Recent Advances in Phytochemistry', ed. T. J. Mabry, R. E. Alston, and V. C. Runeckles, Appleton-Century-Crofts, New York, 1968, vol. 1, p. 58.

(71) (72)

Group or Derived from Diacids') comprise {71; R = Bu^n; R = Me·CH(OH)·[CH$_2$]$_2$; and R = HO·[CH$_2$]$_4$},[168] and the long expected (71; R = Bu^i).[169] Group (II) ('ω-Methylthioalkylglucosinolates and Derivatives') has been supplemented with: {71; R = MeS·[CH$_2$]$_2$·CH(OH)·[CH$_2$]$_2$; R = MeS(O)·[CH$_2$]$_2$··CH(OH)·[CH$_2$]$_2$; and R = MeSO$_2$·[CH$_2$]$_2$·CH(OH)·[CH$_2$]$_2$};[170] {71; R = MeS·[CH$_2$]$_3$·CH(OH)·[CH$_2$]$_2$; R = MeS(O)·[CH$_2$]$_3$·CH(OH)·[CH$_2$]$_2$; and R = MeSO$_2$·[CH$_2$]$_3$·CH(OH)·[CH$_2$]$_2$};[171] {71; R = MeS·[CH$_2$]$_7$};[172] {71; R = MeS(O)·[CH$_2$]$_7$};[173] {71; R = MeS·[CH$_2$]$_8$; R = MeS·[CH$_2$]$_5$·CO·[CH$_2$]$_2$; and R = MeS(O)·[CH$_2$]$_5$·CO·[CH$_2$]$_2$};[172] and {71; R = MeS(O)·[CH$_2$]$_{11}$}.[174]

Additions to group (III) ('Arylmethylglucosinolates') comprise: [71; R = 3,4-(HO)$_2$·C$_6$H$_3$·CH$_2$];[175] [71; R = 3,4,5-(MeO)$_3$·C$_6$H$_2$·CH$_2$];[176] and [72; R = SO$_3$H].[177] {A single reported case of the natural occurrence of an arylglucosinolate, [71; R = 3,4-(MeO)$_2$C$_6$H$_3$],[175] uniquely lacking an obvious protein amino-acid progenitor, occasions surprise and deserves confirmation.} Group (IV) ('2-Arylethyglucosinolates and Derivatives') musters two additions: [71; R = (R)-Ph·CH(OH)·CH$_2$],[173] and [71; R = (R)-p-MeO·C$_6$H$_4$·CH(OH)].[172]

Taken together, the recent findings have added confidence to the previously discussed protein amino-acid derivation of all natural glucosinolates.[167] Moreover, the results suggest that future explorers within this area should hardly anticipate real structural caprices but rather be content with unexceptional modifications on biogenetically transparent frameworks.

Biosynthesis.—Experimental work aimed at clarifying the biosynthesis of glucosinolates was initiated about 10 years ago. By 1968, when a detailed review

[168] A. Kjær and A. Schuster, *Phytochemistry*, 1971, **10**, 3155.
[169] E. W. Underhill and D. F. Kirkland, *Phytochemistry*, 1972, **11**, 2085.
[170] A. Kjær and A. Schuster, *Acta Chem. Scand.*, 1970, **24**, 1631.
[171] A. Kjær and A. Schuster, *Phytochemistry*, 1973, **12**, 929.
[172] A. Kjær and A. Schuster, *Acta Chem. Scand.*, 1972, **26**, 8.
[173] R. Gmelin, A. Kjær, and A. Schuster, *Acta Chem. Scand.*, 1970, **24**, 3031.
[174] A. Kjær and A. Schuster, *Phytochemistry*, 1972, **11**, 3045.
[175] R. Danielak and B. Borkowski, *Diss. Pharm. Pharmacol.*, 1970, **22**, 143.
[176] A. Kjær and M. Wagnières, *Phytochemistry*, 1971, **10**, 2195.
[177] M. C. Elliott and B. B. Stowe, *Phytochemistry*, 1970, **9**, 1629.

$R \cdot [CH_2]_n \cdot CH(NH_2)CO_2H \rightarrow \rightarrow R \cdot [CH_2]_n \cdot CH{=}NOH \rightarrow \rightarrow R \cdot [CH_2]_n \cdot C(SGlc){=}NOSO_3^-$

(73)   (74)   (75)

appeared,[167] the following principal contours of a complete picture had emerged: α-amino-acids (73) are converted, through nitrogenous intermediates, into glucosinolates (75), with loss of the carboxy-carbon atom. Aldoximes (74) appear to be true intermediates on the pathway. A chain-elongation mechanism may operate, by which amino-acids (73; $n = i$), via 2-oxo-acids, addition of acetate, and loss of the carboxy-carbon atom of the latter, are converted into the higher homologues (73; $n = i + 1$), and thence into the corresponding glucosinolates (75; $n = i + 1$). A considerable amount of information, collected after 1967 and discussed in the sequel, has helped to sharpen the details of this picture.

Glucosinolates from Amino-acids, with Preservation of Nitrogen. DL-*N*-Hydroxyphenalanine (76), a conceivable though unproven natural intermediate, but not 2-oximino-3-phenylpropionic acid (77), was converted into benzylglucosinolate in *Tropaeolum majus* shoots better than phenylalanine, and, in fact, as efficiently as phenylacetaldoximine (79). Enzymic studies, conducted in parallel, demonstrated the conversion of (76), but not of (77), into the aldoximine (79),

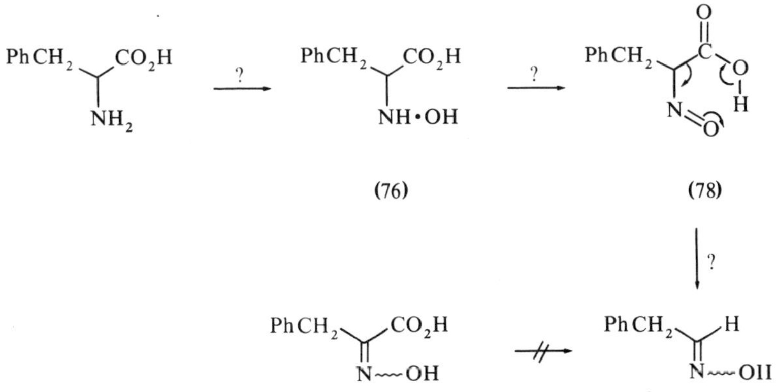

yet without disclosing many details.[178] Passage through the α-nitroso-acid (78), proposed several years ago,[167] remained a real, though experimentally still untested, possibility. The next well-defined landmark on the route to benzylglucosinolate in *T. majus* is phenylacetothiohydroximic acid (80), the role of which as a true intermediate gains credence from an observed 40% conversion

---

[178] K. Kindl and E. W. Underhill, *Phytochemistry*, 1968, **7**, 745.

into benzylglucosinolate,[179] with support from (i) its demonstrated *in vivo* formation from the aldoxime (79), (ii) its occurrence in *T. majus*, established by trapping experiments, and (iii) its enzymic glucosylation, in a cell-free system, to give desulphobenzylglucosinolate (81).[180] In this light, the route from aldoxime (79) to thiohydroximic acid becomes a matter of concern.

$$PhCH_2\text{-C(H)=N-OH} \quad (79) \longrightarrow \cdots \longrightarrow PhCH_2\text{-C(SH)=N-OH} \quad (80) \longrightarrow PhCH_2\text{-C(SGlc)=N-OH} \quad (81)$$

Miscellaneous feeding experiments, some previously reviewed[167] and others more recently described,[181] point to cysteine and, in some cases, methionine (conceivably the biochemical equivalent of cysteine), but not 1-thioglucose, as efficient supplier(s) of reduced sulphur in the glucosinolates of *T. majus* and horseradish plants, though less convincingly so in other plant systems tested.[181] However attractive, the participation of cysteine, or another mercaptide, X·SH, as a general obligatory unit in the *in vivo* synthesis of thiohydroximic acids, *e.g.* by the route (82) → → (83) → (84), is still awaiting experimental

$$R\text{-C(H)=N-OH} \quad (82) \longrightarrow \cdots \longrightarrow R\text{-C(S·X)=N-OH} \quad (83) \longrightarrow R\text{-C(SH)=N-OH} \quad (84)$$

confirmation, preferably at the enzymic level. Assuming its correctness, however, the question arises as to the identity of a biologically reasonable, oxime-derived mercaptide acceptor bridging (82) and (83).

In this context the recently documented, specific incorporation of 1-nitro-2-phenylethane (85) into benzylglucosinolate in *T. majus*, comparable with that of DL-phenylalanine, deserves attention.[182] Definite proof that the nitro-compound, or maybe rather its *aci*-tautomer (86), plays the role as a true intermediate, foreshadowed several years ago,[167] is still outstanding and will supposedly require studies at the enzymic level. If confirmed, the subsequent steps, (86) → (87) → (88), appear unexceptional and not without *in vitro* analogies. What remains a fact is that phenylacetaldoxime in the intact plant gives rise to the production of 1-nitro-2-phenylethane, present in shoots of *T. majus*, albeit in concentrations as low as 50 μg per 100 g of fresh weight.[182]

---

[179] E. W. Underhill and L. R. Wetter, *Plant Physiol.*, 1969, **44**, 584.
[180] M. Matsuo and E. W. Underhill, *Biochem. Biophys. Res. Comm.*, 1969, **36**, 18.
[181] L. R. Wetter and M. D. Chisholm, *Canad. J. Biochem.*, 1968, **46**, 931.
[182] M. Matsuo, D. F. Kirkland, and E. W. Underhill, *Phytochemistry*, 1972, **11**, 697.

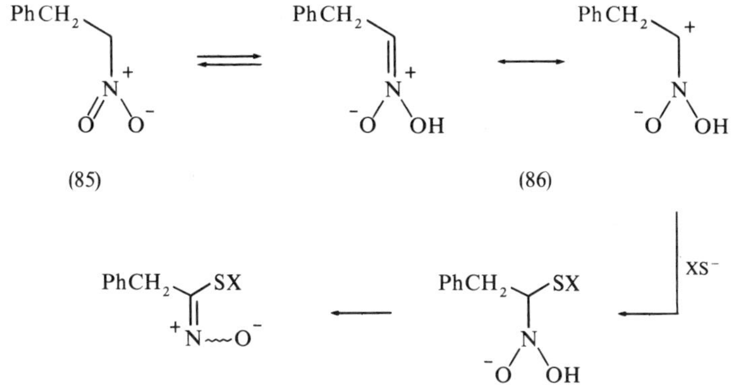

(85) ⇌ (86) → [structures]

(88) ← (87) ← XS⁻

Turning to the late stages in the biosynthetic chain, the isolation from *T. majus* of an *S*-glucosylating enzyme, catalysing transglucosylation from UDPG to phenylacetothiohydroximic acid (89) to give desulphobenzylglucosinolate (90), suggests this step to be one on the true anabolic pathway.[180] Partial purification of the enzyme, combined with a scrutiny of its specificity in terms of donor and acceptor characteristics, affords convincing evidence for (89) and (90) being true intermediates on the indicated pathway. Cell-free extracts of a

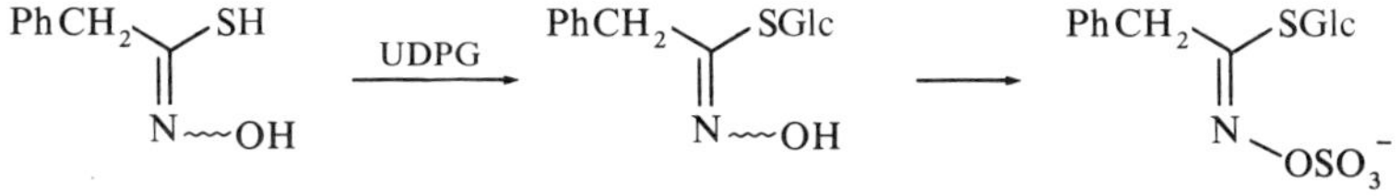

(89) —UDPG→ (90) → (91)

series of glucosinolate-containing plants of the crucifer family exhibited a similar glucosyl-transferring activity.[183] The observed, solitary failure of acetothiohydroximic acid, out of seven thiohydroximic acids studied, to function as a substrate may reflect the discontinuous natural distribution of methylglucosinolate, limited, as far as we know,[184] to the family Capparidaceae, and hence conceivably requiring a different glucosyltransferase.

The terminal step, *O*-sulphonation of desulphobenzylglucosinolate (90), is catalysed by a sulphotransferase, isolated from *T. majus*, utilizing 3'-phosphoadenosine 5'-phosphosulphate (PAPS) as the sulphate donor and possessing very little specificity for the type of desulphoglucosinolate used as the substrate.[185] This finding calls for caution, in addition to that previously ex-

[183] M. Matsuo and E. W. Underhill, *Phytochemistry*, 1971, **10**, 2279.
[184] R. Gmelin and A. Kjær, *Phytochemistry*, 1970, **9**, 569.
[185] E. W. Underhill, personal communication.

pressed,[186] in accepting desulphobut-3-enylglucosinolate rather than, say, desulpho-4-methylthiobutylglucosinolate, as the true natural intermediate on the route to but-3-enylglucosinolate in various cultivars of *Brassica napus*, solely on the basis of an observed, very high *in vivo* incorporation.[186]

The generality of the discussed pathway from α-amino-acids to glucosinolates, deduced mainly on the basis of the *in vivo* synthesis of benzylglucosinolate (91) in *T. majus*, finds support by work in other systems. Thus, a careful search for isobutylglucosinolate (75; $n = 1$, R = Pr^i) in the crucifer *Conringia orientalis*, long recognized as a good source of 2-hydroxy-2-methylpropyl-glucosinolate [75; $n = 1$, R = Me$_2$C(OH)],[167] the latter biosynthesized from L-leucine,[169] has recently been rewarding.[169] Again, retention of the entire amino-acid carbon skeleton, save for the α-carboxy-carbon atom, has been observed on feeding α-aminoadipic acid or, even better, its δ-methyl ester to an *Erysimum* species, synthesizing 3-methoxycarbonylpropylglucosinolate (75; $n = 3$, R = MeO·CO).[187]

In the aromatic series, tyrosine seems capable of traversing the same path. Thus, in *Sinapis* seedlings, *p*-hydroxyphenylacetaldoxime (74; $n = 1$, R = *p*-HO·C$_6$H$_4$) is incorporated into *p*-hydroxybenzylglucosinolate (75; $n = 1$, R = *p*-HO·C$_6$H$_4$) with high efficiency through an intermediate, tentatively identified as desulpho-*p*-hydroxybenzylglucosinolate.[188] Previously reported utilization of *p*-coumaric acid and phenylalanine,[189] though real, is now shown to be marginal in *Sinapis* under normal conditions. To the extent it is observed, the pathway is likely to include tyrosine as an obligatory intermediate.[188] Hence, it causes little surprise that neither phenylacetaldoxime nor benzylglucosinolate is converted into *p*-hydroxybenzylglucosinolate in the *Sinapis*

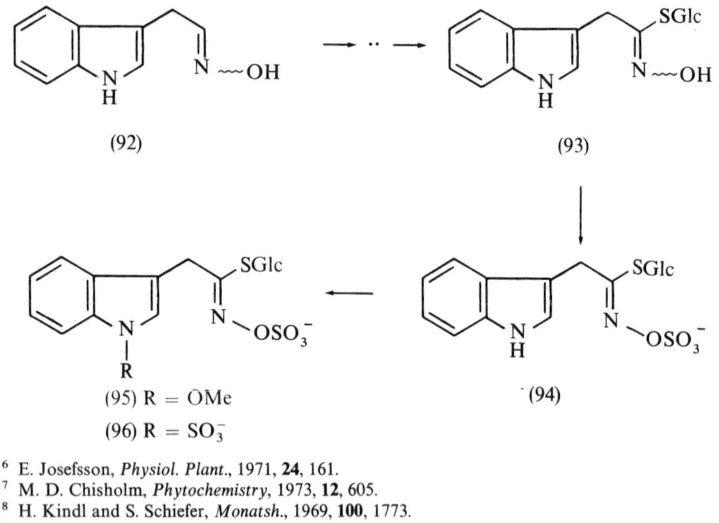

(92)

(93)

(95) R = OMe
(96) R = SO$_3^-$

(94)

[186] E. Josefsson, *Physiol. Plant.*, 1971, **24**, 161.
[187] M. D. Chisholm, *Phytochemistry*, 1973, **12**, 605.
[188] H. Kindl and S. Schiefer, *Monatsh.*, 1969, **100**, 1773.
[189] H. Kindl, *Monatsh.*, 1964, **95**, 439.

system.[188] Much attention has been given to the conversion of L-tryptophan into 3-indolylmethylglucosinolate (94) and two closely related natural products, (95) and (96). Incorporation of 3-^{14}C,^{15}N-labelled L-tryptophan into (94) in rape seedlings proceeded without significant change in the isotope ratio, supposedly by a pathway parallel to that established for benzylglucosinolate in *Tropaeolum majus*.[190] Again, in *Sinapis alba* seedlings, DL-tryptophan was an excellent precursor for 3-indolylmethylglucosinolate (94), clearly surpassed though by 3-indolylacetaldoxime (92), most likely a natural intermediate, converted into (94) *via* the desulphoglucosinolate (93).[190a] In woad (*Isatis tinctoria*), a rich source of the indoleglucosinolates (94),[191] (95),[191] and (96),[177, 191] experiments suggested the function of (94) as an obligatory intermediate for the N-substituted indoleglucosinolates (95) and (96).[192] [^{3}H]Indolylacetaldoxime (92) and L-[^{35}S]cystine both carried their labels efficiently into the indoleglucosinolates during the first eight hours of metabolism, chiefly though into the bivalent sulphur moiety.[193] A trapping technique, involving feeding of labelled oxime (92) and non-labelled cystine, or *vice versa*, permitted isolation of a true intermediate, converted into the indoleglucosinolates to the extent of nearly 60% and tentatively identified as desulphoindolylmethylglucosinolate (93).[193]

Glucosinolates from Amino-acids, without Preservation of Nitrogen. Numerous natural glucosinolates containing side-chains longer than those of the protein amino-acids, frequently structurally modified, evidently pose a challenge to biosynthetic exploration. Knowledge acquired prior to 1968 has been reviewed and discussed.[167] Hence, only more recent evidence will be considered here, uniformly supporting the originally proposed scheme (Scheme 2).

Operation of this sequence in the natural synthesis of 2-phenylethylglucosinolate (100; R = PhCH$_2$) in watercress (*Nasturtium officinale*), proposed on the basis of the efficient precursor function of phenylalanine (97; R = PhCH$_2$), acetate, 2-amino-4-phenylbutyric acid (99; R = PhCH$_2$), and 3-phenylpropionaldehyde oxime,[167] has been supported by demonstration of the presence of small quantities of 2-amino-4-phenylbutyric acid in watercress, and by the observed efficient and specific incorporation of 2-benzylmalic acid (98; R = PhCH$_2$) into 2-amino-4-phenylbutyric acid and 2-phenylethylglucosinolate (100; R = PhCH$_2$).[194] An oxidized counterpart of the latter, (2S)-hydroxy-2-phenylethylglucosinolate [100; R = (S)-PhCH(OH)], present in *Reseda luteola*, is efficiently biosynthesized in this species from 2-amino-4-phenylbutyric acid or, alternatively, phenylalanine and acetate.[167] Recent studies, involving feeding of doubly labelled precursors (^{14}C, ^{15}N) to *R. luteola* shoots, have helped

[190] M. Kutáček and M. Králová, *Biol. Plant.*, 1972, **14**, 279.
[190a] E. Josefsson, *Physiol. Plant.*, 1972, **27**, 236.
[191] M. C. Elliott and B. B. Stowe, *Plant Physiol.*, 1971, **47**, 366.
[192] M. C. Elliott and B. B. Stowe, *Plant Physiol.*, 1971, **48**, 498.
[193] S. Mahadevan and B. B. Stowe, *Plant Physiol.*, 1972, **50**, 43.
[194] E. W. Underhill, *Canad. J. Biochem.*, 1968, **46**, 401.

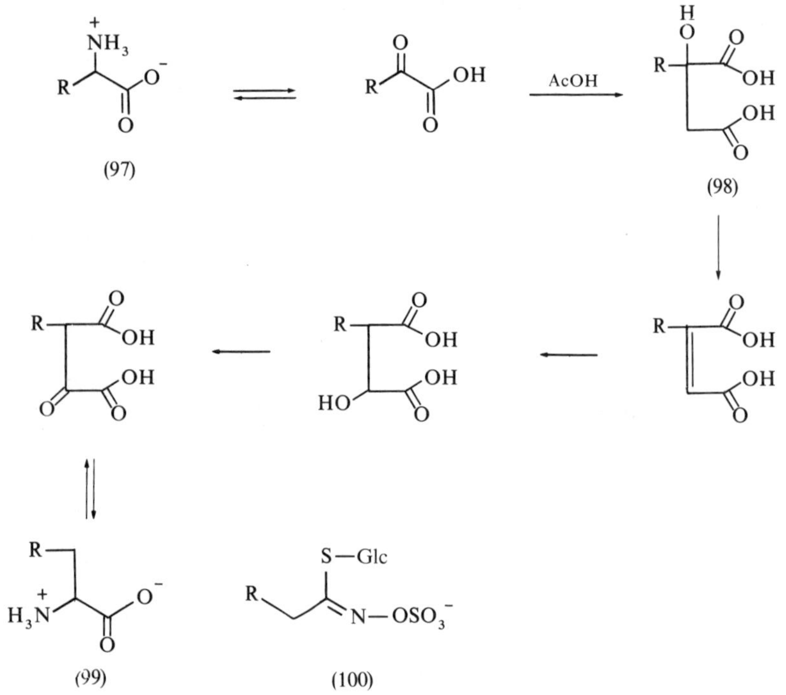

Scheme 2

to clarify the detailed biosynthetic pathway.[195] The precursor efficacy of the appropriate stereoisomers of phenylalanine, 2-benzylmalic acid, 2-amino-4-hydroxy-4-phenylbutyric acid [99; R = PhCH(OH)], and 2-amino-4-phenylbutyric acid, increasing in the order given, supplemented with an observed very low dilution in the biosynthesis of the hydroxylated glucosinolate from its deoxy-analogue, point to hydroxylation as the final step in a sequence otherwise identical with that prevailing in watercress. Uncertainty remains as to the path from the hydroxy-amino-acid, surely not a natural intermediate, to the final product, but the conversion may merely reflect low enzyme specificity.[195] A similar, stereoselective enzymic oxidation could conceivably be involved in the *in vivo* synthesis of the epimeric (2R)-hydroxy-2-phenylethylglucosinolate [100; R = (R)-PhCH(OH)], a natural constituent of the crucifer *Sibara virginica*.[173] The origin of (2R)-hydroxy-2-(4-methoxyphenyl)glucosinolate[100; R = (R)-MeOC$_6$H$_4$CH(OH)], present in seens of *Arabis hirsuta*,[172] is unknown. Tyrosine, O-methyltyrosine, chain-elongation, glucosinolate formation, and, eventually, side-chain oxidation seems a plausible course of

[195] E. W. Underhill and D. F. Kirkland, *Phytochemistry*, 1972, **11**, 1973.

Non-protein Amino-acids, Cyanogenic Glycosides, and Glucosinolates

events, in view of the reported O-methylation of tyrosine preceding incorporation into 4-methoxyphenylacetaldehyde oxime in the crucifer genus *Aubrietia*.[196]

Allylglucosinolate (106) is a classical and important natural compound. Already prior to 1968, when data on its biosynthesis were reviewed,[167] 2-amino-5-methylthiovaleric acid (101) had been traced as an efficient precursor and likely intermediate in the synthesis of allylglucosinolate in horseradish leaves. Subsequently, efficient and specific incorporation of DL-2-[^{15}N]amino-5-methylthio[2-^{14}C]valeric acid – but, significantly, not of doubly labelled allylglycine – into allylglucosinolate in horseradish leaves,[197] experimental evidence for efficient and C—N-intact utilization of 4-methylthiobutyraldoxime (102),[198] and recognition of glucosyl transferase activity in the test plant,[183] together provided support for the operation of an *in vivo* pathway similar to that obtaining with benzylglucosinolate.

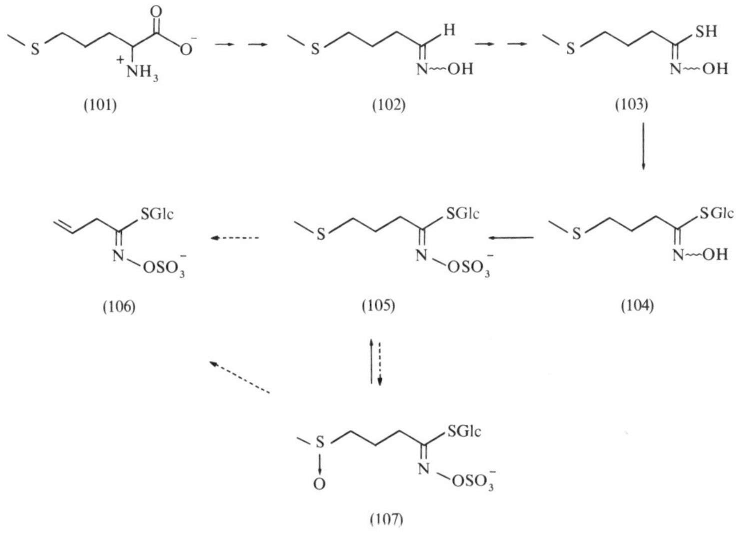

The extent of community is further apparent from the observed incorporation in horseradish leaves of 4-methylthiobutyrothiohydroximic acid (103), desulpho-3-methylthiopropylglucosinolate (104), and the corresponding sulphonated product (105) into allylglucosinolate (106) to the extent of 50–73%,

---

[196] H. Kindl and S. Schiefer, *Phytochemistry*, 1971, **10**, 1795.
[197] M. Matsuo and M. Yamazaki, *Chem. and Pharm. Bull. (Japan)*, 1968, **16**, 1034.
[198] M. Matsuo, *Tetrahedron Letters*, 1968, 4101.

in the order given.[199] Significantly, the sulphide-glucosinolate (105), a long-known substituent of other crucifers[167] and postulated here as an intermediate, has now been encountered in horseradish, clearly deriving from 2-amino-5-methylthiovaleric acid (101).[199] Intriguing, though as yet of limited conclusive value, is the observed biological conversion of extraneous 3-methylsulphinylpropylglucosinolate (107) into allylglucosinolate (106) comparable in degree with, but not exceeding that of, the parent amino-acid (101). The sulphoxide-glucosinolate, well-known from other crucifer sources,[167] but, to be sure, as yet an unproven constituent of horseradish, is also reduced *in vivo* to the sulphide stage (105), although the amino-acid (101) is a three-fold more efficient precursor of the latter. Hence, in accord with the above observations, though insufficiently supported by experiment, the natural pathway may include the sulphoxide-glucosinolate (107), a sizable natural pool of which many undergo formal elimination of MeSOH to give allylglucosinolate, the final product.[199] Experimental results, attesting to uniformity in type and sequence of the main reactions, were obtained from studies conducted on plants of wallflower (*Cheiranthus kewensis*), an established source of 3-methylthiopropylglucosinolate (105) and 3-methylsulphinylpropylglucosinolate (107).[200] Methionine and 2-amino-5-methylthiovaleric acid (101) again served eminently as precursors of both the sulphoxide-glucosinolate (107), the major product, and the sulphide analogue (105). Likewise, sulphoxide reduction, of (107) to (105), the reciprocal of the anabolic course, was shown to be an efficient reaction.[200]

2-Methylthioethylglucosinolate, or its S-oxidized analogues, the expected products from direct methionine incorporation, are curiously absent from our present list of natural glucosinolates.[167] On the other hand, the list abounds in compounds with unbranched chains containing three or more carbon atoms, terminating in olefinic or methylsulphur functionalities. A communal origin of these, involving repeated homologizations *via* the sequence (97) → (99), appears reasonable. In fact, scattered experimental evidence favours this view. Thus, precursor studies performed with plants of the *Brassica* group, notably rape and rutabaga, rich and diverse in their glucosinolate contents[201,202] have shown methionine and 2-amino-5-methylthiovaleric acid to be efficiently converted into glucosinolates with but-3-enyl and pent-4-enyl side-chains, as well as their 2-hydroxylated congeners.[186] Spurred by the desire to minimize the total glucosinolate contents of rape seed grown for industrial purposes, plant breeders have produced cultivars, *e.g.* Bronowski, with greatly diminished contents. The biochemical rationale for this has been traced to an enzymic block on the assumed pathway, subsequent to 2-amino-5-methylthiovaleric acid but prior to desulphobut-3-enylglucosinolate.[186] Experimental results, gained from work with a 'normal' cultivar, suggest, but do not prove, that 2-hydroxylation of desulphobut-3-enylglucosinolate may be a step on the

---

[199] M. D. Chisholm and M. Matsuo, *Phytochemistry*, 1972, **11**, 203.
[200] M. D. Chisholm, *Phytochemistry*, 1972, **11**, 197.
[201] E. Josefsson, *Phytochemistry*, 1967, **6**, 1617.
[202] E. Josefsson and C. Mühlenberg, *Acta Agric. Scand.*, 1968, **18**, 97.

natural pathway, i.e. oxidation after elimination of the methylsulphur fragment, but prior to the final sulphonation. In the Bronowski cultivar, however, hydroxylation is greatly suppressed.[186]

Detailed studies have been directed towards clarification of the pathway to (R)-2-hydroxybut-3-enylglucosinolate in rutabaga plants. Methionine and acetate were efficient precursors,[203] interacting to produce 2-amino-5-methylthiovaleric acid, a natural intermediate, carried through to the hydroxybutenylglucosinolate with integrity of the C—N linkage. The application of a racemic amino-acid precursor was made responsible for a significant, observed increase in the $^{14}C:^{15}N$ ratio on going from precursor to product.[204] Attempts to trace the hydroxylation step included feeding of doubly labelled 5-methylthiovaleraldehyde oxime to rutabaga plants. Efficient and specific incorporation, without severance of the C—N linkage, into the hydroxybutenylglucosinolate permitted the conclusion that hydroxylation occurs at an unknown point beyond the intermediate oxime,[205] a result in line with knowledge acquired from other systems. Most likely, a similar, general pathway is utilized by other higher plants in assembling the many additional ω-methylsulphur-substituted or terminally unsaturated side-chains, including their hydroxylated congeners, present in long-known glucosinolates,[167] but also in more recent additions listed in the beginning of this section.

Since this field was previously reviewed,[167] a vastly more detailed picture has emerged of the beautiful and simple relationship between amino-acids, cyanogenic glycosides, glucosinolates, and, perhaps less apparent, other constituents of higher plants. It is a task for the future to fill in still more details, so that we may ultimately comprehend the full biological significance and consequences of this relationship.

---

[203] G. S. Serif and L. A. Schmotzer, *Phytochemistry*, 1968, **7**, 1151.
[204] C.-J. Lee and G. S. Serif, *Biochemistry*, 1970, **9**, 2068.
[205] C.-J. Lee and G. S. Serif, *Biochim. Biophys. Acta*, 1971, **230**, 462.

4
Biosynthesis of Alkaloids

BY E. LEETE

1 Introduction

This review attempts to cover all material related to the biosynthesis of alkaloids published during 1972, plus a few isolated papers which were not available when last year's chapter[1] was written. Most of the work in this area continues to involve the feeding of labelled putative precursors to alkaloid producing plants, in the hope that they will result in the formation of labelled alkaloids. Tracer work of this type is summarized at the end of this chapter. In 1972 the complete proceedings appeared of the Fourth International Symposium on the Biochemistry and Physiology of Alkaloids held in Halle (Germany) in June 1969.[2] Much of the material presented at this meeting has been published in the interim, so not all the papers appearing in this monograph have been abstracted.

Some of the more significant advances which have been made during 1972 appear in Section 2 of this review. In these yearly reviews it is not possible to discuss the biosynthesis of all classes of alkaloids in depth. It is the plan to select different topics each year so that a more complete story can be told on the individual alkaloids. In Section 3 the biosynthesis of tropic acid is reviewed. Some progress is being made on the isolation of enzymes responsible for the biosynthesis of alkaloids and this appears in Section 4. The use of tissue cultures of various plant organs in studying alkaloid biosynthesis is covered in Section 5.

Discussion of the synthesis of alkaloids in the laboratory under so-called 'physiological conditions', also referred to as 'biomimetic synthesis', has been omitted.

2 Highlights of 1972

Origin of Vasicine.—This alkaloid (2), also known as peganine, is found in *Adhadota vasica* and *Peganum harmala*, and it is generally accepted that it is formed in part from anthranilic acid.[3] The origin of C-1, C-2, C-3, and C-10 is the subject of controversy.[4] In 1971, Liljegren[5] reported the relatively specific

---

[1] E. Leete, in 'Biosynthesis', ed. T. A. Geissman (Specialist Periodical Reports), The Chemical Society, London, 1972, Vol. 1, Chap. 5, p. 158.
[2] *Biochem. Physiol. Alkaloide, Int. Symp.,* 4th. 1969 (Pub. 1972), ed. K. Mothes, Akad.-Verlag: Berlin, E. Germany.
[3] D. Gröger, S. Johne, and K. Mothes, *Experientia,* 1965, **21**, 13.
[4] D. Gröger, in 'Biosynthese der Alkaloide', ed. K. Mothes and H. R. Schütte, Deut. Verlag der Wissen. Berlin, 1969, p. 556.
[5] D. J. Liljegren, *Phytochemistry,* 1971, **10**, 2261.

Biosynthesis of Alkaloids

incorporation of [2-^{14}C]ornithine into this unit, the activity at C-1 and C-10 being 42.7 and 41.6%, respectively. This result suggested that the ornithine was being decarboxylated affording putrescine, a symmetrical molecule, which would then be incorporated resulting in equal labelling at C-1 and C-10. However, the incorporation of [1,4-^{14}C]putrescine was not especially specific: C-1 27.7%, C-2 2.7%, C-3 24.3%, C-10 41.8%. Gröger and co-workers[6] have now discovered that N-acetylanthranilic acid (1) is a specific precursor of C-3, C-10, and the anthranilic acid-derived portion of the alkaloid (see Table under *Adhatoda vasica*). The results of Liljegren can be rationalized by postulating that [2-^{14}C]ornithine serves as a source of [1-^{14}C]acetate which is then incorporated *via* N-acetylanthranilic acid. This metabolic route is illustrated in Scheme 1. The relatively high specific incorporation of [2-^{14}C]glycine into C-3 of the alkaloid is consistent with the formation of [2-^{14}C]acetate from this glycine, a route which was discussed in detail last year.[7] The origin of C-1, C-2, and N-11 remains obscure.

Formation of Pipecolic Acid from D-Lysine.—Gilbertson[8] has compared the

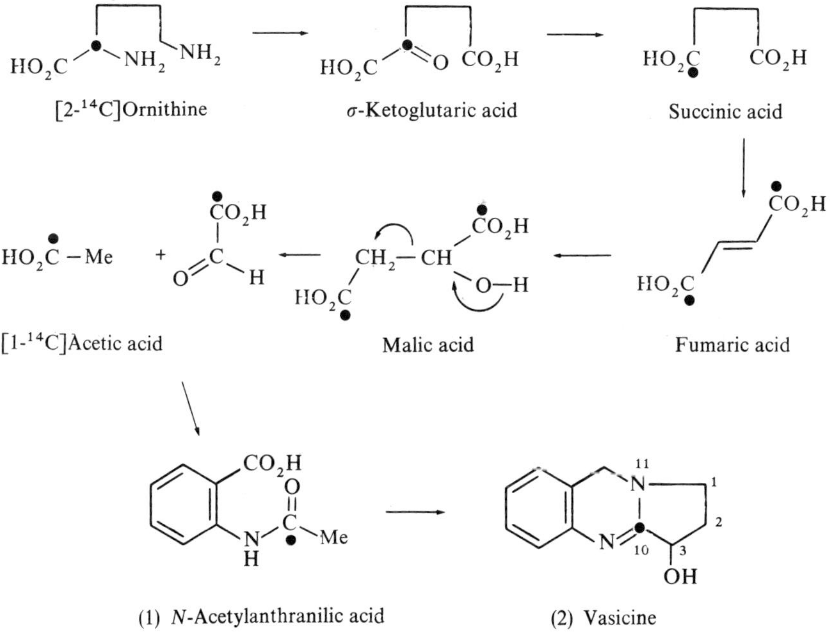

Scheme 1 *Biosynthesis of vasicine*

---
[6] K. Waiblinger, S. Johne, and D. Gröger, *Phytochemistry*, 1972, **11**, 2263.
[7] Ref. 1, p 161.
[8] T. J. Gilbertson, *Phytochemistry*, 1972, **11**, 1737.

incorporation of D- and L-[2-^{14}C]lysine into the piperidine rings of anabasine (4) and pipecolic acid (5) in *Nicotiana glauca*. As expected, only L-lysine served as a precursor of the piperidine ring of anabasine, although D-lysine was approximately eight times more efficient than L-lysine as a precursor of pipecolic acid. The rate of absorption of the D-lysine from the hydroponic nutrient solution in which the roots of the plants were growing was much slower than that of the L-lysine. The Reporter considers that the formation of pipecolic acid from L-lysine in this species may be simply a detoxification mechanism whereby the plant gets rid of the undesirable (?) D-lysine. Previous work[9] strongly favours α-keto-ε-aminocaproic acid (3) as an intermediate between lysine and pipecolic acid.

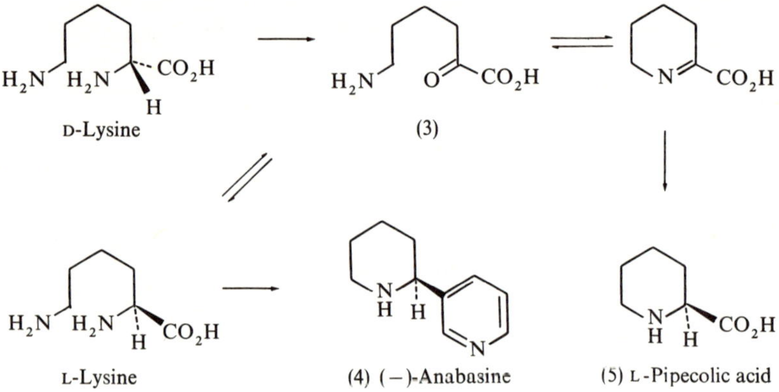

Scheme 2 *Biosynthesis of pipecolic acid*

Betanine Biosynthesis.—Dreiding and co-workers have reported definitive work on the origin of betanine (9), the red colouring matter of beets (*Beta vulgaris*). Some tracer experiments were carried out with beets; however, more significant results with good incorporations were obtained on feeding various labelled compounds to *Opuntia* species.[10] It has been established that [1-^{14}C]-dopa (6) yields betanine labelled at C-10 and C-19.[11] Tritium was retained in the betalamic acid (7) portion of the molecule when L-[3′,5′-^{3}H$_2$]tyrosine was administered, and degradations were consistent with all the tritium being present at C-11.[12] This result supports the biosynthetic pathway illustrated in Scheme 3, where an oxidative cleavage occurs between carbons 4′ and 5′ of

[9] R. N. Gupta and I. D. Spenser, *Phytochemistry*, 1970, **9**, 2329.
[10] H. E. Miller, H. Rösler, A. Wohlport, H. Wyler, M. E. Wilcox, H. Frohofer, T. J. Mabry, and A. S. Dreiding, *Helv. Chim. Acta*, 1968, **51**, 1470.
[11] E. Dunkelblum, H. E. Miller, and A. S. Dreiding, *Helv. Chim. Acta*, 1972, **55**, 642.
[12] N. Fischer and A. Dreiding, *Helv. Chim. Acta*, 1972, **55**, 649.

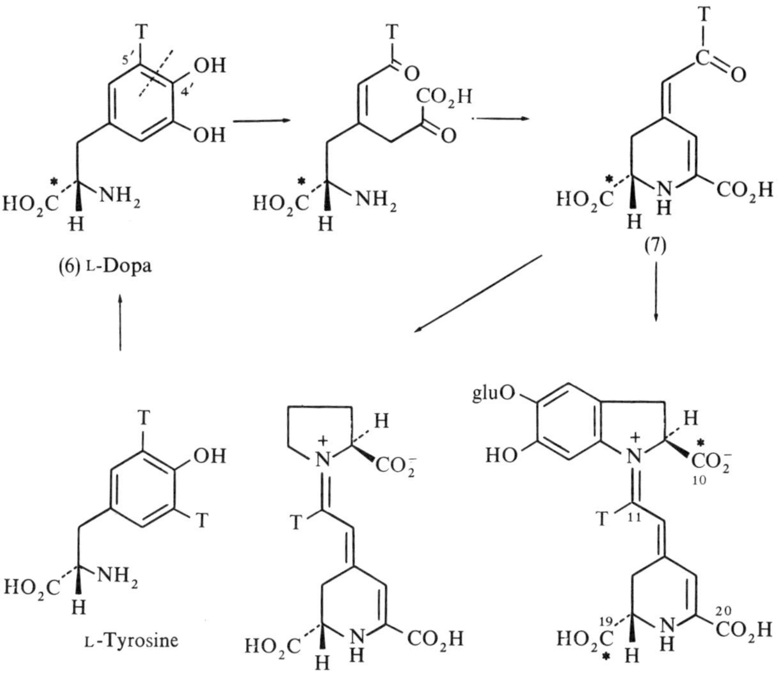

Biosynthesis of Alkaloids

Scheme 3 *Biosynthesis of indicaxanthine and betanine*

dopa. Complementary work has been carried out on the biosynthesis of indicaxanthine (8) by Piattelli and co-workers.[13]

Aberrant Biosynthesis of Unnatural Alkaloids.—Kirby[14] has demonstrated that the opium poppy (*Papaver somniferum*) is capable of carrying out transformations of 'unnatural' precursors to unnatural alkaloids. The O-demethylation of various codeine derivatives to unnatural derivatives of morphine was observed (see Table under *P. somniferum*). The efficiency of the unnatural transformation was compared with the natural conversion of codeine into morphine by feeding to the plant a mixture of a [2-^{3}H]codeine derivative and [*N*-methyl-^{14}C]codeine. One unnatural transformation which was more efficient than the natural one was the conversion of dihydrodeoxycodeine (10) into dihydrodeoxymorphine (11) (Scheme 4). Kirby concluded that neither the 6-hydroxy-group nor the 7,8-double bond in codeine is important for binding to the enzyme responsible for demethylation of the 3-methoxy-group of codeine.

[13] G. Impellizzeri and M. Piatelli, *Phytochemistry*, 1972, **11**, 2499.
[14] G. W. Kirby, S. R. Massey, and P. Steinreich, *J.C.S. Perkin I*, 1972, 1642.

Since other methyl ethers, *e.g.* reticuline and its structural isomers, are not demethylated in *P. somniferum*,[15,16] it is concluded that only one enzyme is responsible for the O-demethylation of codeine and its derivatives.

The ability of higher plants to carry out unnatural or aberrant syntheses is a relatively unexplored area of research. Some other examples are illustrated in Scheme 4. 5-Fluoronicotinic acid (12) is converted into 5-fluoronicotine (13) in the tobacco plant (*Nicotiana tabacum*).[17] Rapoport[18] has found that methylated analogues of 1-methyl-1-pyrrolinium chloride yield methylated analogues

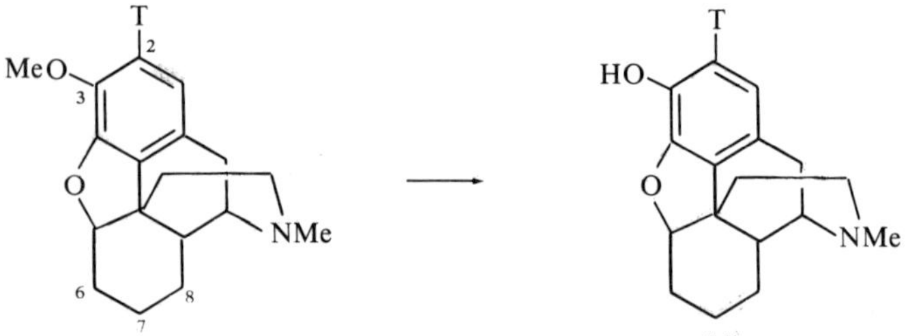

(10) (11)

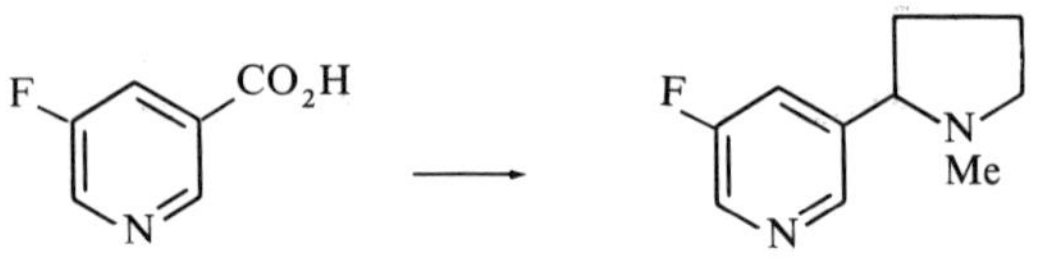

(12) 5-Fluoronictonic acid (13) 5-Fluoronicotine

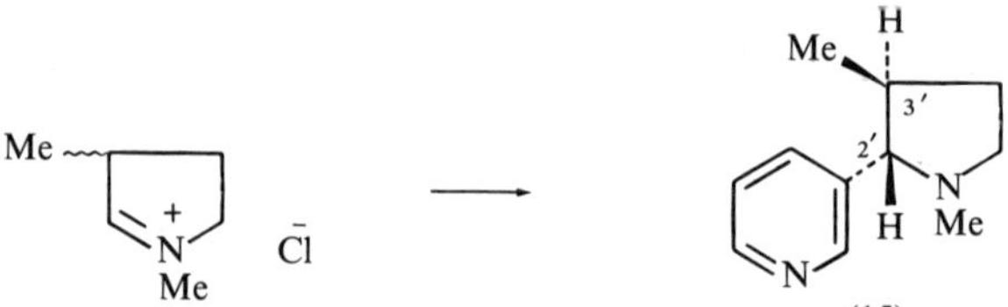

(14) (15)

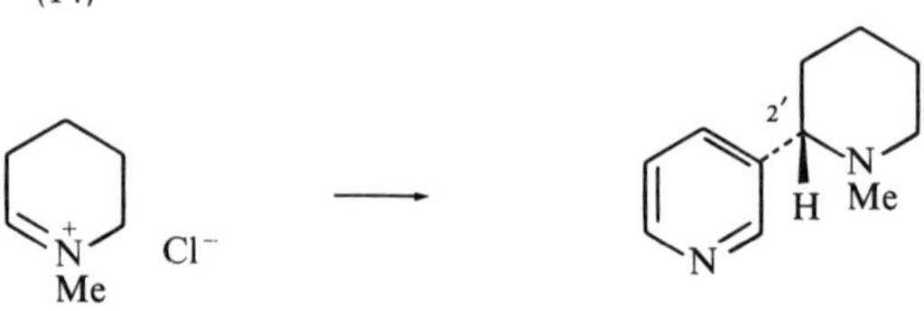

(16) (17) (−)-*N*-Methylanabasine

Scheme 4 *Aberrant syntheses*

[15] D. H. R. Barton, G. W. Kirby, W. Steglich, G. M. Thomas, A. R. Battersby, T. A. Dobson, and H. Ramuz, *J. Chem. Soc.*, 1965, 2423.

[16] A. R. Battersby, D. M. Foulkes, and R. Binks, *J. Chem. Soc.*, 1965, 3323.

[17] E. Leete, G. B. Bodem, and M. F. Manuel, *Phytochemistry*, 1971, **10**, 2687.

[18] M. L. Rueppel and H. Rapoport, *J. Amer. Chem. Soc.*, 1970, **92**, 5528; 1971, **93**, 7021.

of nicotine. Thus the (RS)-1,3-dimethyl-1-pyrrolinium chloride (14) afforded 2'-(S)-*trans*-3'-methylnicotine (15). N-Methyl-Δ'-piperideinium chloride (16) was converted in *N. glauca* and *N. tabacum* into the higher homologue of nicotine, *i.e.* (−)-N-methylanabasine (17).[19] These homologues of nicotine had the same chirality at C-2' as natural (−)-nicotine, strongly suggesting that the enzymes involved in nicotine biosynthesis were being used in these unnatural reactions.

Biotransformations of Alkaloids.—For a long time it was considered that alkaloids were end-products of metabolism. However, it is now being realized that many of them are rapidly metabolized, either to other alkaloids, or to non-alkaloidal compounds. Waller and co-workers have carried out extensive work in the past few years on ricinine metabolism.[20] In senescent leaves of *Ricinus communis*, ricinine (18) is demethylated to N-demethylricinine (19). In

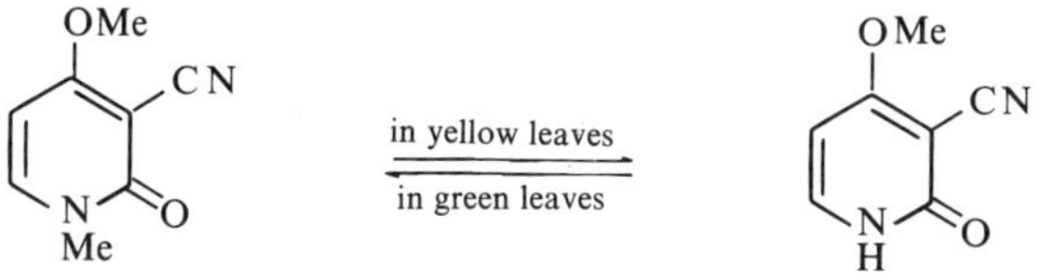

(18) Ricinine

in yellow leaves
in green leaves

(19) N-Demethylricinine

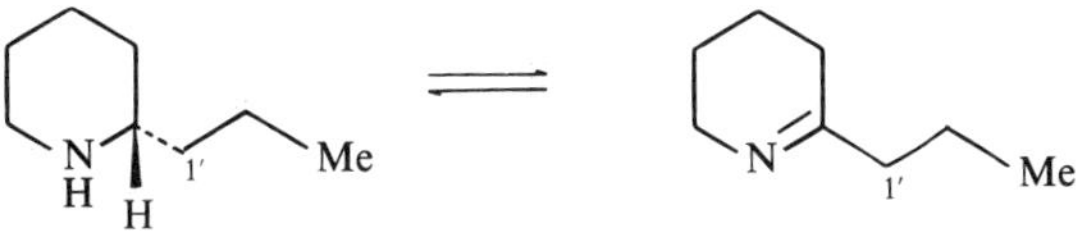

(20) (+)-Coniine

(21) γ-Coniceine

fresh green leaves this compound is methylated back to ricinine. The N-demethylation does not seem to occur in green leaves.

Another reversible reaction is the interconversion of (+)-coniine (20) and γ-coniceine (21) which occurs in hemlock (*Conium maculatum*).[21-23] The reaction is stereospecific, only the natural (+)-coniine being dehydrogenated to γ-coniceine.[21] γ-Coniceine is formed in hemlock from 5-oxoöctanal by a transamination involving L-alanine[24]. By administering [1'-^{14}C,^{15}N]coniine and [1'-^{14}C,^{15}N]-γ-coniceine to hemlock, and then re-isolating the alkaloids

---

[19] E. Leete and M. R. Chedekel, *Phytochemistry*, 1972, **11**, 2751.
[20] (a) G. R. Waller and J. L.-C. Lee, *Plant Physiol.*, 1969, **44**, 522; (b) L. Skursky, D. Burleson, and G. R. Waller, *J. Biol. Chem.*, 1969, **244**, 3238; (c) G. R. Waller and L. Skursky, *Plant Physiol.*, 1972, **50**, 622.
[21] E. Leete and J. O. Olson, *J. Amer. Chem. Soc.*, 1972, **94**, 5472.
[22] J. W. Fairbairn and P. N. Suwal, *Phytochemistry*, 1961, **1**, 38.
[23] J. W. Fairbairn and A. A. E. R. Ali, *Phytochemistry*, 1968, **7**, 1599.
[24] M. F. Roberts, *Phytochemistry*, 1971, **10**, 3057.

1 and 9 days later, it was established that there was little, if any, loss of the ^{15}N relative to the ^{14}C. It is thus apparent that the nitrogen of these hemlock alkaloids is not removed by a reversal of their biosynthesis, or by other metabolic reactions.

Very little is known about the non-alkaloidal metabolites of alkaloids in the species where they are formed. Fairbairn[25] has shown that morphine is rapidly metabolized in the latex of the opium poppy to polar non-alkaloidal compounds of unknown structure. Iljin[26] found that the administration of [^{14}C]nicotine to tobacco led the production of many radioactive amino-acids.

Digenis[27] has claimed that gramine (22), an alkaloid formed in the leaves of germinating barley, is transformed to 3-hydroxymethylindole (24) (10.1% incorporation), indole-3-carboxylic acid (23) (6.2% inc.), and tryptophan (0.84% inc.). [*methylene*-^{14}C,^{3}H]gramine (^{3}H:^{14}C = 0.5) was fed to 60-day-old

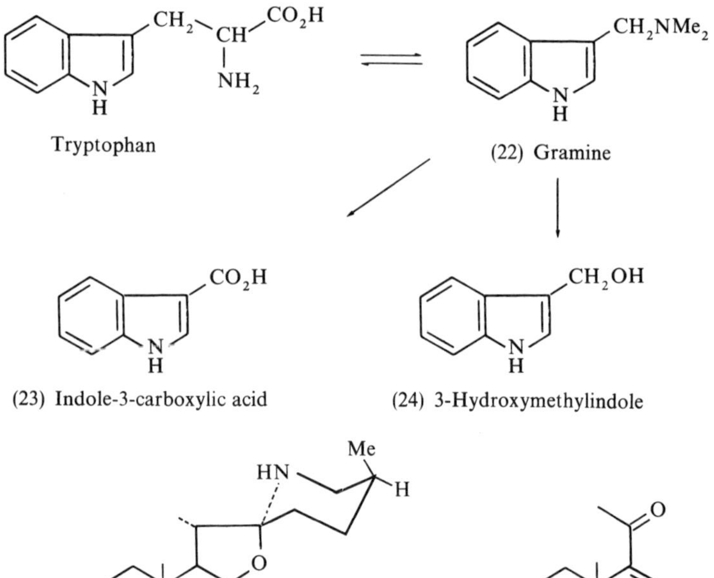

Tryptophan

(22) Gramine

(23) Indole-3-carboxylic acid

(24) 3-Hydroxymethylindole

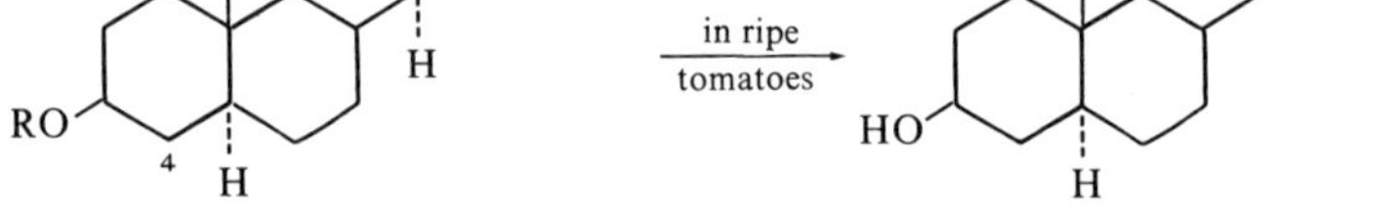

(25) Tomatine

(26) Allopregnenolone

[25] J. W. Fairbairn and S. El-Masry, *Phytochemistry*, 1967, **6**, 499.
[26] G. S. Iljin and M. Y. Lovkova, ref. 2, p. 207.
[27] G. A. Digenis, *J. Pharm. Sci.*, 1969, **58**, 39, 42; G. A. Digenis, B. A. Faraj, and C. I. Abou-chaar, *Biochem. J.*, 1966, **101**, 27C.

Biosynthesis of Alkaloids

barley shoots. The tryptophan isolated 7 days later contained the same $^3\text{H}:^{14}\text{C}$ ratio. Surprisingly it was reported that tryptophan was the only radioactive amino-acid isolated from the alkaline hydrolysis of the plant proteins. This result strongly suggested that gramine was serving as a direct precursor of tryptophan, a reversal of its biosynthesis.[28] The radioactive tryptophan derived from the gramine was not degraded to determine the location of the ^3H and ^{14}C, and it is hoped that this will be done in order to substantiate this remarkable biotransformation.

One transformation which has economic importance is the degradation of the toxic alkaloid tomatine (25) in ripe tomatoes.[29] Radioactive allopregnenolone (26) was isolated from tomatoes which had been fed [4-^{14}C]tomatine. Unripe green tomatoes do not apparently degrade the alkaloid.[30]

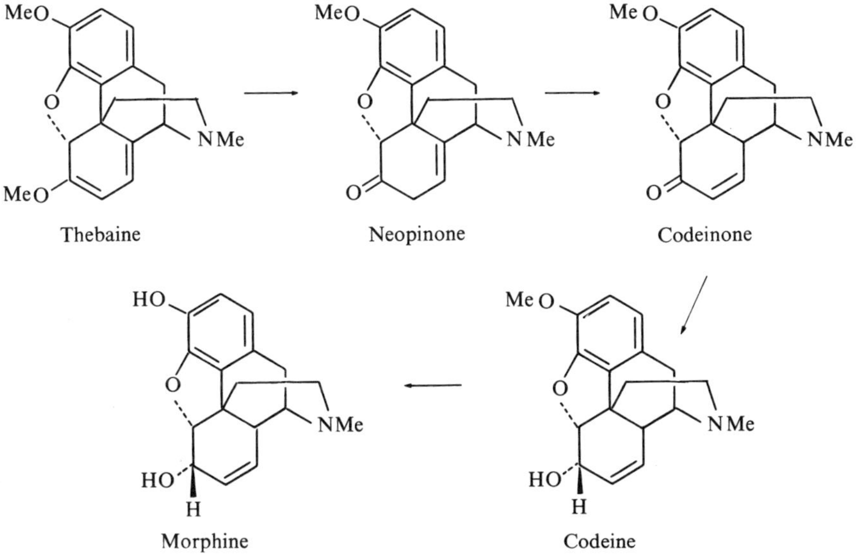

Thebaine → Neopinone → Codeinone

Morphine ← Codeine

Scheme 5

Details of the transformation of thebaine to morphine (Scheme 5) have been worked out by studying the metabolism of potential intermediates in *Papaver somniferum*.[31] The results obtained (see Table under *P. somniferum*) are consistent with the sequence: thebaine→neopinone→codeinone→codeine→morphine.

---

[28] D. O'Donovan and E. Leete, *J. Amer. Chem. Soc.*, 1963, **85**, 461, and ref. cited therein.
[29] E. Heftmann and S. Schwimmer, *Phytochemistry*, 1972, **11**, 2783.
[30] H. Sander, *Planta*, 1956, **47**, 374.
[31] H. I. Parker, G. Blaschke, and H. Rapoport, *J. Amer. Chem. Soc.*, 1972, **94**, 1276.

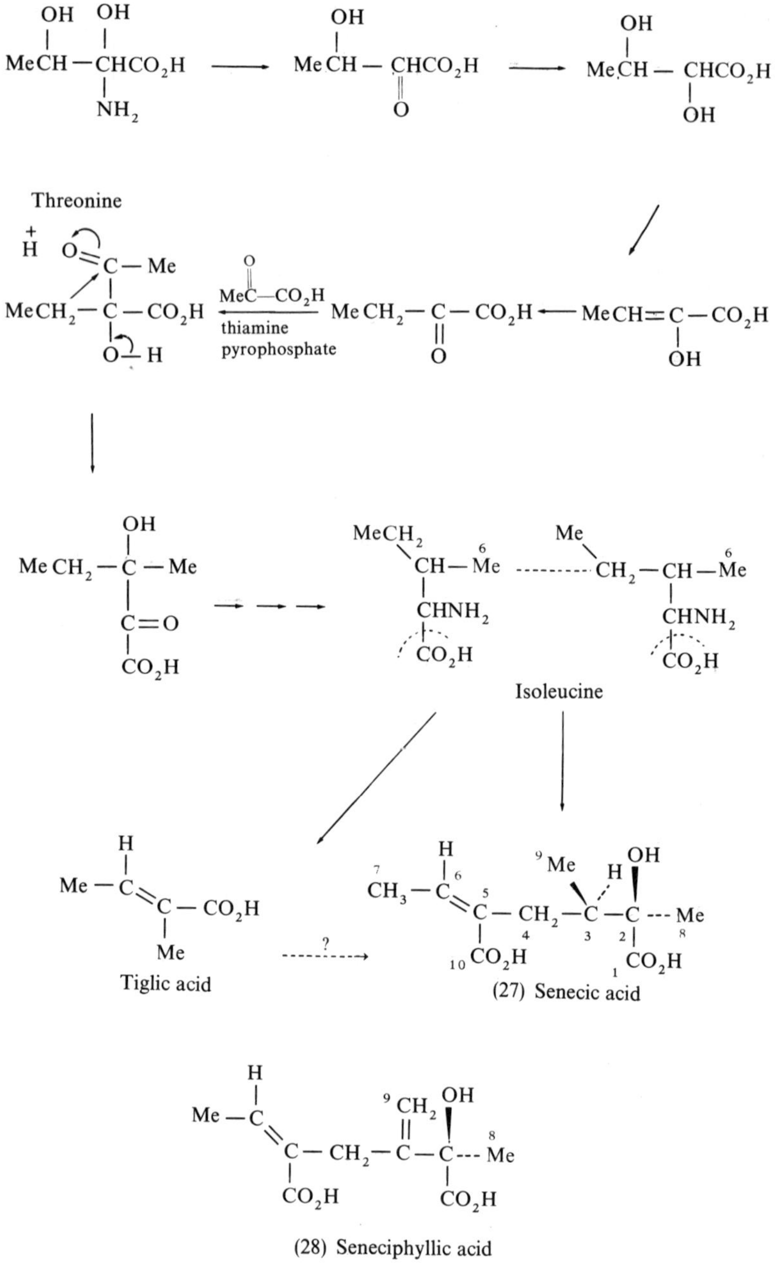

Biosynthesis

Threonine

(27) Senecic acid

(28) Seneciphyllic acid

Scheme 6 *Biosynthesis of senecic acid*

Biosynthesis of Alkaloids

Senecic Acid Biosynthesis.—This necic acid (27) is esterified with the pyrrolizine base retronecine in the alkaloid senecionine. It has now been established[32] that it is formed from two molecules of isoleucine, as illustrated in Scheme 6. Threonine is also incorporated specifically into senecic acid, and this result is readily rationalized since threonine serves as a precursor of isoleucine *via* α-ketobutyric acid. Tiglic acid is also formed from isoleucine with loss of the carboxy-group of the amino-acid,[33,34] and examination of this compound (or its *cis*-isomer angelic acid) as an intermediate between isoleucine and senecic acid would be worthwhile. In earlier work[35] it was found that the administration of L-[*methyl*-^{14}C]methionine to *Senecio douglasii* yielded seneciphylline with preferential (73—83%) incorporation of activity into seneciphyllic acid (28). Furthermore, 22—26% of the activity was found at C-8. It is considered that the methyl group of methionine is incorporated into C-3 of serine which then affords [3-^{14}C]pyruvate by established metabolic pathways. The methyl group of pyruvate ultimately becomes C-6 of isoleucine, resulting in labelling at C-4 and C-8 of seneciphyllic acid. Crout[32] showed that the serine obtained from pea seedlings which had been fed L-[*methyl*-^{14}C]methionine was labelled solely at C-3 (0.14% incorporation).

3 Biosynthesis of Tropic Acid

Phenylalanine as a Precursor.—Tropic acid is found in Nature as the acid moiety of the ester alkaloids hyoscyamine and hyoscine. A large amount of information, some conflicting, has been published in the past decade on the origin of this relatively simple molecule. We will review these data critically in the hope that it will aid work on the currently unsolved problem of its biosynthesis. It was discovered in 1960[36] that the administration of [3-^{14}C]phenylalanine to intact *Datura stramonium* plants yielded tropic acid having essentially all its activity at C-2. Later workers confirmed this result in *D. stramonium* (intact plants)[37] and *D. metel* (sterile root cultures).[38] It was then established that the other carbons of the phenylalanine side-chain were used for the production of the side-chain of tropic acid.[39-41] The pattern of labelling found in the tropic acid after feeding variously labelled phenylalanines indicated that a migration of the carboxy-group from C-2 to C-3 occurs to form the carbon skeleton of tropic acid as illustrated in Scheme 7. The tropic acid formed from [1,3-^{14}C]-phenylalanine had the same ratio of activity at C-1/C-2 as at C-1/C-3 in the

---

[32] D. H. G. Crout, N. M. Davis, E. H. Smith, and D. Whitehouse, *J.C.S. Perkin I*, 1972, 671.
[33] W. C. Evans and J. G. Woolley, *J. Pharm. Pharmacol.*, 1965, **17**, Suppl. 37 S.
[34] E. Leete and J. B. Murrill, *Tetrahedron Letters*, 1967, 1727.
[35] D. H. G. Crout, H. M. Benn, H. Imaseki, and T. A. Geissman, *Phytochemistry*, 1966, **5**, 1.
[36] E. Leete, *J. Amer. Chem. Soc.*, 1960, **82**, 612.
[37] E. W. Underhill and H. W. Youngken, *J. Pharm. Sci.*, 1962, **51**, 121.
[38] D. Gross and H. R. Schütte, *Arch. Pharm.*, 1963, **296**, 1.
[39] E. Leete and M. L. Louden, *Chem. and Ind.*, 1961, 1405.
[40] M. L. Louden and E. Leete, *J. Amer. Chem. Soc.*, 1962, **84**, 1510.
[41] M. L. Louden and E. Leete, *J. Amer. Chem. Soc.*, 1962, **84**, 4507.

administered phenylalanine,[42,43] indicating that a molecular rearrangement of some type was occurring. The use of [U-^{14}C]phenylalanine as a precursor established that the phenyl group was incorporated intact into tropic acid.[43] Shikimic acid, a precursor of the phenyl group of phenylalanine, also served as a precursor of the benzene ring of tropic acid.[43] Before we discuss the mechanism of this rearrangement the status of other compounds which have been suggested as precursors of tropic acid should be evaluated.

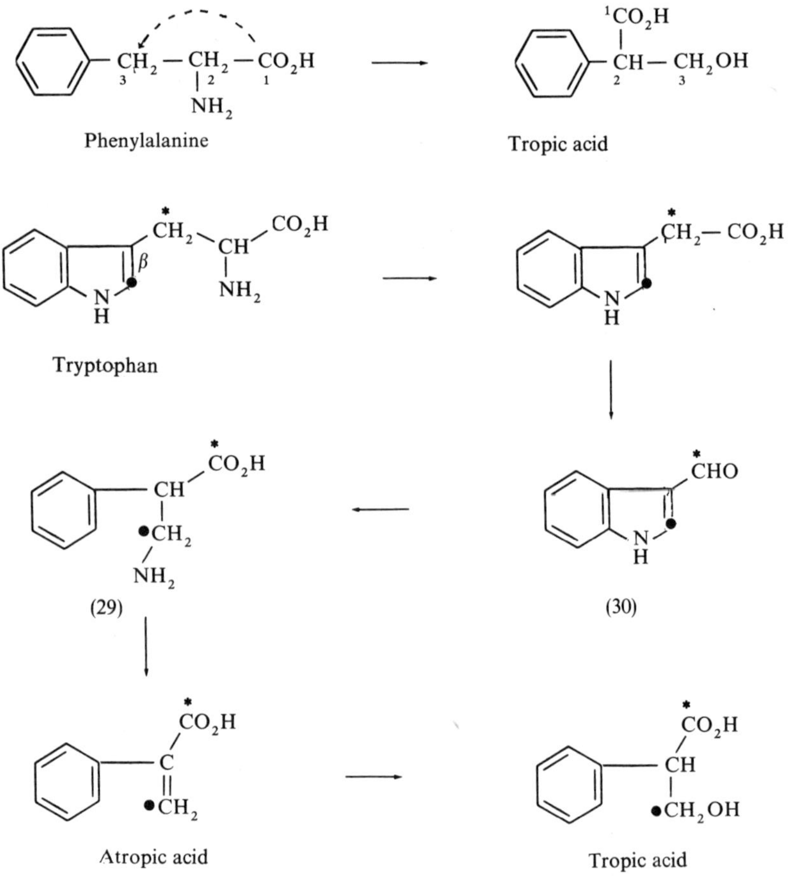

Scheme 7 *Biosynthetic hypotheses for tropic acid*

[42] E. Leete and M. L. Louden, *Abh. dtsch.·Akad. Wiss. Berlin, Kl. Chem., Geol., Biol.*, Nr. 3, 1966, 538.
[43] C. A. Gibson and H. W. Youngken, *J. Pharm. Sci.*, 1967, **56**, 854.

Biosynthesis of Alkaloids

Tryptophan and Phenylacetic acid as Precursors.—In 1961 it was reported[44] that the administration of [β-^{14}C]tryptophan to *D. stramonium* yielded tropic acid labelled on its carboxy-group. The incorporation of activity was extremely low (0.003 %),[45] but this result was so remarkable that this Reporter repeated the feeding of [β-^{14}C]tryptophan to *D. stramonium*.[46] The resultant tropic acid was completely inactive. Interest in the conversion of tryptophan into tropic acid has now been revived by the recent work of Hamon and Eyolfson,[47] reported in detail in the Table (see under *D. innoxia*) at the end of this chapter. The tracer results which they obtained support the biosynthetic route illustrated in Scheme 7. Some of the steps in the sequence have no biochemical or chemical analogy, especially the conversion of indole-3-aldehyde (30) into α-phenyl-β-alanine (29). The tropic acid derived from [aryl-U-^{14}C]tryptophan had most of its activity (94 %) located in its benzene ring. It was also claimed that tropic acid derived from [2-^{14}C]tryptophan had most of its activity (61 %) residing on the hydroxymethyl group (C-3). This publication contains other surprising and irrational results which will be discussed later. Serious questions can be raised concerning the validity of the degradations, which were carried out on tropic acid samples of relatively low activity. One can also be concerned about the radiochemical purity of the isolated alkaloids, and the tropic acid obtained on hydrolysis. Thus the alkaloids (hyoscyamine and hyoscine) derived from [aryl-U-^{14}C]tryptophan had apparently 17 % of their activity located in their tropane moieties. There is no reasonable way in which the benzene ring of tryptophan could provide such a substantial amount of activity in the tropane nucleus of the alkaloids. It seems highly unlikely that tropic acid would be produced by two quite different biochemical mechanisms in the same species. Repetition of this work using the feeding methods of Hamon is certainly required.

Phenylacetic acid is another compound which has served as a precursor of tropic acid.[37,47] Good incorporations (compared with phenylalanine) of [1-^{14}C]phenylacetic acid into tropic acid have been reported. Degradations carried out on this tropic acid were consistent with all the activity being located at C-3. In view of the work with the doubly labelled phenylalanine, it is unlikely that phenylacetic acid is an intermediate between phenylalanine and tropic acid. The incorporation of phenylacetic acid can be rationalized if it is assumed that it undergoes a carboxylation leading to phenylpyruvic acid and thence to phenylalanine by transamination. This reaction has been observed by Allison in ruminal bacteria[48] and photosynthetic anaerobic bacteria.[49] He showed that phenylalanine derived from [1-^{14}C]phenylacetic was labelled solely at C-2.[49]

The Reporter, in his early work,[36] found that none of the usual one-carbon

---

[44] A. M. Goodeve and E. Ramstad, *Experientia*, 1961, **17**, 124.
[45] Reported by S. Agurell, *Abh. dtsch. Akad. Wiss. Berlin, Kl. Chem., Geol., Biol.*, Nr. 3, 1966, 538.
[46] E. Leete, unpublished work, reported in ref. 42.
[47] N. W. Hamon and J. L. Eyolfson, *J. Pharm. Sci.*, 1972, **61**, 2006.
[48] M. J. Allison, *Biochem. Biophys. Res. Comm.*, 1965, **18**, 30.
[49] M. J. Allison and I. M. Robinson, *J. Bacteriol.*, 1967, **93**, 1269.

sources (DL-[*methyl*-¹⁴C]methionine, [¹⁴C]formate, [¹⁴C]formaldehyde) served as a precursor of any of the carbons of tropic acid. Activity was found in the alkaloids, but it was located on the *N*-methyl group of the tropane bases. Underhill and Youngken[37] also found that [¹⁴C]formate and [3-¹⁴C]serine yielded radioactive hyoscyamine and hyoscine which had insignificant amounts of activity in their tropic acid moieties. Hamon and Eyolfson now report[47] that [3-¹⁴C]serine afforded tropic acid with significant activity at C-1 (58%) and C-3 (29%). The administration of [¹⁴C]formate resulted in a similar distribution of activity: C-1 (41%), C-3 (21%). These mysterious and conflicting results of Hamon may possibly be explained on the basis of the transformation of the administered compounds by micro-organisms prior to incorporation into tropic acid. This is suggested because the amounts of compounds actually administered to the *D. innoxia* roots were extremely small (0.004 mg of [3-¹⁴C]-serine, 0.002 mg of [1-¹⁴C]phenylacetic acid, 0.005 mg of DL-[*aryl-U*-¹⁴C]-tryptophan.)

Mechanism of the Rearrangement of the Phenylalanine Side-chain.—[2-¹⁴C,-*aryl*-³H]Phenylalanine was incorporated into tropic acid with retention of all the tritium,[50] indicating that no change or substitution occurs in the benzene ring during the rearrangement. However, loss of tritium takes place when it is located on the side-chain of phenylalanine. When [1-¹⁴C,2-³H]phenylalanine was fed to a sterile culture of *D. metel* roots the resultant tropic acid had lost 93% of the ³H relative to ¹⁴C.[50,51] The tropic acid derived from [2-¹⁴C,2,3-³H₃]-phenylalanine retained 39% of the ³H relative to ¹⁴C. This result is consistent with the retention of one of the hydrogens at C-3, and loss of the hydrogen at C-2, if it is assumed that the three hydrogens at C-2 and C-3 of phenylalanine were equally labelled with ³H (this was not stated and the method of synthesis of the labelled phenylalanine not described). The loss of ³H from C-2 of phenylalanine is consistent with the formation of phenylpyruvic acid by a transamination reaction. Phenylpyruvic acid certainly serves as a precursor of tropic acid, and in competition with phenylalanine (see Table under *D. stramonium*) it is almost as efficient a precursor of tropic acid.[50] However, the expected facile interconversion of these two compounds does not enable us to make a decision as to which is biosynthetically closer to tropic acid. The same remarks apply to phenyl-lactic acid, the reduction product of phenylpyruvic acid. This acid is esterified with tropine in the alkaloid littorine (36), and phenylalanine has been established as its precursor in *D. sanguinea*[52] and *D. innoxia*.[53] Phenyl-lactic acid is a good precursor of tropic acid,[50,53] and was reported to be superior to phenylalanine in competitive feeding experiments.

Cinnamic acid, a metabolite of phenylalanine in many plant species, failed

---

[50] (a) H. W. Liebisch, G. C. Bhavsar, and H. J. Schaller, ref. 2, p. 233; (b) H. W. Liebisch, Abstracts of the Seventh International Symposium on the Chemistry of Natural Products, Riga, 1970, p. 557.
[51] H. R. Schütte and H. W. Liebisch, *Z. Pflanzenphysiol.*, 1967, **57**, 440.
[52] W. C. Evans and V. A. Woolley, *Phytochemistry*, 1969, **8**, 2183.

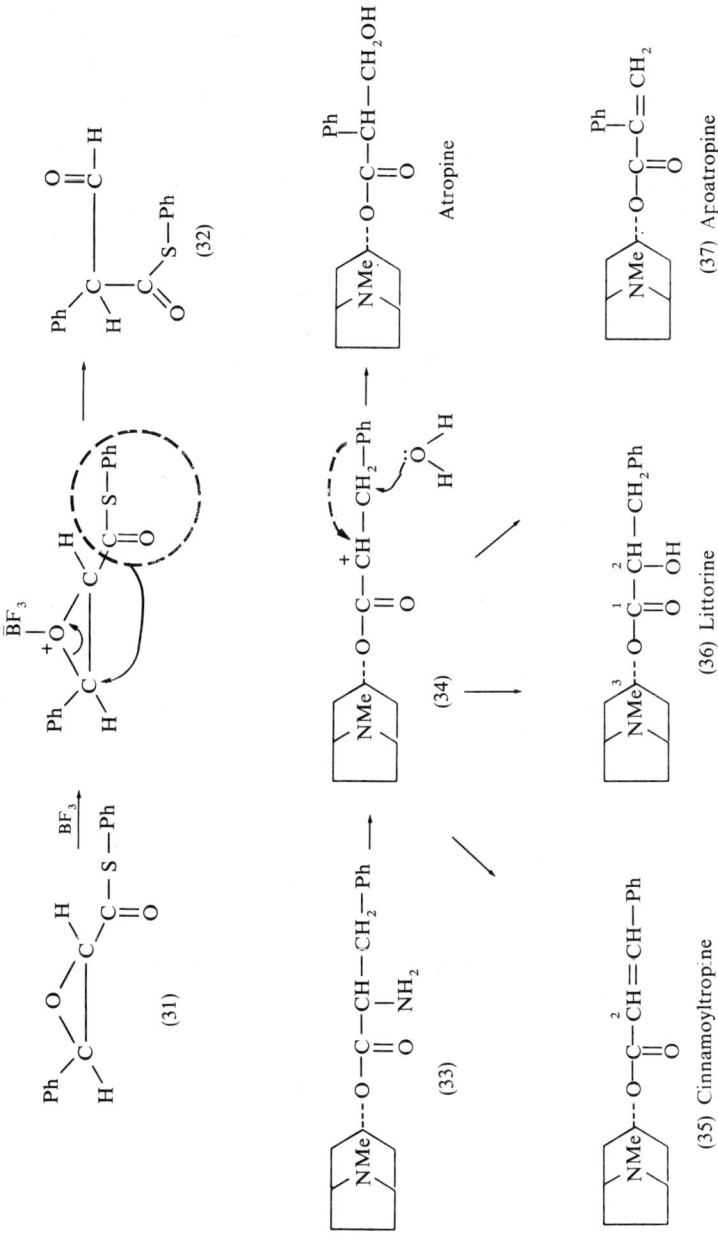

Scheme 8 In vitro transformations as models for tropic acid biosynthesis

to serve as a precursor of tropic acid.[50b, 53] This compound was a prime candidate as an intermediate between phenylalanine and tropic acid since phenyl *trans*-3-phenylthiolglycidate (31) on treatment with boron trifluoride rearranges to phenyl 2-formylphenylthiolactate (32)[54] as illustrated in Scheme 8. This reaction involves the 1,2-migration of a thiol ester group,[55] analogous to the rearrangement which occurs in the biosynthesis of tropic acid. However, negligible activity was detected in the tropic acid moiety of hyoscyamine and hyoscine isolated from *D. stramonium* plants which had been fed potassium [2-^{14}C]β-phenylglycidate.[56] It was considered that perhaps rearrangement of the C_6–C_3 skeleton occurs after esterification of tropine with an acid derived from phenylalanine. Thus [2-^{14}C]cinnamoyl-[*N-methyl*-^{14}C]tropine (35) was fed to *D. stramonium* plants.[57] Activity was found in both hyoscyamine (0.94% inc.) and hyoscine (0.91% inc.). However, it was found that all the activity was located on the *N*-methyl groups of the alkaloids, indicating that the administered ester had undergone hydrolysis in the plant, affording tropine and cinnamic acid, the latter not being used at all for the biosynthesis of tropic acid. Littorine has recently been shown to be widely distributed in *Datura* species.[58] This ester was also tested as a precursor of hyoscyamine.[57] Dual labelled littorine {3α-([1-^{14}C]-2-hydroxy-3-phenylpropionyloxy)-[3-^{3}H]tropane, ^{3}H:^{14}C = 6.8} was fed to *D. stramonium* plants which yielded radioactive hyoscyamine (^{14}C inc. 0.06%) (^{3}H:^{14}C = 31). Both halves of the littorine molecule were incorporated into hyoscyamine, but the drastic change in the ^{3}H:^{14}C ratio indicated that the ester had been hydrolysed to tropine and phenyl-lactic acid, the latter undergoing a rearrangement to tropic acid prior to esterification with tropine.

It has been demonstrated that DL-phenylalanine-3α-tropanyl ester (33) is converted into atropine (DL-hyoscyamine), littorine, apoatropine (37), and cinnamoyltropine, on treatment with nitrous acid.[59] This reaction presumably involves the intermediate formation of the carbonium ion (34). The products obtained are derivable from this carbonium ion either by loss of a proton (cinnamoyltropine), hydration (littorine), migration of the phenyl group and loss of a proton (apoatropine), or migration of the phenyl group and hydration of the new carbonium ion (atropine). This reaction thus has no analogy to the formation of atropine *in vivo* where a migration of the carboxy-group occurs. The nitrous acid deamination of the ethyl ester of phenylalanine which yields tropic acid,[60] also involves a phenyl migration.[60b, 61]

[53] W. C. Evans, J. G. Woolley, and V. A. Woolley, ref. 2, p. 227.
[54] J. N. Wemple, *J. Amer. Chem. Soc.*, 1970, **92**, 6694.
[55] J. Domagala and J. Wemple, *Tetrahedron Letters*, 1973, 1179.
[56] E. Leete and J. D. Braunstein, unpublished work, *cf.* J. D. Braunstein, Ph.D. dissertation, Univ. of Minnesota, 1971.
[57] E. Leete and E. P. Kirven, unpublished work.
[58] W. C. Evans, A. Ghani, and V. A. Woolley, *Phytochemistry*, 1972, **11**, 2527.
[59] Y. Takeuchi, K. Koga, T. Shiori, and S. Yamada, *Chem. and Pharm. Bull. (Japan)*, 1971, **19**, 2603.
[60] (*a*) S. Yamada, T. Kitagawa, and K. Achiwa, *Tetrahedron Letters*, 1967, 3007; (*b*) K. Koga, C. C. Wu, and S. Yamada, *ibid.*, 1971, 2283, 2287; (*c*) K. Koga, C. C. Wu, and S. Yamada, *Chem. and Pharm. Bull. (Japan)*, 1972, **20**, 1272, 1282.
[61] E. Leete, *Tetrahedron Letters*, 1968, 5793.

4 Enzyme Studies

Nicotiana **Species.**—Some progress is being made in isolating the enzymes responsible for the biosynthesis of nicotine. The immediate precursor of the pyrrolidine ring of nicotine is the 1-methyl-1-pyrrolinium salt (42).[62,63] This is formed from putrescine [(38), the decarboxylation product of ornithine] by the route illustrated in Scheme 9. A putrescine-*N*-methyltransferase was

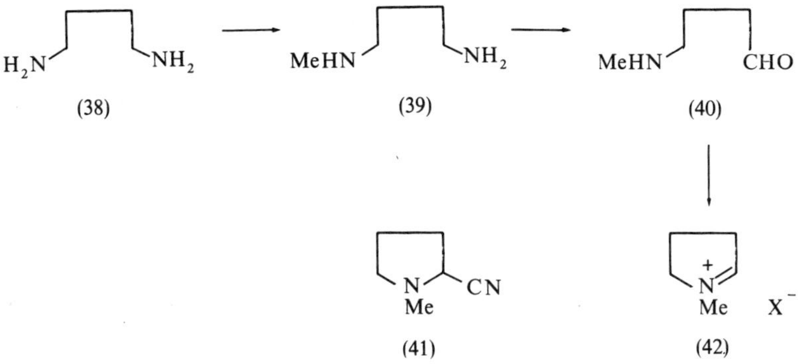

H_2N⌐⌐NH_2 ⟶ MeHN⌐⌐NH_2 ⟶ MeHN⌐⌐CHO

(38)   (39)   (40)

⌐⌐$\underset{Me}{N}$–CN   ⌐⌐$\underset{Me}{\overset{+}{N}}$ X^-

(41)   (42)

Scheme 9 *Enzymatic reactions in tobacco*

isolated from tobacco roots,[64] and this enzyme catalysed the formation of *N*-methylputrescine (39) from putrescine and *S*-adenosyl-L-methionine. The next step is catalysed by an *N*-methylputrescine oxidase, also isolated from *Nicotiana tabacum* roots.[65] The product of this oxidation is 4-methylaminobutanal (40) which cyclizes in acidic media to the pyrrolinium salt. The enzyme was assayed by using as a substrate [*N-methyl*-^{14}C]methylputrescine, and then measuring the radioactivity found in 2-cyano-1-methylpyrrolidine (41), obtained by the reaction of potassium cyanide with the pyrrolinium salt. A 152-fold purification of the enzyme was achieved by ammonium sulphate fractionation, chromatography on DEAE-cellulose, and gel filtration on Sephadex G-200 gel. The pH optimum for the enzyme was 8.0, activity being reduced by half at pH 7.2 and 8.7. Complete inhibition of the enzyme occurred on treatment with diethyldithiocarbamate, a reagent which is usually considered to be an inhibitor of copper-containing enzymes. The enzyme was found only in the roots of *N. tabacum*, the level of activity being increased by prior decapitation of the tobacco plants. Putrescine and cadaverine were also oxidized by this enzyme, whereas histamine, spermine, tyramine, n-hexylamine, and phenethylamine were unaffected.

[62] E. Leete, *J. Amer. Chem. Soc.*, 1967, **89**, 7081.
[63] S. Mizusaki, T. Kisaki, and E. Tamaki, *Plant Physiol.*, 1968, **43**, 93
[64] S. Mizusaki, Y. Tanabe, M. Noguchi, and E. Tamaki, *Plant Cell Physiol.*, 1971, **12**, 633.
[65] S. Mizusaki, Y. Tanabe, M. Noguchi, and E. Tamaki, *Phytochemistry*, 1972, **11**, 2757.

Long ago it was shown that the β-carbon of serine[66] and the α-carbon of glycolic acid[67] and glycine[66] served as precursors of the N-methyl group of nicotine, in addition to the usual methyl donors such as methionine. Glycine and glyoxalate decarboxylation have been studied in isolated mitochondria obtained from *N. rustica* roots.[68] Glycine was decarboxylated yielding principally carbon dioxide, formaldehyde, and formic acid. In the presence of tetrahydrofolic acid, serine was produced in large amounts. The formic acid appeared to be formed by oxidation of formaldehyde and was not an immediate product of glycine decarboxylation. It was not possible to obtain an active soluble preparation from the roots, and mitochondrial integrity is apparently essential for glycine decarboxylation. Pyridoxal phosphate was a required cofactor. Glyoxalate in the presence of pyridoxamine phosphate also underwent decarboxylation yielding formaldehyde.

Nicotinic acid is a precursor of the pyridine ring of nicotine and related alkaloids such as anabasine. The carboxy-group is lost at some stage in the biosynthetic sequence leading to nicotine. A particulate fraction has been isolated from *N. rustica* roots which catalyses the decarboxylation of nicotinic acid.[69] This particulate fraction which contained mitochondria was assayed by measuring the release of $^{14}CO_2$ from [7-^{14}C]nicotinic acid. Oxygen was required for enzymatic activity. Quinolinic acid (pyridine-2,3-dicarboxylic acid) was not decarboxylated in this system. No cofactors could be established. Crude extracts of the leaves and stems of the tobacco plant yielded fractions which had only 1—5% of the activity of the root extract. No comparable activity was found in an extract of *Ricinus communis* cotyledons [this species produces ricinine (44) which is formed from nicotinic acid without loss of the carboxy-group].

Ricinus communis.—The specific intermediates between nicotinic acid and ricinine are unknown; a reasonable biosynthetic pathway is illustrated in Scheme 10. Robinson and co-workers[70] have obtained a crude enzyme from *R. communis* seedlings, and resolved it by chromatography on DEAE-cellulose into three components, all of which catalysed the oxidation of 3-cyano-1-methylpyridinium perchlorate (43) to the pyridones (46) and (47). The optimum activity for all these fractions was between pH 9.5 and 10.5. The enzymes were relatively non-specific. All the following pyridinium salts were oxidized: 3-formyl-1-methylpyridinium iodide, 3-nitro-1-methylpyridinium iodide, 3-acetyl-1-methylpyridinium iodide, 3-cyano-1-ethylpyridinium iodide, and 1-benzyl-3-cyanopyridinium chloride. N-Methylnicotinamide, trigonelline sulphate, 1-methylpyridinium iodide, nicotinic acid, 1-methylquinolinium iodide, and 3-cyanopyridine, were not oxidized to any appreciable extent.

---

[66] R. U. Byerrum, R. L. Hamill, and C. D. Ball, *J. Biol. Chem.*, 1954, **210**, 645.
[67] R. U. Byerrum, L. J. Dewey, R. L. Hamill, and C. D. Ball, *J. Biol. Chem.*, 1956, **219**, 345.
[68] C. W. Prather and E. C. Sisler, *Phytochemistry*, 1972, **11**, 1637.
[69] J. L. R. Chandler and R. K. Gholson, *Phytochemistry*, 1972, **11**, 239.
[70] P. Fu, J. Kobus, and T. Robinson, *Phytochemistry*, 1972, **11**, 105.

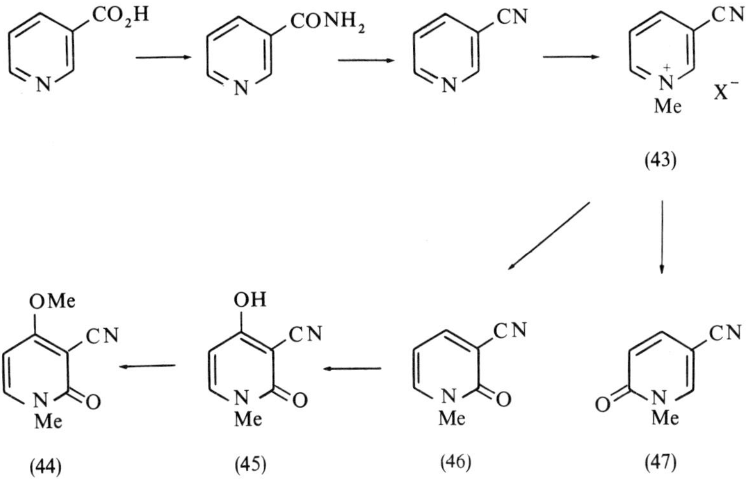

Scheme 10 *Enzymatic reactions in* Ricinus communis

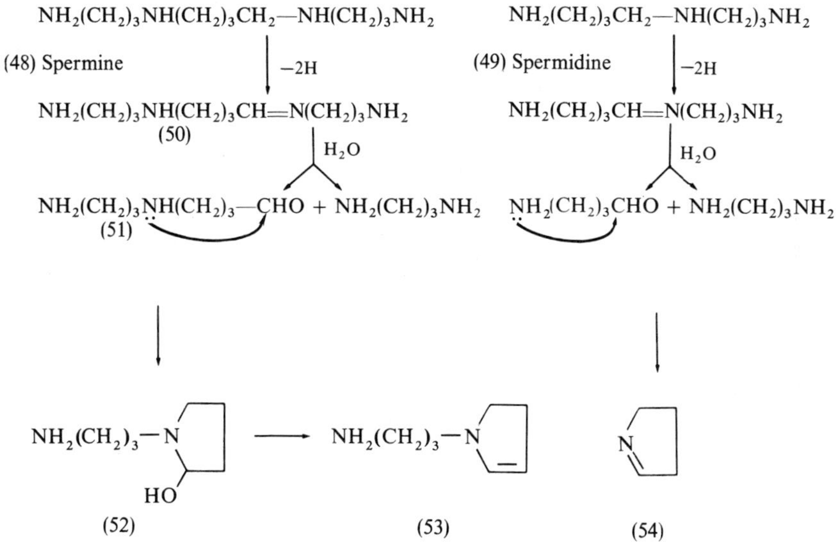

Scheme 11 *Enzymatic oxidation of spermine and spermidine*

Phalaris tuberosa.—A tryptophan decarboxylase, purified 20-fold, was isolated from seedlings of this species, a member of the Gramineae.[71] The enzyme activity was highest from 4-day-old seedlings, and was assayed by measuring the amount of [β-^{14}C]tryptamine formed from L-[β-^{14}C]tryptophan, or by the production of $^{14}CO_2$ from L-[*carboxyl*-^{14}C]tryptophan. The enzyme was specific for L-tryptophan, the D-isomer being unaffected. DL-5-Hydroxytryptophan was decarboxylated yielding 5-hydroxytryptamine. *NN*-Dimethyltryptamine, one of the major alkaloids of this species, inhibited the enzyme. This observation may indicate that the tryptamine is translocated away from the site of tryptophan decarboxylation prior to N-methylation.

Hordeum (Barley).—Smith[72] has isolated from barley shoots a particulate fraction which contains an amine-oxidase. This enzyme converts spermine (48) into 1,3-diaminopropane and 1-(3-aminopropyl)-Δ^2-pyrroline (53). The spermine is presumably dehydrogenated, affording the Schiff base (50), which on hydrolysis yields 1,3-diaminopropane and the amino-aldehyde (51). Cyclization of this compound yields the methanolamine (52) and thence the pyrroline by dehydration (Scheme 11). Spermidine (49) yields Δ^1-pyrroline (54) and 1,3-diaminopropane by a similar series of reactions. This amine-oxidase was also found in oats (*Avena sativa*), maize (*Zea mays*), wheat (*Triticum vulgare*), and rye (*Secale cereale*).

Mercurialis perennis.—When aliphatic aldehydes are fed to this plant they are converted into the corresponding primary amines. A transaminase has been isolated[73] which catalyses the transamination between α-amino-acids and aldehydes. L-Alanine was the most efficient amino donor, and all aldehydes in the homologous series from ethanal to undecanal were active amino acceptors. No requirement for pyridoxal phosphate could be demonstrated. In a later publication[74] it was shown that this reaction takes place in many flowering plants.

5 Alkaloid Production in Tissue Cultures

It was thought that tissue cultures derived from various parts of whole plants would be an ideal system in which to study alkaloid biosynthesis, since sterile conditions and uniform conditions of growth are readily maintained. However, work in this area has been disappointing. In general, tissue cultures produce much less alkaloid than an intact plant, or intact organs. For example, cultures of excised roots of tomato (*Lycopersicum esculentum*) synthesized the steroidal alkaloid tomatine; however, the amount produced was 40 times less than that found in the intact seedling radicles.[75] The addition of established precursors

---

[71] C. Baxter and M. Slaytor, *Phytochemistry*, 1972, **11**, 2763.
[72] (a) T. A. Smith, *Xenobiotica*, 1971, **1**, 449; (b) T. A. Smith, *Phytochemistry*, 1972, **11**, 899.
[73] T. Hartmann, D. Dönges, and M. Steiner, *Z. Pflanzenphysiol.*, 1972, **67**, 404.
[74] T. Hartmann, H.-I. Ilert, and M. Steiner, *Z. Pflanzenphysiol.*, 1972, **68**, 11.
[75] J. G. Roddick and D. N. Butcher, *Phytochemistry*, 1972, **11**, 2019.

of tomatine, acetic acid or cholesterol, did not increase the amount of alkaloid produced.[76] Surprisingly, mevalonic acid reduced the level of tomatine. Tris-(2-diethylaminoethyl)phosphate, a compound which blocks the conversion of lanosterol into zymosterol, decreased the amount of tomatine produced; however, the overall root growth was also inhibited, and thus the concentration of tomatine remained the same.

A cell suspension culture of *Datura ferox*, when supplied with DL-[2-^{14}C]-ornithine, yielded radioactive putrescine, citrulline, arginosuccinate, arginine, γ-aminobutyric acid, glutamic acid, aspartic acid, α-keto-δ-guanidovalerate, and α-keto-δ-aminovalerate. However, none of the tropane alkaloids produced by this species was radioactive.[77] It was suggested that this *in vitro* cell culture lacks the enzyme which catalysed the reaction between Δ^1-pyrroline-2-carboxylic acid (55) and acetoacetyl coenzyme A. The product of this condensation ultimately yields hygrine (57) and then tropine (56) as illustrated in Scheme 12.

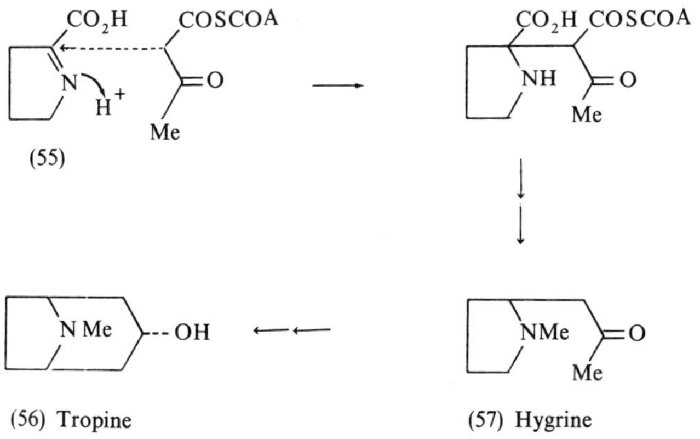

(55)

(56) Tropine

(57) Hygrine

Scheme 12

A tissue culture of *Scopolia parviflora* produced an extremely small amount of atropine (0.003 %) compared with intact rhizomes (0.34 %).[78] The addition of tropic acid increased the yield of alkaloid to 0.08 %, suggesting that the tissue culture lacked the capability of synthesizing tropic acid. It has been previously shown that the level of tropane alkaloids in *Hyoscyamus niger*[79] and *Datura stramonium*[80] is lower in callus tissue cultures than in intact plants.

[76] J. G. Roddick and D. N. Butcher, *Phytochemistry*, 1972, **11**, 2991.
[77] H. Elze and E. Teuscher, ref. 2, p. 239.
[78] M. Tabata, *Phytochemistry*, 1972, **11**, 949.
[79] J. Telle and R. J. Gautheret, *Compt. rend.*, 1947, **224**, 1653.
[80] T. M. Steinstra, *Proc. K. Ned. Acad. Wet.*, 1954, **57**, 584.

Callus and suspension cultures of *Nicotiana tabacum* have been studied.[81] Stem callus tissues did not produce nicotine until roots were developed. The fungicide cycloheximide (58) initiated the synthesis of nicotine. The only known precursor of nicotine which increased the yield of the alkaloid was *N*-methylputrescine.

A cell suspension culture established from the roots of *Phaseolis vulgaris* produced no alkaloids when the culture medium contained only inorganic nitrogen (ammonium and nitrate). Addition of tryptophan resulted in the formation of harman (60) and norharman (59).[82]

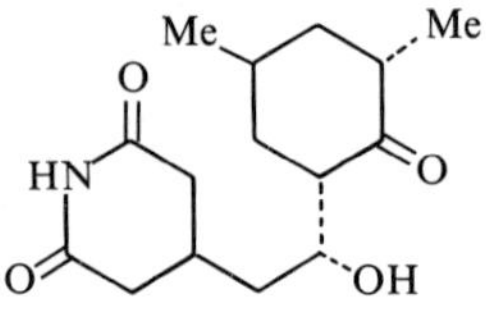

(58) Cycloheximide 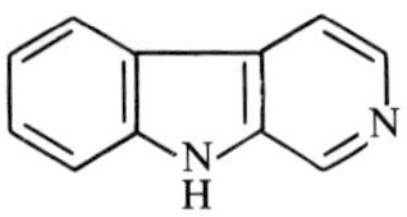 (59) Norharman 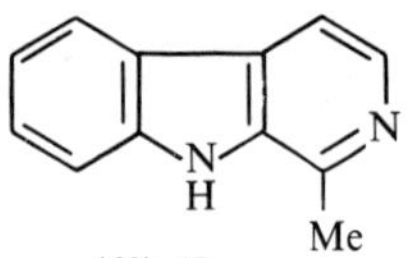 (60) Harman

6 Table of Tracer Work Relating to Alkaloid Biosynthesis

Species are listed in alphabetical order. The position of isotopes in the precursors fed and in the labelled alkaloids formed are indicated with a heavy dot or an asterisk. Incorporations are reported either as *Absolute incorporations* (the total radioactivity found in the isolated alkaloid divided by the total radioactivity fed to the plant) or *Specific incorporations* (the specific activity of the isolated alkaloid divided by the specific activity of the administered compound). Specific incorporations are recorded in square brackets [ ].

References are recorded at the end of the Table unless the data have been discussed earlier in the chapter.

---

[81] D. Neumann and E. Müller, *Biochem. Physiol. Pflanz.*, 1971, **162**, 503.
[82] I. A. Veliky, *Phytochemistry*, 1972, **11**, 1405.

Table*

Species Compound administered	Alkaloids labelled	Incorporation/%	Ref.
Adhatoda vasica (all compounds fed to roots for 2 days)	Vasicine (Peganine)		6

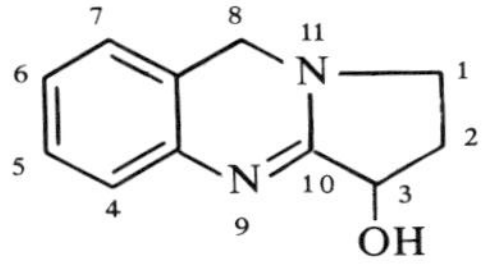

Vasicine structure with positions labelled: 1, 2, 3-OH, 4, 5, 6, 7, 8, 9, 10, 11 (N at 10 and 11)

Distribution of activity (%)

	C-1	C-2	C-3	C-10
Sodium [1-^{14}C]acetate	4.6	2.4	23.9	21.4	[0.51]
Sodium [2-^{14}C]acetate	14.4	3.4	55.7	0	[0.23]
Sodium [3-^{14}C]pyruvate	2		84	6.8	[0.046]
[2-^{14}C]Glycine	7.6	nd	47.2	nd	[0.12]
N-[2'-^{14}C]Acetyl[^{15}N]anthranilic acid	0	0	97.8	0	^{14}C: [0.32] ^{15}N: [0.34]

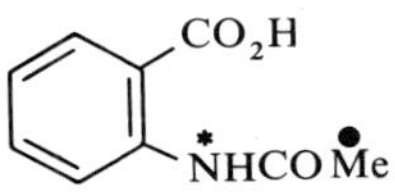

[structure: 2-(N-acetylamino)benzoic acid with CO$_2$H and *NHCOṀe labels; * marks ^{14}C, ● marks ^{15}N]

| [4-^{14}C]4-Amino-2-hydroxybutanoic acid | | 10.7 | 1.3 | 19 | 1.7 | [0.15] |

[nd—not determined]

All Table references are on p. 182.

Species

Compound administered	Alkaloids labelled	Incorporation/%	Ref.
Anabasis aphylla [2-^{14}C]Lysine (fed in October) (The results of a feeding carried out in July are also reported, with the same trends in activity in the various alkaloids, except that the high lupinine activity was not found in the lower parts of the plant)	Anabasine — Distribution of activity in alkaloids: lower parts of plant 17%, higher parts 28%; Anabasamine 21 / 15; Lupinine 40 / 18		83

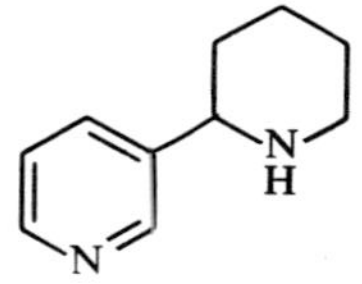

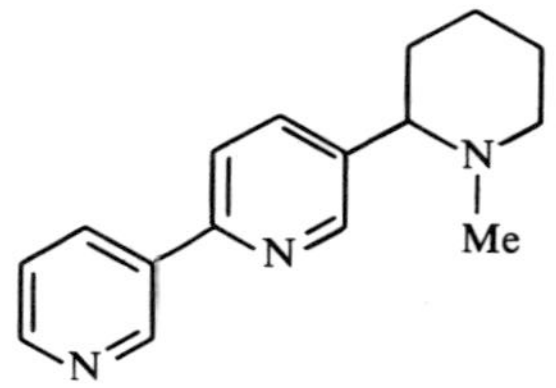

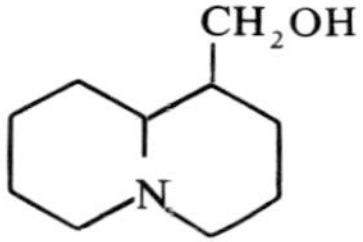

Anabasine

Anabasamine

Lupinine

Aphylline 11 20

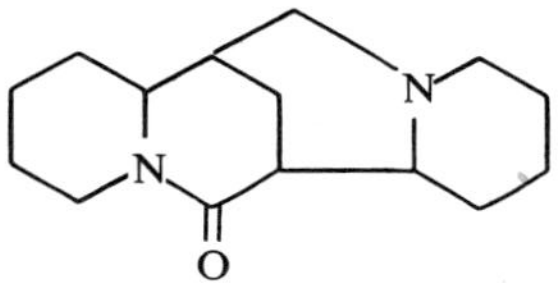

[structure of aphylline: bicyclic with N, N, C=O]

[2-¹⁴C]Lysine
(fed for 3 days to shoots of the plant which were pruned and dipped into a solution of the lysine)

Unidentified alkaloids	10	22
Anabasine (the most active)	+++	84
Anabasamine		
Aphylline } Lesser activity		
Lupinine		

(A higher incorporation was obtained using plants pruned in July compared with those fed in October.)

Magnoflorine 12.1 85

[structure of magnoflorine with NMe₂⁺ X⁻, MeO, HO, HO, MeO groups]

(>90% on the N-Me group)

Aquilegia species
('McKana hybrid')
(RS)-[N-methyl-1-¹⁴C]Reticuline
(fed by wick method)

[structure of reticuline with N—Me (labelled), MeO, HO, HO, MeO groups]

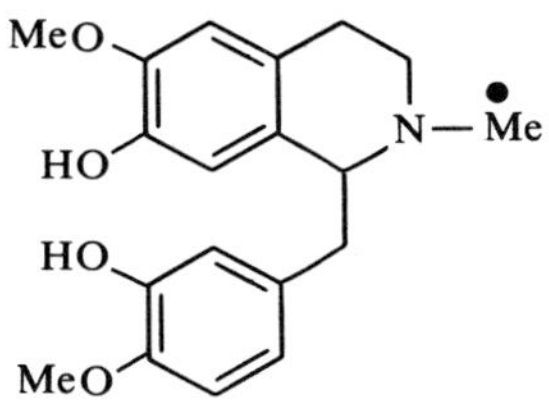

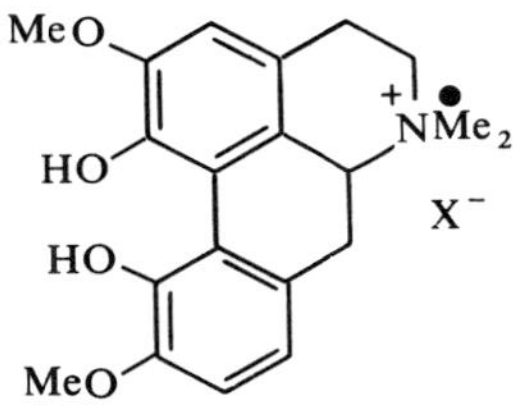

Species	Compound administered	Alkaloids labelled	Incorporation/%	Ref.
Aquilegia—contd.	(RS)-[3-^{14}C,7-O-methyl-^{3}H]Norprotosinomenine	Magnoflorine	none	
Cephaelis acuminata (all compounds fed 10 days to 4—5-year-old plants in summer except where indicated)	DL-[3-^{14}C]Tyrosine	Cephaeline	0.0064	86

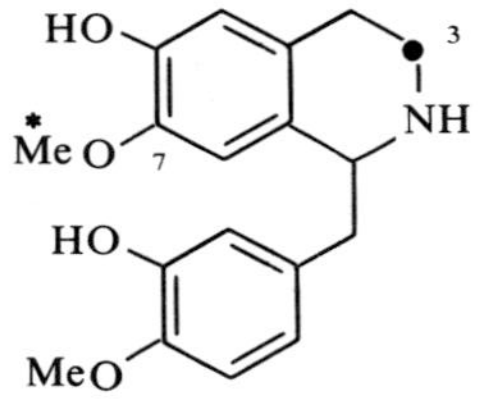

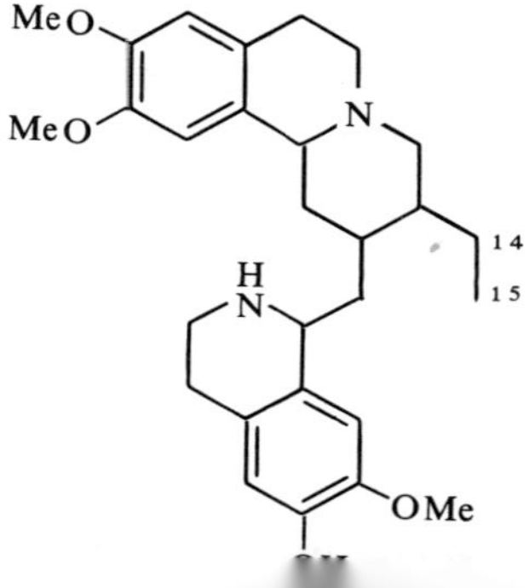

Precursor	Product	Incorporation
[1,4-^{14}C]Succinic acid	Cephaeline	inactive
[2,3-^{14}C]Succinic acid	Cephaeline	0.0006
[1,3-^{14}C]Glycerol	Cephaeline	0.0004
Sodium [2-^{14}C]pyruvate	Cephaeline	0.0002
	Ipecoside	small activity
Sodium [3-^{14}C]pyruvate	Cephaeline	0.0002
L-[U-^{14}C,4,5-^{3}H$_2$]Leucine	Cephaeline	inactive
Sodium [1-^{14}C]acetate (fed 2 days)	Cephaeline	inactive
Sodium [1-^{14}C]acetate (fed 4 days)	Cephaeline	0.0001
Sodium [2-^{14}C]acetate	Cephaeline	0.0001
	(The cephaeline from the second feeding experiment had C-14: 6.3%, C-15: 7.0%)	
Sodium [1-^{14}C]glycolate	Cephaeline	0.0008
Calcium [2-^{14}C]glycolate	Cephaeline	0.0025
	(C-14: 1.2%, C-15: 1.3%)	0.0216
[1-^{14}C]Glycine	Cephaeline	0.0003
[2-^{14}C]Glycine	Cephaeline	0.022
	(C-14: 0%, C-15: 16%)	
[2-^{14}C]Glycine	Cephaeline	0.0077
	(C-14: 0.7%, C-15: 17.0%)	
[2-^{14}C]Glycine (fed to 2-year-old plant in winter)	Cephaeline	0.012
	(C-14 + C-15: 1.2%)	
	Ipecoside	+
	(C-15 + C-16: 0.6%)	

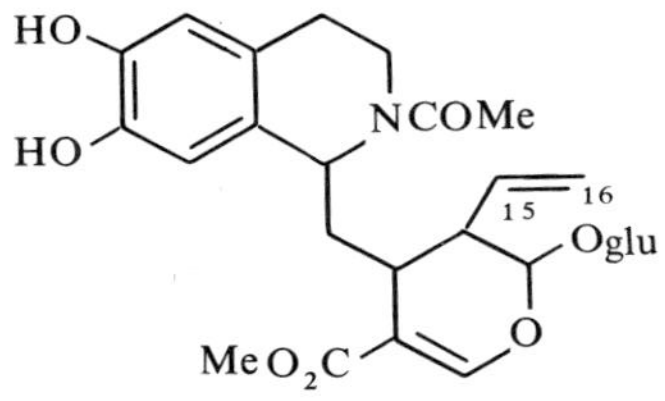

[Structure: cephaeline — dihydroxy-tetrahydroisoquinoline with NCOMe group linked to a dihydropyran bearing MeO$_2$C, vinyl at C-15, and OGlu at C-16]

Species Compound administered	Alkaloids labelled	Incorporation/%	Ref.
Cephaelis acuminata—contd.			
[2-^{14}C]Glycine (fed 21 days)	Cephaeline	0.0004	
Sodium [2-^{14}C]glyoxalate	Cephaeline	0.0061	
	(C-14 + C-15: 0.4%)		
	Ipecoside	+	
	(C-15 + C-16: 2.3%)		
	Activity was also detected in the emetine (OMe ether of cephaeline) in several of the experiments where activity was found in cephaeline.		
Choisya ternata			
[3-^{14}C]Dictamnine 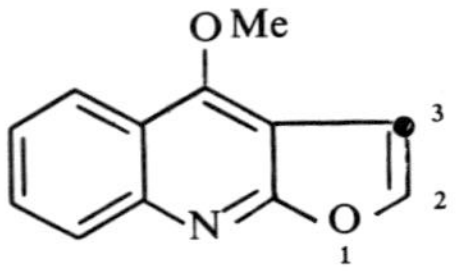	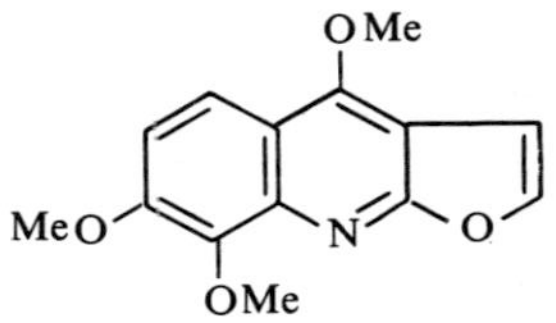 Skimmianine	1.2—3.5	87

(no degradations carried out)

Clivia miniata			
[*O*-methyl-^{14}C,3′,5′-^{3}H$_2$]-*O*-Methylnorbelladine 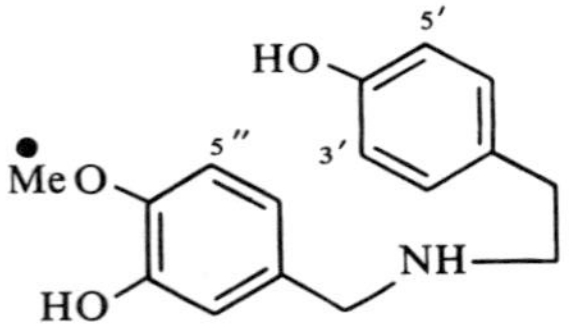	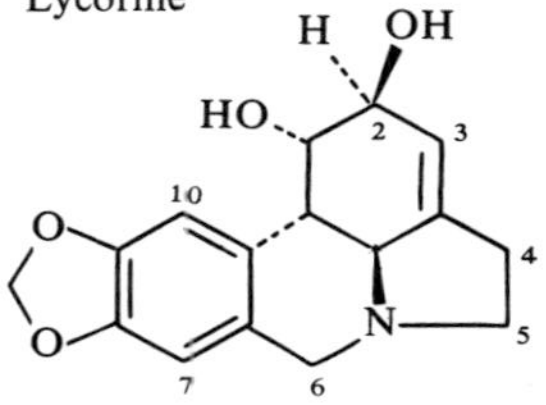 Lycorine	0.7(^{14}C)	88

(no ^{3}H retained)

Clivonine   0.01

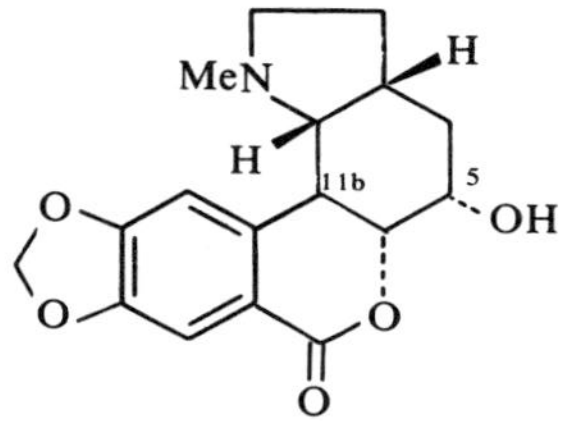

Biosynthesis of Alkaloids

[1-^{14}C,3',5',5'',-^{3}H$_3$]-O-Methylnorbelladine
(equal ^{3}H at the 3 positions)

Lycorine
(37% retention of ^{3}H, all at C-10)
Clivonine
(32% retention of ^{3}H, none at C-5 or C-11b)   +
Lycorine
(92% retention of ^{3}H, all at C-2)   +

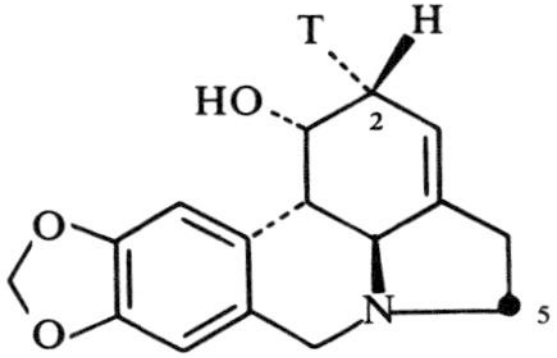

[5-^{14}C,2α-^{3}H]Caranine

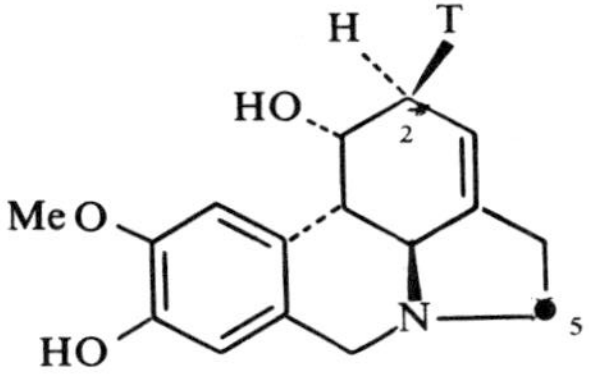

Lycorine
(20% retention of ^{3}H)   8.0 (^{14}C)
Clivonine
(Caranine and norpluvine are thus intermediates between
O-methylnorbelladine and lycorine, no oxo-derivative is
involved as an intermediate and hydroxylation at C-2 occurs
with retention of configuration)

[5-^{14}C,2β-^{3}H]Norpluvine   3.4 (^{14}C)
                                                  inactive

Species Compound administered	Alkaloids labelled	Incorporation/%	Ref.

Coffea arabica
Sodium [^{14}C]bicarbonate
	Caffeine
	
	          (in seeds)
	          (in pericarp)
	        C-2 C-6 N-Me
	seed: 0.7 6.7 12.2
	pericarp: 1.9 7.4 19.4
		0.0076
		0.003
			89

L-[*methyl*-^{14}C]Methionine	Caffeine (not degraded) (in seeds) (in pericarp) (in pericarp)	0.19 3.0—4.7 0.29	
Colchicum autumnale (all compounds fed in spring) [2-^{14}C]Tyramine	Colchicine	0.4	90
[*aryl*-^{3}H]Dopamine	Colchicine	1.1	

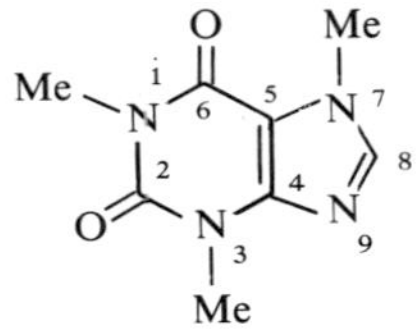

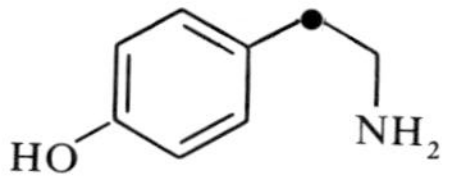

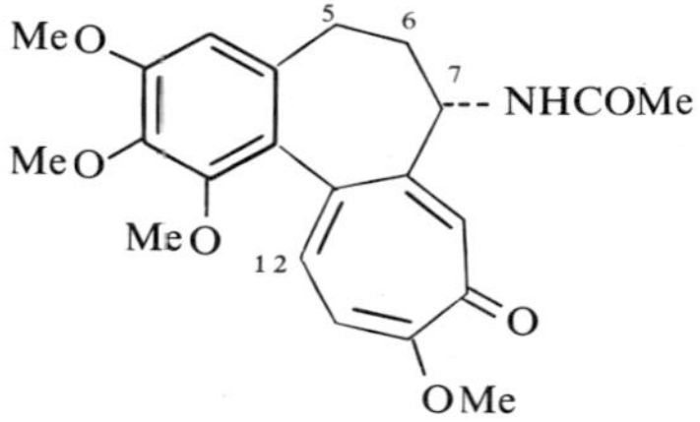

Precursor	Product	Incorporation (%)
(RS)-[8-³H]-1,2,3,4-Tetrahydro-7-hydroxy-6-methoxy-1-phenethyl-2-methylisoquinoline (isoquinoline 1)	Colchicine	0.007
(RS)-[9-¹⁴C]-4'-Hydroxyisoquinoline 1	Colchicine	1.4
(RS)-[9-¹⁴C]-4'-Hydroxy-3'-methoxyisoquinoline 1	Colchicine	3.8
(RS)-[9,¹⁴C]-3',4'-Dihydroxy-5-methoxyisoquinoline 1	Colchicine	0.52
(RS)-[9-¹⁴C]Autumnaline (3'-hydroxy-4',5'-dimethoxyisoquinoline 1)	(C-6: 100%)	9.6
N-Nor[aryl-³H]autumnaline	Colchicine	0.04
O-[¹³H]Methylandrocymbine	Colchicine	15.2

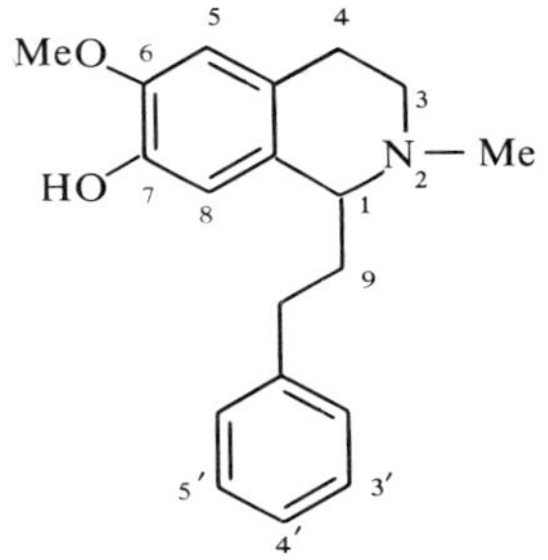

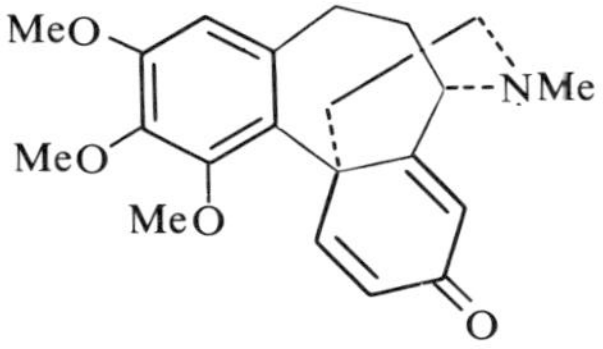

Biosynthesis

Species			
Compound administered	Alkaloids labelled	Incorporation/%	Ref.
Colchicum autumnale—contd.			
DL-[3-^{14}C]Tyrosine	Colchicine (About 85% of the activity at C-12 with only minor activity at other positions)	+	91
	Colchicine	inactive	
1-{[6-^{3}H]-5-Hydroxy-2-hydroxymethyl-4-methoxyphenyl-3-{[2,6-^{3}H]-3-hydroxy-4,5-dimethoxyphenyl}propylamine	Colchicine (90 ± 4% retention of ^{3}H)	0.8 (^{14}C)	92
	Demecolcine (80 ± 9% retention of ^{3}H)	0.011 (^{14}C)	
(2S,3R)-[1-^{14}C,3-^{3}H$_1$]Phenylalanine			

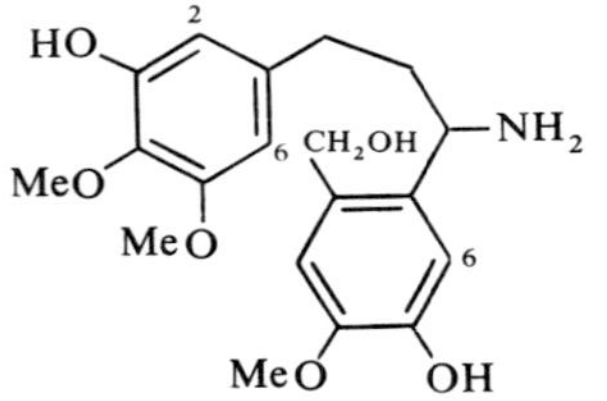

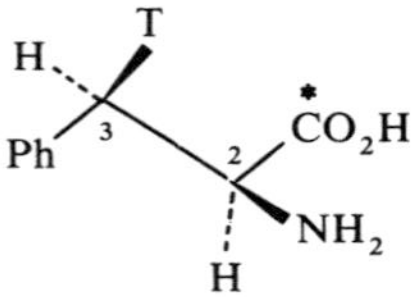

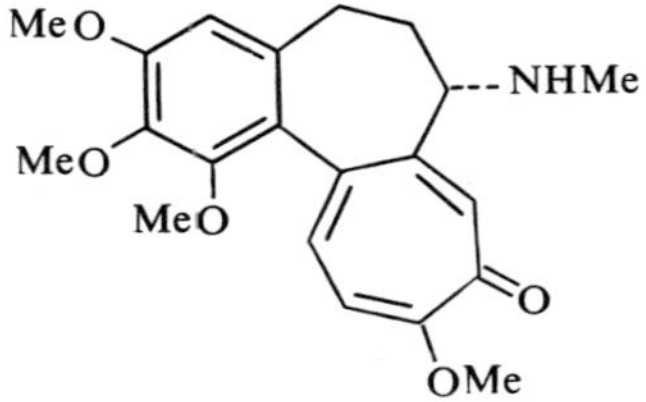

Biosynthesis of Alkaloids

Compound fed	Alkaloid	Incorporation (%)
$(2RS,3S)$-[1-^{14}C,3-^{3}H$_1$]Phenylalanine	Colchicine (only 12 ± 2% retention of ^{3}H)	1.14 (^{14}C)
	Demecolcine (10 ± 3% retention of ^{3}H)	0.008 (^{14}C) 90
	(Therefore it is the 3-*pro-S* hydrogen of phenylalanine which is eliminated in the formation of cinnamic acid which is then incorporated into colchicine and demecolcine)	
Colchicum byzantinum (compounds fed in autumn) [2-^{14}C]*p*-Coumaric acid HO–⟨●CH=CH–CO$_2$H⟩	Colchicine	<0.01
[2-^{14}C]Ferulic acid MeO, HO–⟨●CH=CH–CO$_2$H⟩	Colchicine	<0.01
[2-^{14}C]Isoferulic acid HO, MeO–⟨●CH=CH–CO$_2$H⟩	Colchicine	0.001
	Demecolcine	0.003
3-Hydroxy-4,5-dimethoxy[2-^{14}C]cinnamic acid	Colchicine	0.003
	Demecolcine	0.006
(RS)-[9-^{14}C]-4′-Hydroxyisoquinoline 1 (*cf.* under *C. autumnale*)	Demecolcine	0.30

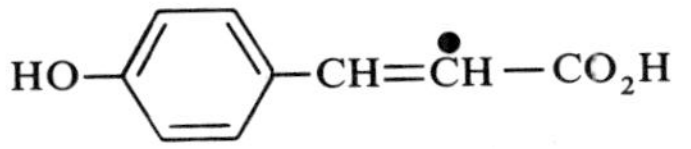

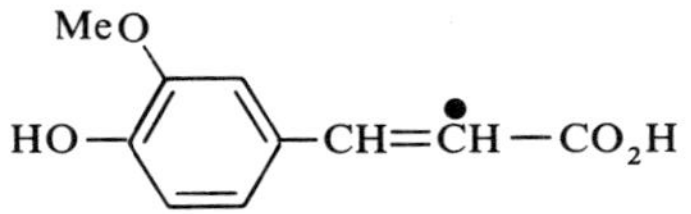

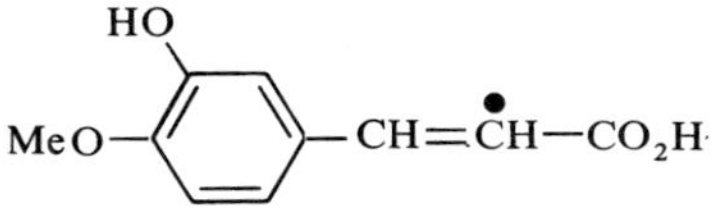

Species / Compound administered	Alkaloids labelled	Incorporation/%	Ref.
Colchicum byzantinum—contd.			
(RS)-[9-^{14}C]-4'-Hydroxy-3'-methoxyisoquinoline 1	Colchicine	0.030	
	Demecolcine	0.22	
(RS)-[9-^{14}C]-3'-Hydroxy-4'-methoxyisoquinoline 1	Colchicine	0.012	
	Demecolcine	0.06	
[9-^{14}C]Autumnaline	Colchicine (all activity at C-6)	0.26	
(RS)-[aryl-^{3}H]-4'-Hydroxy-3',5'-dimethoxyisoquinoline 1	Demecolcine	1.22	
	Colchicine	0.002	
	Demecolcine	0.02	
O-Methyl-[aryl-3H]autumnale	Colchicine	not determined	
	Demecolcine	0.005	
O-[^{3}H]Methylandrocymbine	Colchicine	0.65	
Conium maculatum (Minnesota variety unless otherwise stated)			
Sodium [1-^{14}C]acetate (fed for 1 day)	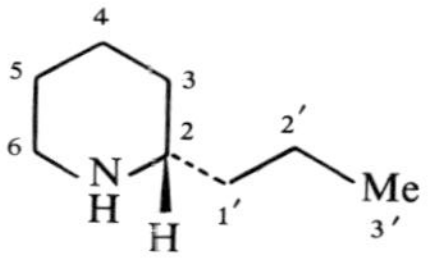 Coniine	[0.009]	21
Sodium [2-^{14}C]acetate (fed for 1 day)	Coniine (C-6: 24.5%, C-2: 24.4%, C-3': 0.7%) Coniine (C-6: 10.1%, C-2: 9.8%, C-3': 15.1%)	[0.04]	
Sodium [2-^{14}C]acetate (fed 14 days)	Coniine (C-4: 12.5%, C-5: 13.0%, C-6: 12.3%, C-2: 12.2%, C-3': 13.2%)	[0.035]	

Coniine structure: piperidine ring with positions labelled 2,3,4,5,6 and N–H, with substituent –CH(H)–CH$_2$–CH$_3$ (1', 2', 3') at C-2.

Biosynthesis of Alkaloids

Precursor	Product (distribution)	Incorporation
Sodium [1-¹⁴C]hexanoate (fed 7 days)	Coniine (C-4: 92%, C-5: 0.9%, C-6: 1.4%, C-2′: 1.6%, C-3′: 0.9%)	[0.024]
Sodium [1-¹⁴C]hexanoate (fed 1 day to Chelsea variety)	γ-Coniceine	[0.05]
Sodium [1-¹⁴C]octanoate (fed 1 day)	Coniine (C-4: 95%, C-5: 0.7%, C-6: 0.6%, C-2′: 1.0%, C-3′: 0.9%)	[0.01]
Sodium [1-¹⁴C]octanoate (fed 7 days)	Coniine (C-6: 93%, C-2′: 0.6%, C-3′: 0.3%)	[0.01]
Sodium [7-¹⁴C]octanoate (fed 1 day)	Coniine (C-6: 85%, C-2′: 3.2%, C-3′: 1.1%)	[0.07]
Sodium [8-¹⁴C]octanoate (fed 1 day)	Coniine (C-6: 0.6%, C-2′: 93%, C-3′: 1.0%)	[0.01]
Sodium [8-¹⁴C]octanoate (fed 7 days)	Coniine (C-5: 9%, C-6: 6%, C-2′: 6%, C-3′: 48%)	[0.01]
Sodium [6-¹⁴C]-5-oxo-octanoate (fed 1 day)	Coniine (C-5: 9%, C-6: 10%, C-2′: 9%, C-3′: 28%)	[0.61]
Sodium [6-¹⁴C]-5-oxo-octanoate (fed 5 days to Chelsea variety)	Coniine (C-1′: 95%)	[0.06]
	γ-Coniceine (C-1′: 94%)	[0.05]
[6-¹⁴C]-5-Oxo-octanal (fed 1 day)	Coniine (C-1′: 98%)	[1.1]
(+)-[2′-¹⁴C]Coniine hydrochloride (fed to Chelsea variety 7 days) 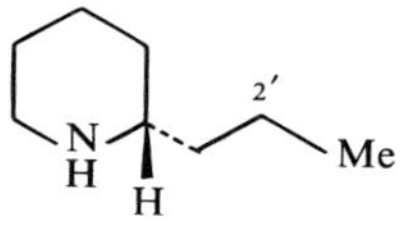	Coniine (C-1′: 95%, C-2: 2%)	[0.27]
	γ-Coniceine (C-2′: 99%) (C-2′: 98%)	[0.23]

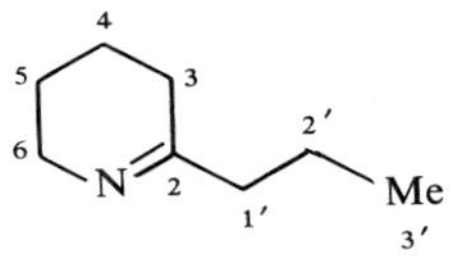

Species / Compound administered	Alkaloids labelled	Incorporation/%	Ref.
Conium maculatum—contd.			
(−)-[2′-^{14}C]Coniine hydrochloride (fed to Chelsea variety 7 days)	γ-Coniceine (C-2: 93%)	[0.019]	
(+)-[1′-^{14}C,^{15}N]Coniine HCl (fed 1 day)	Coniine (C-1′: 99%)	^{14}C: [5.03] ^{15}N: [5.04]	
(+)-[-^{14}C,^{15}N]Coniine HCl (fed 9 days)	Coniine (not degraded)	^{14}C: [4.58] ^{15}N: [3.07]	
(+)-[1′-^{14}C,^{15}N]Coniine HCl (fed 9 days to Chelsea variety)	Coniine (not degraded)	^{14}C: [9.43] ^{15}N: [9.11]	
[1′-^{14}C,^{15}N]-γ-Coniceine (fed 1 day)	Coniine (not degraded)	^{14}C: [1.44] ^{15}N: [1.48]	
[1′-^{14}C,^{15}N]-γ-Coniceine (fed for 9 days)	Coniine (not degraded)	^{14}C: [3.28] ^{15}N: [3.32]	
	Conhydrinone 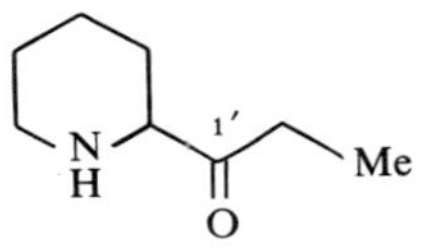	^{14}C: [3.50] ^{15}N: [3.36]	
[1′-^{14}C,^{15}N]-γ-Coniceine (fed 9 days to Chelsea variety)	(C-1′: 98%) γ-Coniceine (not degraded)		
Datura innoxia [1′-^{14}C,^{15}N]-γ-Coniceine (fed to the roots for 9 days)		^{14}C: [1.99] Roots ^{15}N: [2.00] Aerial parts	

[Structure: 2-propanoyl piperidine, with position 1′ labelled]

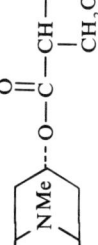

Compound fed					
DL-[2-^{14}C] Phenylalanine	Apohyoscine	—	[0.053]	53	
	(no degradations) Hyoscyamine	[0.3]	[0.077]		
	Hyoscine	[0.17]	[0.009]		
	Littorine	[1.86]	[0.22]		
DL-[2-^{14}C]-2-Hydroxy-3-phenylpropanoic acid (phenyl-lactic acid)	Apohyoscine Hyoscyamine Hyoscine	— [0.45] [0.55]	[0.12] [0.2] [0.11]		
[2-^{14}C]trans-Cinnamic acid (the following compounds fed to excised roots for 4 h)	No incorporation into any of the alkaloids				

Species	Alkaloids labelled	Incorporation/%	Ref.
Compound administered			
Datura innoxia—contd.			
L-[3-^{14}C]Serine	Hyoscine + hyoscyamine (in tropic acid moiety: 80%)	2.6	47

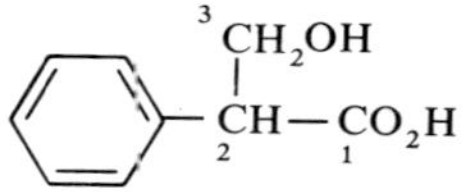

$$\underset{2}{\text{CH}}-\underset{1}{\text{CO}_2\text{H}} \qquad \overset{3}{\text{CH}_2\text{OH}}$$ (on phenyl ring)

DL-[2-^{14}C]Tryptophan	(C-1: 58%, C-2: 3%, C-3: 29%, Ph: 10%) Hyoscine + hyoscyamine (in tropic acid: 76%) (C-1: 20%, C-2: 4%, C-3: 61%, Ph: 15%)	2.7	
DL-[*aryl-U*-^{14}C]Tryptophan	Hyoscine + hyoscyamine (in tropic acid: 83% C-1: 2%, C-2: 1%, C-3: 3%, Ph: 94%)	0.71	
[^{14}C]Formic acid	Hyoscine + hyoscyamine (in tropic acid: 42%) C-1: 41%, C-2: 8%, C-3: 21%, Ph: 30%)	1.7	
[1-^{14}C]Phenylacetic acid	Hyoscine + hyoscyamine (in tropic acid: 91%) C-1: 12%, C-2: 6%, C-3: 64%, Ph: 18%)	2.5	
DL-[1-^{14}C]Phenylalanine	Hyoscine + hyoscyamine (in tropic acid: 83%) C-1: 72%, C-2: 6%, C-3: 12%, Ph: 10%)	2.0	

Datura metel
(fed to isolated root cultures)

Sodium [2-^{3}H]acetate + Sodium [3-^{14}C]acetoacetate	Hyoscyamine (The significantly higher incorporation of the ^{14}C labelled acetoacetate favours this compound as the immediate precursor of the tropane moiety of the alkaloid)	^{14}C: [0.39] ^{3}H: [0.086]	93

Biosynthesis of Alkaloids

Precursor	Product	Incorporation	Ref.
Sodium [2-^{14}C,^{3}H]acetate	Hyoscyamine (The loss of ^{3}H is probably due to metabolism of the acetate into the Krebs cycle intermediates which could exchange off the ^{3}H)	^{14}C: [0.18]	50a
[2-^{14}C,p-^{3}H]Phenylalanine (^{3}H:^{14}C = 6.9)	Tropic acid (from the hyoscyamine) (^{3}H:^{14}C = 6.9)	+	
[2-^{14}C,m-^{3}H]Phenylalanine (^{3}H:^{14}C = 1.5)	Tropic acid (from hyoscyamine) (^{3}H:^{14}C = 1.9)	+	
[2-^{14}C,o-^{3}H]Phenylalanine (^{3}H:^{14}C = 3.0)	Tropic acid (from hyoscyamine) (^{3}H:^{14}C = 6.0*) *This surprising result was not confirmed on repetition of this feeding experiment,[50b] i.e. tropic acid having the same ^{3}H:^{14}C ratio as the administered [2-^{14}C,o-^{3}H]phenylalanine was obtained.	+	
[1-^{14}C,2-^{3}H]Phenylalanine	Hyoscyamine (The tropic acid moiety retained only 7% of the ^{3}H relative to ^{14}C)	^{14}C: [3.41]	
[2-^{14}C,2,3-^{3}H$_3$]Phenylalanine (^{3}H:^{14}C = 4.1)	Hyoscine Tropic acid (from hyoscyamine) (^{3}H:^{14}C = 1.6)	^{14}C: [5.96] +	50, 51
Datura meteloides [N-methyl-^{14}C,3β-^{3}H]Tropine (fed for 7 days to intact plant)	Meteloidine	2.87 [3.08]	94

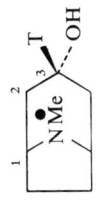

(102% retention of ^{3}H, all located at C-3, all ^{14}C on N-Me)

Species	Alkaloids labelled	Incorporation/%	Ref.
Compound administered			
Datura meteloides—contd.			
	Hyoscine (94% retention of ^{3}H, no degradations)	0.85 [0.31]	
	Hyoscyamine (103% retention of ^{3}H, no degradations)	0.36 [1.36]	
	7β-Hydroxy-3α,6β-ditigloyltropane	0.15 [0.18]	
	(96% retention of ^{3}H, no degradations) Tropine (reisolated from plant) (96% retention of ^{3}H)	[9.9]	
Datura sanguinea			
DL-[1-^{14}C]Phenylalanine	Hyoscine	0.0032	52, 53
	Hyoscyamine	0.011	
	Littorine	0.017	
	(86% of the activity located on the carboxy-group of the phenyl-lactic acid moiety)		
DL-[3-^{14}C]Phenylalanine	Hyoscine	0.0043	
	Hyoscyamine	0.023	

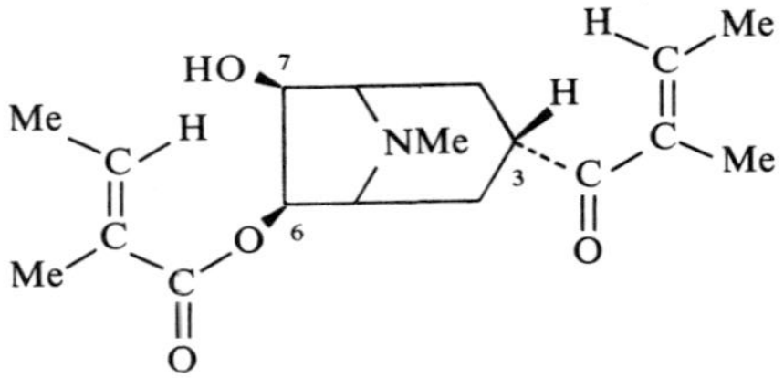

Biosynthesis of Alkaloids

Precursor	Product / Notes	Incorporation	Ref.
[U-^{14}C]-Valine or L-Leucine (it is not clear from the data presented which was actually fed)	Littorine (74% of activity present on the phenyl group and its adjacent carbon, presumably all on the latter)	0.016	95
	3,6-Disenecioyloxytropane + 3,6-di-isovaleroyloxytropane	0.011	
	3,6-disenecioyloxytropane-7-ol + 3,6-di-isovaleroyl derivative	0.046	
	3-Senecioyl(+isovaleroyl)oxytropane	0.0062	
	3-Senecioyl(+isovaleroyl)oxytropane-6,7-diol	0.0048	
	(essentially all activity found in the acid moieties of the above ester alkaloids. The significance of this work is dubious since these alkaloids were isolated by dilution and they may not be normal components of the plant. The apparent conversions may be aberrant syntheses.)		

Datura stramonium

Competition feeding experiments between the following pairs of compounds:

Precursors	Product		Ref.
[U-^{3}H]Phenylalanine + Sodium [1-^{14}C]pyruvate (^{3}H:^{14}C = 12.2)	The tropic acid was obtained by hydrolysis of the resultant hyoscyamine with the indicated ^{3}H:^{14}C ratios: Tropic acid: ^{3}H:^{14}C = 24.8		50a
[U-^{3}H]Phenylalanine + Sodium [3-^{14}C]pyruvate (^{3}H:^{14}C = 14.1)	Tropic acid: ^{3}H:^{14}C = 64.3		
[U-^{3}H]Phenylalanine + [U-^{14}C]Shikimic acid (^{3}H:^{14}C = 28.2)	Tropic acid: ^{3}H:^{14}C = 320		
[U-^{3}H]Phenylalanine + [2-^{14}C]Tyrosine (^{3}H:^{14}C = 1.8)	Tropic acid: ^{3}H:^{14}C = 120		
[U-^{3}H]Phenylalanine + [1-^{14}C]phenylpyruvate (^{3}H:^{14}C = 1.4)	Tropic acid: ^{3}H:^{14}C = 2.0		
[7-^{14}C]Hygrine	Hyoscyamine (>85% at C-3 of tropine moiety)	[0.20]	93
	Hyoscine	[0.18]	

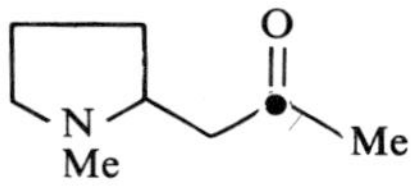

[Structure: N-methylpyrrolidine ring with -CH$_2$-C(=O)-Me substituent (hygrine)]

Species Compound administered	Alkaloids labelled	Incorporation/%	Ref.
Datura stramonium—contd.			
[U-^{14}C]Leucine (fed for 5 days)	Disenecioyl(+isovaleroyl)oxytropane (alkaloid 1)	0.014	95
	Disenecioyl(+isovaleroyl)oxytropane-7-ol (alkaloid 2)	0.088	
	3-Senecioyl(+isovaleroyl)oxytropane (alkaloid 3)	0.015	
	3-Senecioyl(+isovaleroyl)oxytropane-6,7-diol (alkaloid 4)	0.0085	
[U-^{14}C]Valine (fed for 5 days)	Alkaloid 1	0.015	
	Alkaloid 2	0.016	
	Alkaloid 3	0.012	
	Alkaloid 4	0.0037	
	(*cf.* remarks under *D. sanguinea* regarding these experiments)		
Datura stramonium var. *tatula* (compounds fed to sterile root cultures)			
L-[U-^{14}C]Proline	Hyoscyamine	0.37	96
	Hyoscine	0.09	
L-[U-^{14}C]Proline + puromycin (for 1 week)	Hyoscyamine	1.4	
	Hyoscine	0.24	
L-[U-^{14}C]Proline + puromycin (for 2 weeks)	Hyoscyamine	0.43	
	Hyoscine	0.04	
	(No degradations carried out on the labelled alkaloids. The puromycin, an inhibitor of protein synthesis, caused an increase in the conversion of proline into hyoscyamine, presumably by suppressing the incorporation of proline into proteins)		
Datura (unidentified species)			
Sodium [2-^{14}C]acetate	Hyoscyamine and hyoscine	+	97
[1,3-^{14}C]Acetone	Hyoscyamine and hyoscine	+	
[^{14}C]Methylamine	Hyoscyamine and hyoscine	+	
[1,4-^{14}C]Succinic acid	Hyoscyamine and hyoscine	+	
[2,3-^{14}C]Succinic acid	Hyoscyamine and hyoscine	+	

Biosynthesis of Alkaloids

Dioscorea hispida
Sodium [1-^{14}C]acetate

(The methylamine was the best precursor of the alkaloids; however, since no degradations were reported the results are of doubtful significance)

Dioscorine 0.2 98

[Structure of Dioscorine with numbered positions: 1, 4, 5, 7, 8, 9, 10, 11, 12, 13, NMe, C=O, Me]

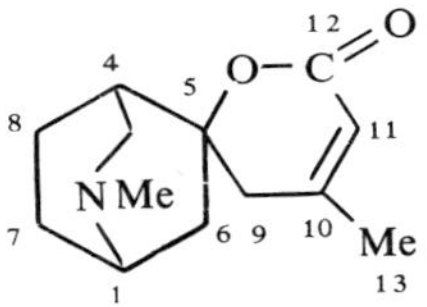

DL-[2-^{14}C]Lysine
DL-[6-^{14}C]Lysine
[6-^{14}C]-Δ^1-Piperideine

(C-5: 31 ± 1%, C-10: 28 ± 2%, C-12: 28 ± 1%)
Dioscorine 0.003 99
Dioscorine 0.007 98
Dioscorine 0.03

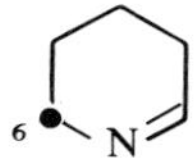

(A partial degradation indicated that the pattern of labelling was the same as that found after feeding [1-^{14}C]acetate, indicating that the Δ^1-piperideine was probably catabolized to acetate prior to incorporation)

[Structure of Δ^1-piperideine, position 6 labeled]

Dolichothele sphaerica
DL-[2-^{14}C]Leucine
(All compounds fed by injection and then allowed to grow 3 weeks)

Dolichotheline 2.98 100

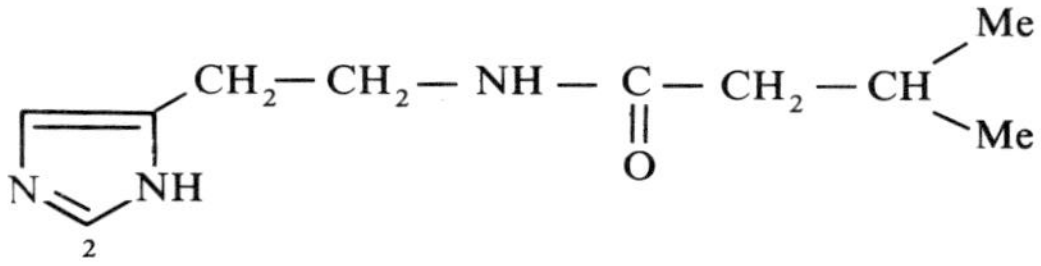

[Structure of Dolichotheline: imidazole-CH$_2$-CH$_2$-NH-C(=O)-CH$_2$-CH(Me)$_2$, NH position 2]

(C=O: 97%)

Species / Compound administered	Alkaloids labelled	Incorporation/%	Ref.
Dolichothele sphaerica—contd.			
L-[*U*-¹⁴C]Valine	Dolichotheline (C=O < 1%, essentially all activity in the isovaleroyl moiety)	1.36	
Sodium [1-¹⁴C]isovalerate	Dolichotheline (C=O 97%)	10.32	
L-[2-¹⁴C]Histidine	Dolichotheline (C-2: 100%)	8.09	
[2-¹⁴C]Histamine	Dolichotheline (C-2: 100%)	15.13	
(These results complement those of Horan and O'Donovan[101])			
Echinocereus merkeri			
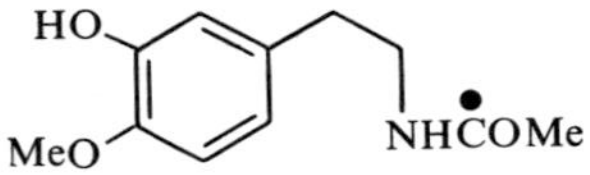 *N*-[1-¹⁴C]Acetyl-3-hydroxy-4-methoxyphenethylamine	Salsoline 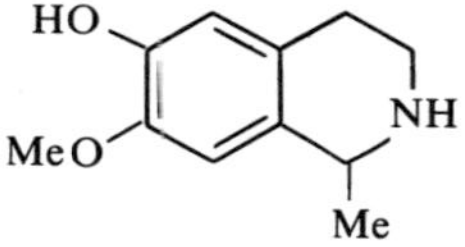	inactive	102
DL-[2-¹⁴C]Tyrosine	Salsoline	inactive	103
	3,4-Dimethoxyphenethylamine (C-α: >99%)	0.54	

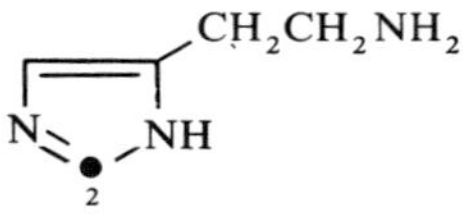

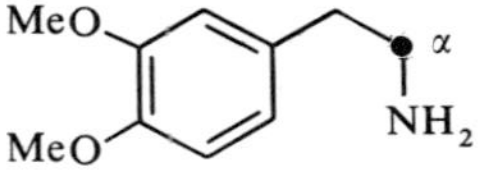

Biosynthesis of Alkaloids

Precursor	Product	Incorporation (%)
Tyramine (structure: HO-C6H4-CH2-CH(α)-NH2)	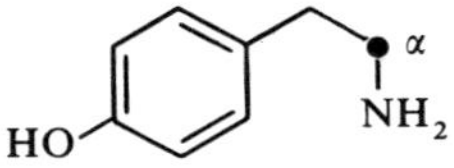	4.9
[α,β-^{3}H$_{2}$]Dopamine hydrochloride	Salsoline (C-α: >99%)	inactive / 0.75
[aryl-^{3}H]-3-Hydroxy-4-methoxyphenethylamine hydrochloride	3,4-Dimethoxyphenethylamine (>98% of the ^{3}H in the side-chain)	9.0
	3,4-Dimethoxyphenethylamine (95% of the ^{3}H in the benzene ring)	3.0
	Salsoline (no degradations)	
DL-[1-^{14}C]-1-Methyl-6-hydroxy-7-methoxy-1,2,3,4-tetrahydroisoquinoline-1-carboxylic acid 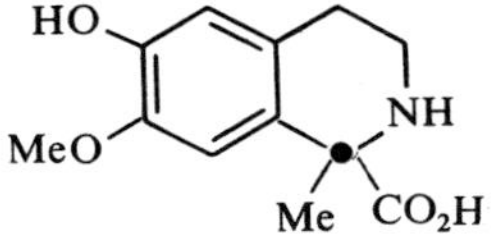	Salsoline (no degradations)	0.14
Elaeagnus angustifolia L-[β-^{14}C]Tryptophan (hydroponic feeding)	Eleagnine	0.012

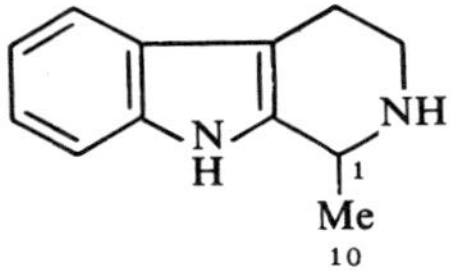

102

Species	Alkaloids labelled	Incorporation/%	Ref.
Compound administered			
Elaeagnus angustifolia—contd.			
[α-¹⁴C]Tryptamine (wick feeding)	Eleagnine (no degradations)	4.5	
[β-¹⁴C,aryl-³H]Tryptamine (hydroponic feeding)	Eleagnine (55% retention of ³H, no explanation given for the loss of ³H)	0.46 (¹⁴C)	
N-[1-¹⁴C]Acetyltryptamine	Eleagnine (This result is apparently in conflict with the work of O'Donovan and Kenneally,¹³⁰ who reported specific labelling at C-2 after feeding [1-¹⁴C]acetate to this species)	inactive	

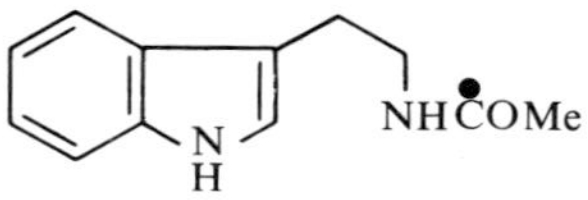

[structure: tryptamine with NHCOMe side chain]

N-[1-¹⁴C]Acetyl-[aryl-³H]tryptamine	Eleagnine	inactive	
[1-¹⁴C]Harmalan hydrochloride	Eleagnine (no degradations)	0.70	

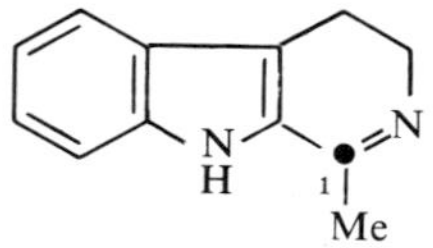

[structure: dihydro-β-carboline with N-Me]

L-[β-¹⁴C,G-³H]Tryptophan (wick and hydroponic feeding)	N-Acetyltryptamine (Cold material was added to trap any that might be formed)	inactive	
Erythroxylon coca			
Sodium [1-¹⁴C]acetate	Cocaine	[0.072]	93

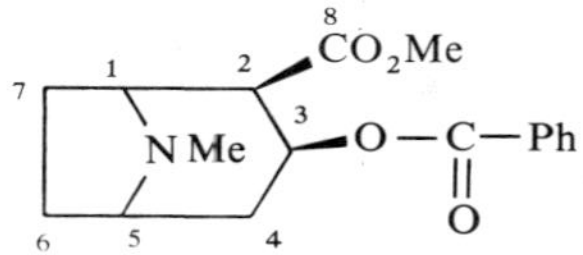

[structure: cocaine with numbered positions 1–7, NMe, CO₂Me at C-8, and O–C(=O)–Ph benzoate]

Heimia salicifolia
Sodium [1-¹⁴C]acetate

(on benzoic acid: 29.4%, essentially all on carboxy-group, C-8: 2.9%, C-3: 3.1%, C-1 + C-5 + C-6 + C-7 + N-Me: 2.0%. It seems that most of the activity is located on the methyl group of the methoxycarbonyl group)

Cryogenine   0.023 [0.4]   104

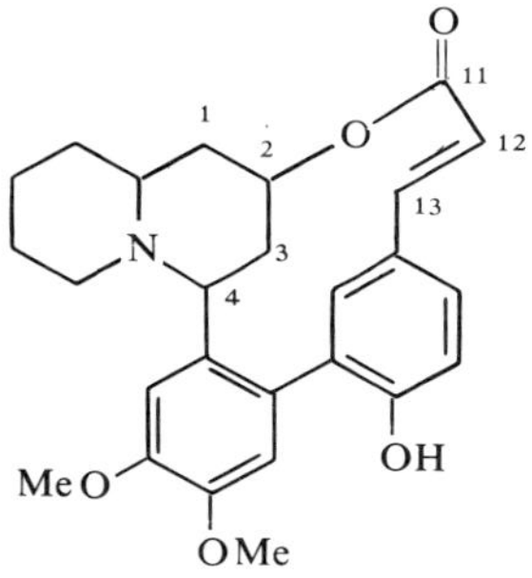

DL-[3-¹⁴C]Phenylalanine

(no degradations)
Cryogenine
(C-13: 46%, degradative products consistent with 33% of activity at C-4)

Hordeum vulgare
[*methylene*-¹⁴C]Gramine

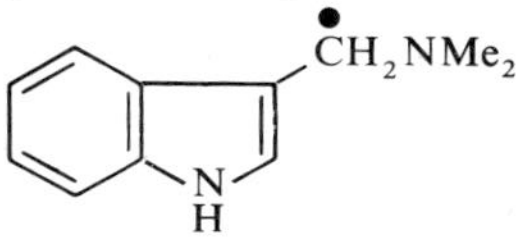

The feeding of gramine to the plant did not result in an increased yield of the alkaloid from the plant. Thus the introduced gramine is degraded and the *de novo* synthesis of alkaloid inhibited.

0.3*[0.5]   105
*in a preliminary feed an incorporation of 0.06% was obtained

Species / Compound administered	Alkaloids labelled	Incorporation/%	Ref.
Hordeum vulgare—contd.			
[β-¹⁴C]Tryptophan	This amino-acid is an established precursor of gramine; however, the addition of gramine to the plant inhibited the formation of [¹⁴C]gramine. The incorporation into gramine also varied in the leaves, 0.068% at top, 1.47% at the base.		
Leucaena leucocephala			
[*U*-¹⁴C]Mimosine	Mimoside (Mimosine-*O*-glucoside)	+	106
	(The formation of mimoside was also accomplished by an extract of *L. leucocephala* seedlings + UDP-glucose. The reverse reaction was also observed)		
Lophophara williamsii			
DL-[2-¹⁴C]Tyrosine	Tyramine	7.2	107
(Various phenethylamines were added to the workup of the cactus, 10 days after feeding the tyrosine, to determine whether they were significant intermediates en route to mescaline)	*N*-Methyltyramine	1.7	
	4-Hydroxy-3-methoxyphenethylamine	0.5	
	N-Methyl-4-hydroxy-3-methoxyphenethylamine	0.4	
	Dopamine	0.07	
	Epinine	0.09	

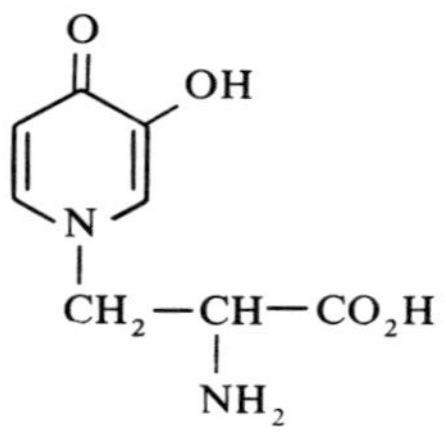

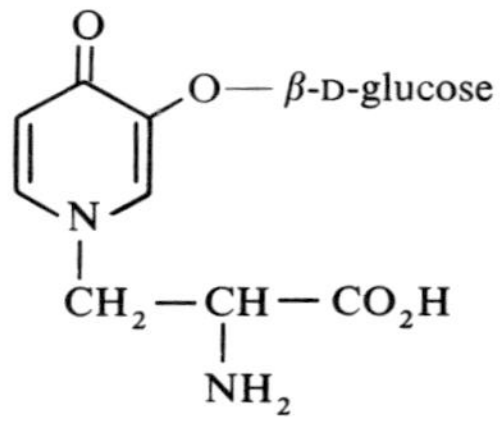

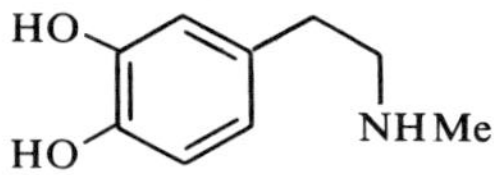

Biosynthesis of Alkaloids

HO-C6H3(OH)-CH2CH2-NHMe

DL-[2-^{14}C]Dopa	
[α-^{14}C]Dopamine	
[α-^{14}C]Dopamine (5-day feed)	
N-[1-^{14}C]Acetyl-3-hydroxy-4,5-dimethoxyphenethylamine	

Dopa (3,4-dihydroxyphenylalanine)	0.005
Dopamine	0.09
Epinine	0.06
4-Hydroxy-3-methoxyphenethylamine (>98% at C-α)	13.5
N-Methyl-4-hydroxy-3-methoxyphenethylamine (C-α: 100%)	3.2
N,N-Dimethyl-4-hydroxy-3-methoxyphenethylamine (C-α: 100%)	1.7
3,4-Dihydroxy-5-methoxyphenethylamine (C-α: 96%)	0.29
3,4,5-Trihydroxyphenethylamine	0.01
Anhalonidine	inactive

102

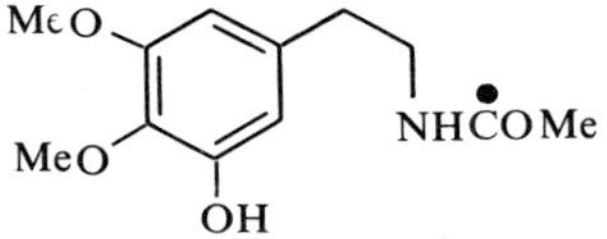

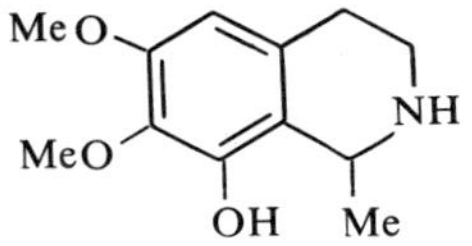

Anhalonidine structure (MeO, MeO, OH, NH, Me)

(confirms the results of Lundström[108] and Kapidia et al.[109])

Species

Compound administered	Alkaloids labelled	Incorporation/%	Ref.

Lycopersicon esculentum
[4-^{14}C]Cholesterol
(The cholesterol, dissolved in 96% ethanol containing DL-α-tocopherol, was applied to the leaves of the plant 3 times/week for 5 weeks, and the plants harvested one week after the final feeding)

Tomatine

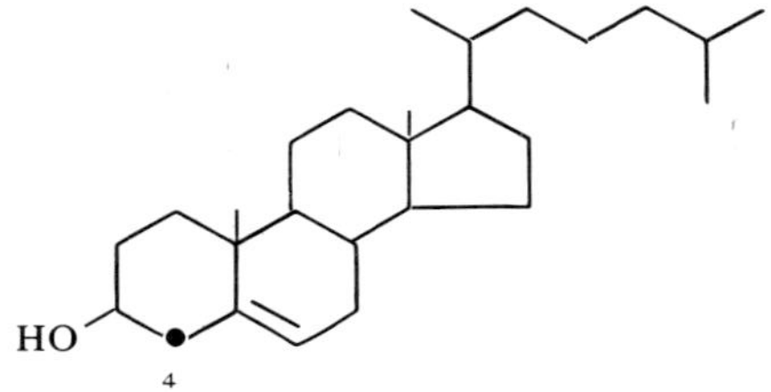

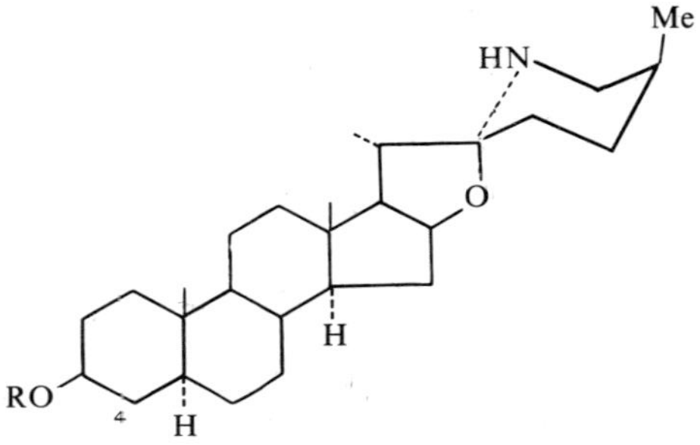

(R = D-xylose-D-glucose-D-glucose—
  D-galactose)

(This [4-^{14}C]tomatine was used in a study of the catabolism of the alkaloid in ripe tomato fruits)

1.0   29

Lycopodine

0.0023 [0.0014]   110

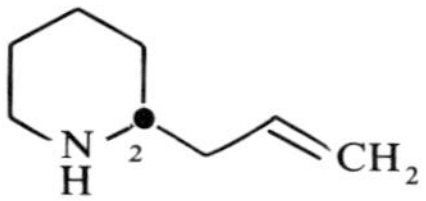

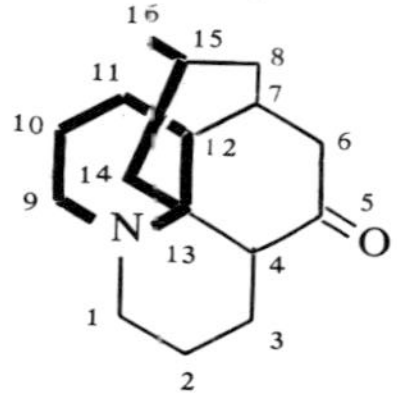

(C-5: 3 ± 1%)

Lycopodium tristachyum
[2-^{14}C]-2-Allylpiperidine

Precursor	Alkaloid	Incorporation (%)	Ref.
[2-14C]-Δ'-Piperideine + inactive pelletierine	Lycopodine (C-5: 88 ± 2%, C-7 to C-16: 9%. The excess cold pelletierine apparently enters that part of the alkaloid indicated with heavy bonds)	0.008 [0.034]	
	Pelletierine 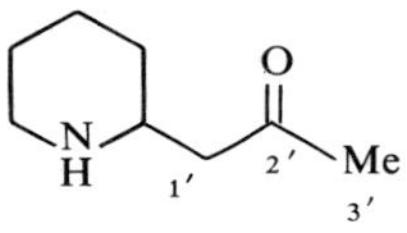	[0.0041]	
[1,5-14C]Cadaverine + inactive pelletierine	Lycopodine (C-5:44 ± 1%. This value would have been 25% if the cold pelletierine had not suppressed the incorporation of the [1,5-14C]cadaverine into atoms C-9 to C-16. The cold pelletierine reduced the incorporation of the cadaverine about 13 times[131]	0.023 [0.0008]	
	Pelletierine (C-2' + C-3': inactive)	[0.00011]	
Mimosa pudica [U-14C]Mimosine (see under *Leucaena leucoephala*)	Mimoside (Mimosine-O-glucoside)	+	106
[side-chain-14C]Mimosine	Mimoside	+	
Narcissus pseudonarcissus [2,4-3H2]Crinine	Narciclasine	inactive	111

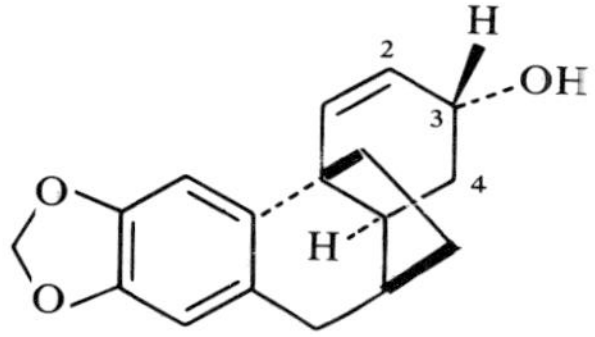

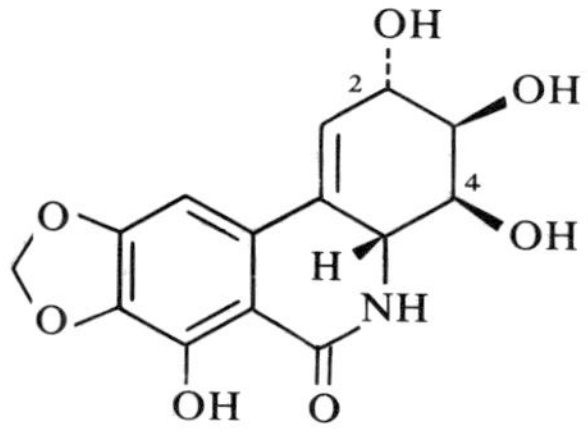

Species

Compound administered	*Alkaloids labelled*	*Incorporation/%*	*Ref.*
Narcissus pseudonarcissus—contd.			
(±)-[3-^{3}H]Crinine	Narciclasine (This result shows that narciclasine has the indicated stereochemistry and is derived from the enantiomer of crinine, *i.e.* vittatine. See under *Pancratium maritimum* for related work)	2.1	
(2*RS*,3*S*)-[3-^{14}C,3-^{3}H$_1$]Phenylalanine	Haemanthamine (no significant retention of ^{3}H)	0.10 (^{14}C)	91
	Oduline (no significant retention of ^{3}H)	0.055 (^{14}C)	
(2*RS*,3*R*)-[3-^{14}C,3-^{3}H$_1$]Phenylalanine	Haemanthamine	0.18 (^{14}C)	
	Oduline (no significant retention of ^{3}H)	0.11 (^{14}C)	

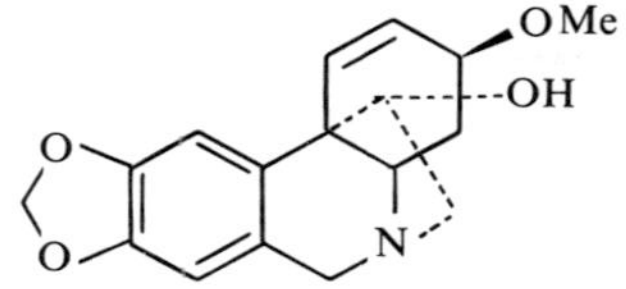

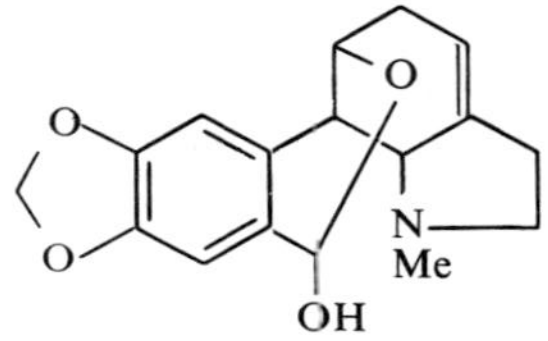

Biosynthesis of Alkaloids

Plant / Precursor	Alkaloid	Incorporation [Dilution]	Ref.
Nicotiana glauca L-[2-¹⁴C]Lysine	Anabasine	0.075 [2.17]	8
	Pipecolic acid (*The author actually recorded the specific incorporation of the pipecolic acid; however, the value reported is meaningless since the amino-acid was isolated by dilution)	0.044*	
D-[2-¹⁴C]Lysine (This compound and the L-isomer were added to the nutrient solution in which the plants were growing hydroponically)	Anabasine Pipecolic acid (no degradations)	0.0004 [0.072] 0.37*	
[U-¹⁴C,¹⁵N]Aspartic acid	Anabasine (The ¹⁵N was distributed as follows: pyridine ring: 43% piperidine ring: 57%) (−)-Anabasine (C-2′: 100%)	¹⁴C: [1.08] ¹⁵N: [0.73] 1.21 [1.4]	112 19
N-Methyl-[2-¹⁴C]-Δ′-piperideinium chloride 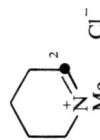	(−)-*N*-Methylanabasine (C-2′: 100%)	0.71 [104]	

Species

Compound administered	Alkaloids labelled	Incorporation/%	Ref.
Nicotiana tabacum			
[^{14}C]Nicotine (fed 3 days) (obtained biosynthetically by feeding [2-^{14}C]acetate to tobacco)	Nicotine (recovered) Activity was detected in the following amino-acids: serine, aspartic acid, glutamic acid, lysine, histidine, arginine, alanine, valine, phenylalanine, and leucine	7.0	26
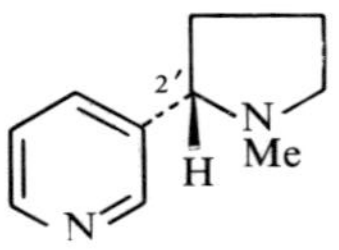 [structure of nicotine with 2' position and N-Me labelled]			
[^{14}C]Nicotine (fed 3 days) (obtained by feeding [^{14}C]aspartic acid to tobacco)	Nicotine (recovered) Activity found in all the above amino-acids, but to a much smaller extent	6.6	
[^{14}C]Nicotine (fed 3 days) (obtained by feeding [^{14}C]glutamic acid to tobacco)	Nicotine (recovered) Low activity in the amino-acid fraction	7.0	
[U-^{14}C,^{15}N]Aspartic acid	Nicotine (The ^{15}N was distributed as follows: pyridine ring: 74% pyrrolidine ring: 26% The nicotinic acid obtained on oxidation of the nicotine contained 67% of the ^{14}C activity)	^{14}C: [1.12] ^{15}N: [0.76]	112
N-Methyl-[2-^{14}C]-Δ^1-piperideinium chloride (fed hydroponically for 5 days)	Anabasine (C-2′: 100%)	1.28 [38]	19
	(−)-N-Methylanabasine (C-2′: 100%)	2.25 [102]	
	Nicotine	0.037 [0.03]	

Biosynthesis of Alkaloids

DL-[2-^{14}C]Lysine
(fed hydroponically for 5 days)

N-Methylanabasine 0.0002
(cold *N*-methylanabasine was added to the harvested plant to trap activity that might be located in this alkaloid)
Nicotine 0.003 [0.006]

Opuntia decumbens

L-[3',5'-^{3}H$_2$]Tyrosine

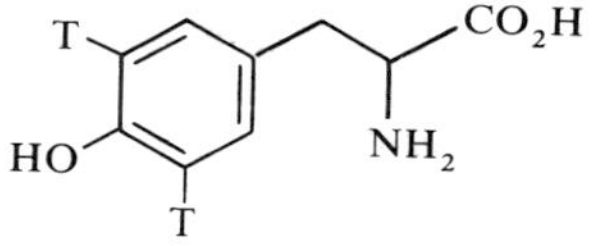

Betanine 0.029—0.085 12

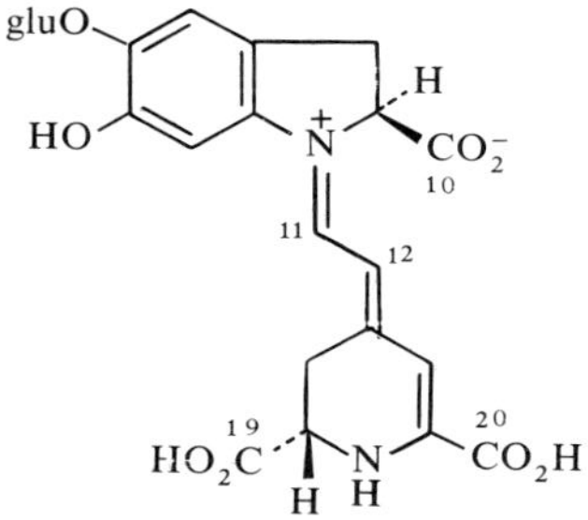

(structure of betanine shown with labels: gluO, HO, H, CO_2^- at position 10, N$^+$, 11, 12, HO_2C at position 19, N, H, H, CO_2H at position 20)

(88—94% of the ^{3}H was present in the lower betalamic acid portion of the molecule. Degradations consistent with all the ^{3}H being located at C-11)

L-[1-^{14}C]Tyrosine Betanine 0.528
L-[2-^{14}C]Dopa Betanine 2.60
DL-[1-^{14}C]Dopa Betanine 2.65 11
(fed to fresh fruits of this species or (essentially all the activity at C-10 and C-19, about 90% at
O. bergeriana) the latter position)

Species Compound administered	Alkaloids labelled	Incorporation/%	Ref.
Opuntia dillenii [10-^{14}C]Betanidine (The administered compound also contained the 2R,15R, 2S,15R, and 2R,15S isomers)	Betanine	4.4 (based on the incorporation of the 2S,15S isomer)	113
(injected by means of the syringe into the fruits and allowed to metabolize for 12 h) L-[2-^{14}C]Tyrosine	Betanine	0.35	
Opuntia ficus-indica L-[1-^{14}C,3′,5′-^{3}H$_2$]Tyrosine (injected directly into fruits)	Indicaxarthine	0.23—0.57 (^{14}C)	13

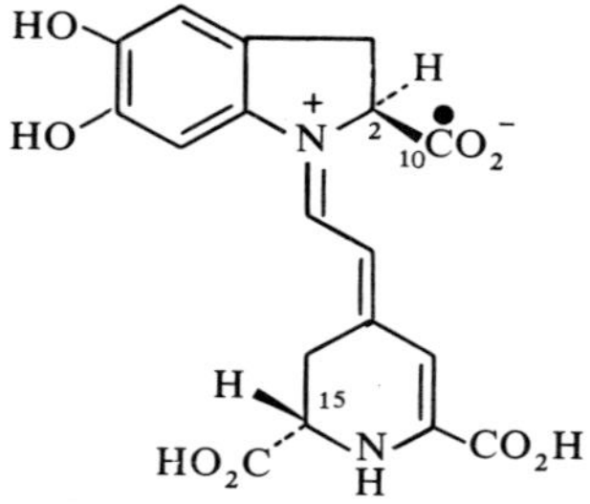

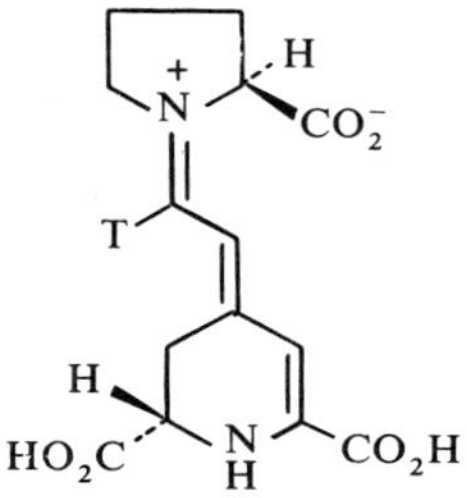

Biosynthesis of Alkaloids

111

+

(Approximately 50% retention of ^{3}H, a result consistent with the intermediate formation of dopa. Degradations consistent with all the ^{3}H being located at the indicated position)

Vittatine

0.8

Narciclasine
(see under *Narcissus pseudonarcissus*) (^{3}H at C-2:C-4 = 2, i.e. half the ^{3}H is lost from C-4 during the hydroxylation at this position)

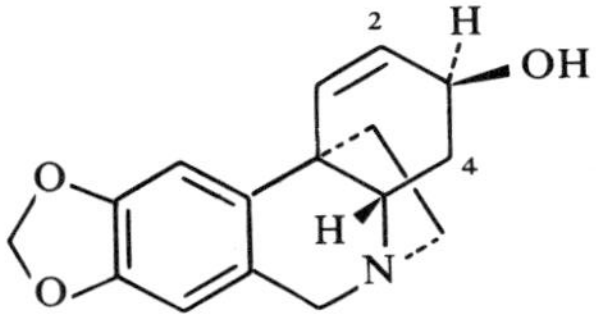

(100% retention of ^{3}H, all located at C-2 and C-4)

Pancratium maritimum
[1-^{14}C,3′,5′-^{3}H$_2$]-*O*-Methylnorbelladine

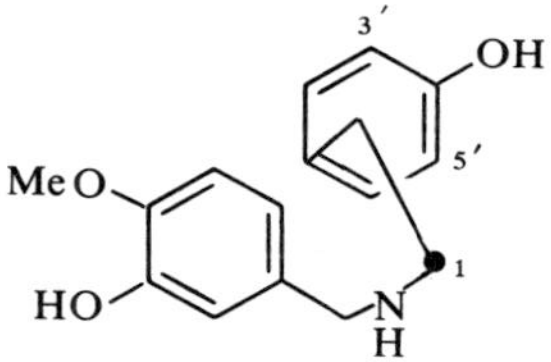

[2,4-^{3}H$_2$]Vittatine
(obtained biosynthetically as above)
(^{3}H at C-2:C-4 = 1)

Species Compound administered	Alkaloids labelled	Incorporation/%	Ref.
Pancratium maritimum—contd.	Haemanthidine	2.3	
	(no degradations reported)		
Papaver bracteatum (RS)-[N-methyl-¹⁴C]Tetrahydropalmatine methiodide	Alpinigenine	0.56 [0.15]	114
	(N-Me: 100%) Alpinigenine (The ratio of activity at C-18:NMe was the same as at C-8:NMe in the administered precursor)		
(RS)-[N-methyl,8-¹⁴C]Tetrahydropalmatine	Alpinigenine (not degraded)	0.52 [0.06]	
(RS)-[8-¹⁴C]Tetrahydropalmitine		14.2 [2.8]	

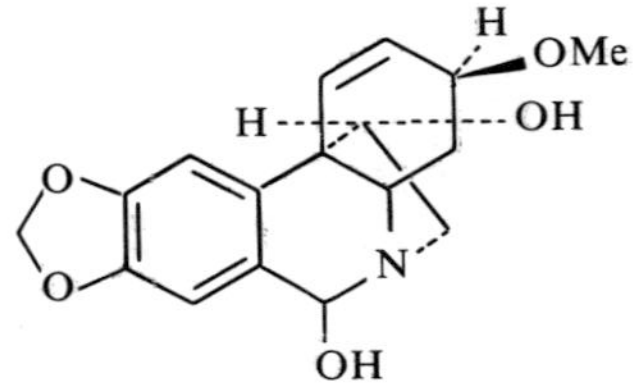

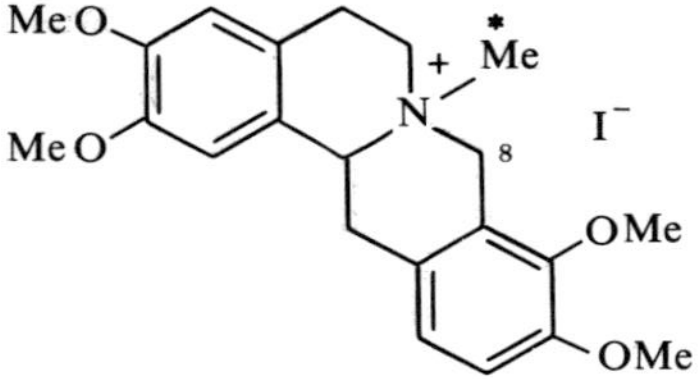

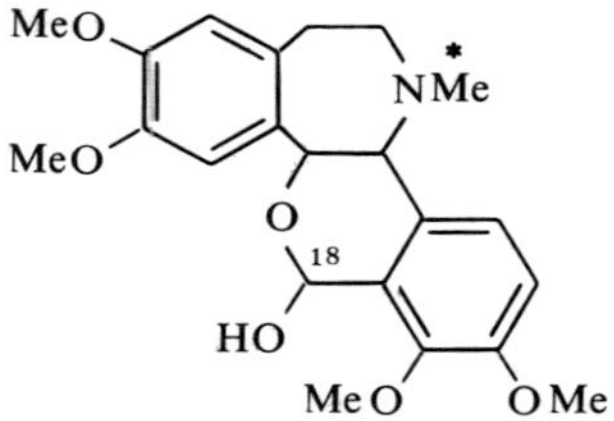

Papaver somniferum
[G-^{14}C]Codeine methyl ether (prepared from generally labelled morphine obtained biosynthetically by feeding $^{14}CO_2$ to *P. somniferum*) (fed for 24 h)

Codeine methyl ether	52	31
Codeine 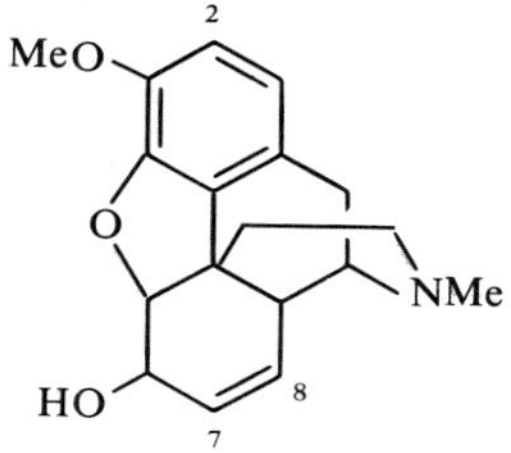	4.7	
Morphine 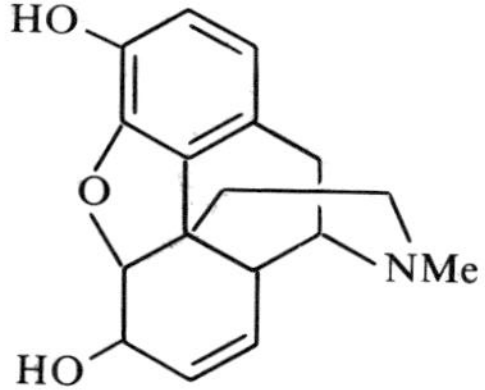	3.0	

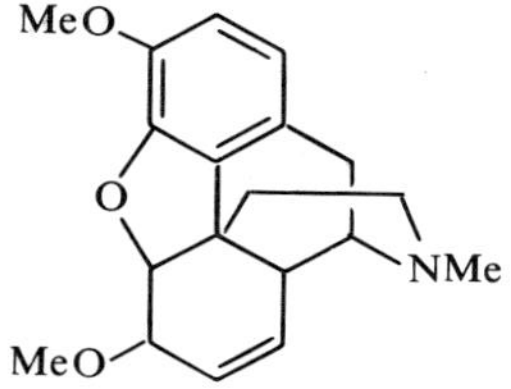

Species / Compound administered	Alkaloids labelled	Incorporation/%	Ref.
Papaver somniferum—contd.			
[G-^{14}C]Codeinone (fed for 24 h) 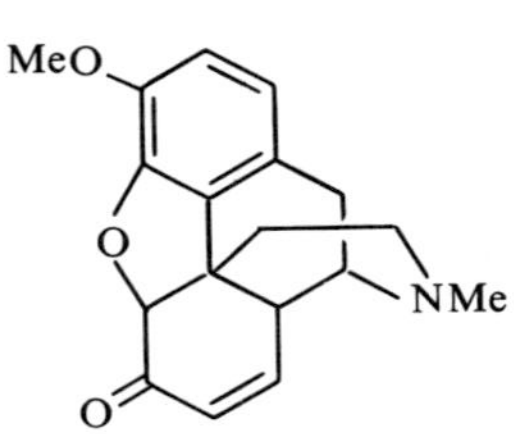	Thebaine 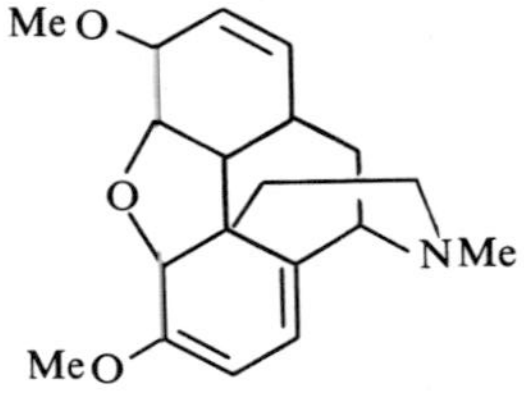	inactive	
	Codeinone	14	
	Codeine	14.5	
	Morphine	2.8	
[G-^{14}C]Codeine (fed for 24 h)	Codeinone	inactive	
	Codeine	54	
	Morphine	13	
	Thebaine, codeinine and codeine were obtained with decreasing specific activities		
[^{14}C]Carbon dioxide (fed under steady state conditions for 3 h, i.e. the atmosphere contained 0.04% CO_2)	Radioactive thebaine and codeine were isolated from the plants. Inactive codeine methyl ether, codeinone, or neopinone were added as carriers to detect any activity in these plausible intermediates between thebaine and morphine. Negligible (<0.02%) of the activity of the thebaine was detected in methyl ether. Significant activity was found in codeinone (5.4%) and neopinone (6.1%)		
[^{14}C]Carbon dioxide (Plants, 53—66 days old, were exposed to 50 mCi of $^{14}CO_2$ for 4 h, during which time 90% of the activity was absorbed)			
[G-^{14}C]Neopinone (It was not possible to remove all codeinone	Codeinone	19	
	Codeine	8.0	

Biosynthesis of Alkaloids

from the administered material, and about 5% of the neopinone is converted into codeinone during the 28 h feeding experiment)	Codeine Morphine	8.0 8.7

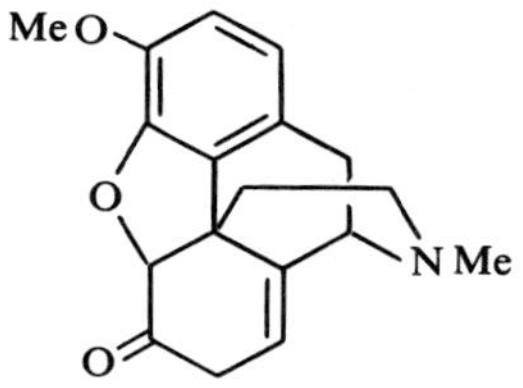

MeO — (structure with N Me, fused ring system with ketone)

[G-^{14}C]Codeinone (fed under the same conditions as the neopinone)	Codeinone Codeine Morphine	2.3 16 4.7
[U-^{14}C]Tyrosine		115
[2-^{14}C]Tyrosine	Morphine was synthesized in isolated latex. The rate of morphine synthesis depended on the stage of development of the plant, and was a maximum at the budding stage	
[^{14}C]Carbon dioxide	The greatest accumulation of labelled alkaloids occurred during the flowering stage. The alkaloids were not final products, but were converted into non-alkaloidal materials	116
[N-methyl-^{14}C,2-^{3}H]Codeine (^{3}H:^{14}C = 5.32)	Morphine (^{3}H:^{14}C = 5.57) Codeine (^{3}H:^{14}C = 5.70) (This slight increase in the ^{3}H:^{14}C ratio may indicate partial N-demethylation occurring in the plant.)	0.99 (^{14}C) 2.69 (^{14}C) 114

Species	Alkaloids labelled	Incorporation/%	Ref.
Compound administered			
Papaver somniferum—contd.			
[N-methyl-^{14}C]Codeine + [2-^{3}H]-7,8-Dihydrocodeine	Codeine	19.4 (^{14}C)	
	Morphine	1.26 (^{14}C)	
	7,8-Dihydrocodeine	31.8 (^{3}H)	
	7,8-Dihydromorphine (C-2: 96% of the ^{3}H)	0.59 (^{3}H)	
[2-^{3}H]-7,8-Dihydrocodeine	7,8-Dihydrocodeine	39	
	7,8-Dihydromorphine	0.51	
	(No significant incorporation into morphine or codeine)		
[N-methyl-^{14}C]Codeine + [2-^{3}H]Isocodeine	Codeine	16.2 (^{14}C)	
		<0.003 (^{3}H)	
	Isocodeine	<0.01 (^{14}C)	
		32 (^{3}H)	
	Morphine	9.1 (^{14}C)	
		<0.01 (^{3}H)	
	Isomorphine	<0.05 (^{14}C)	
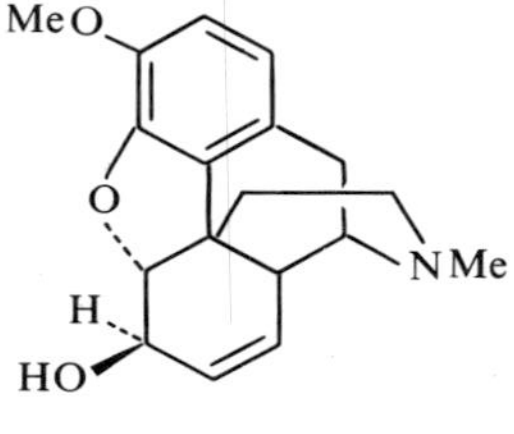		1.38 (^{3}H)	

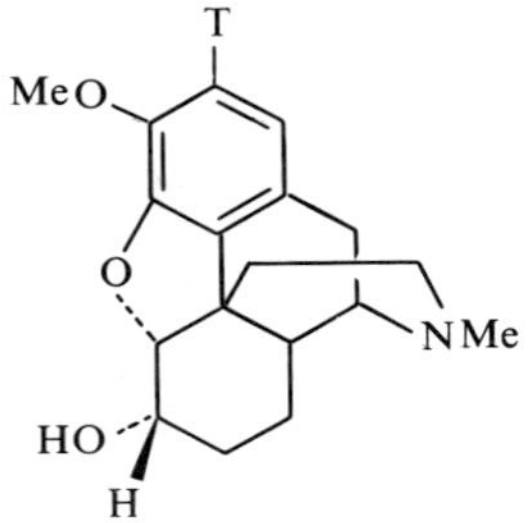

Biosynthesis of Alkaloids

[2-³H]Isocodeine

Compound	Activity
Codeine	0.008
Isocodeine	9.4
Morphine	0.02
Isomorphine	0.27

[N-methyl-¹⁴C]Codeine + [2-³H]Codeine methyl ether

Compound	Activity
Codeine	14.6 (¹⁴C)
	0.49 (³H)

(The presence of a significant amount of ³H in the codeine indicates that the methyl ether can be demethylated in opium poppies)

Compound	Activity
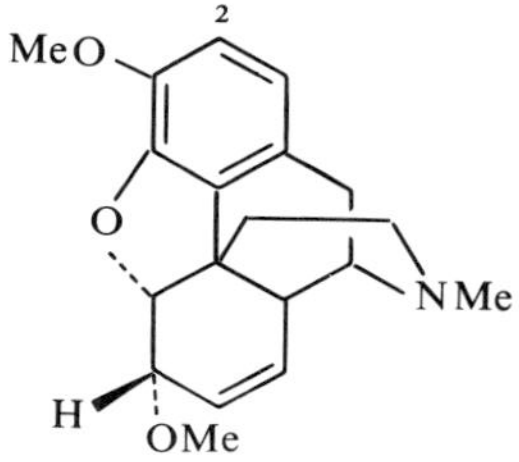 Codeine methyl ether	0.182 (¹⁴C)
	18.4 (³H)

(The presence of ¹⁴C in this compound indicates that it is a natural component of opium and confirms earlier work¹³²)

Compound	Activity
Morphine	0.90 (¹⁴C)
	0.08 (³H)

(The ³H-labelled morphine is presumably derived from the [³H]codeine)

Compound	Activity
Morphine methyl ether	0.012 (¹⁴C)
	0.62 (³H)

(C-2: 100% of the ³H activity)

[2-³H]Codeine methyl ether

Compound	Activity
Codeine	0.88
Codeine methyl ether	16.4
Morphine	0.048
Morphine methyl ether	1.02

[N-methyl-¹⁴C]Codeine + [2-³H]Dihydrodeoxycodeine

Compound	Activity
Codeine	38.4 (¹⁴C)
	0.08 (³H)
Dihydrodeoxycodeine	30.6 (³H)
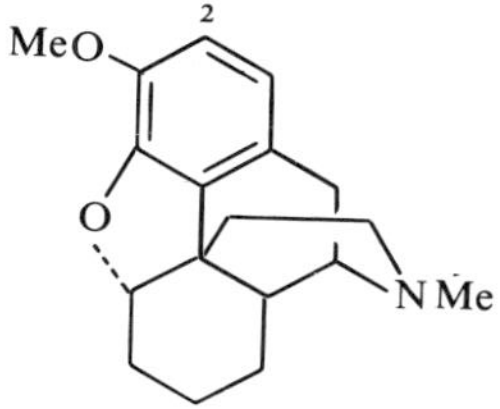 Morphine	8.55 (¹⁴C)
Dihydrodeoxymorphine	0.07 (¹⁴C)
	9.27 (³H)

(C-2: 95.9% of the ³H)

Species / Compound administered	Alkaloids labelled	Incorporation/%	Ref.
Papaver somniferum—contd.			
[*N-methyl*-^{14}C]Codeine + 1-Bromo-[2-^{3}H]Codeine	Codeine	61 (^{14}C)	
		0.02 (^{3}H)	
	1-Bromocodeine	0.03 (^{14}C)	
		35.2 (^{3}H)	
	Morphine	2.57 (^{14}C)	
		0.02 (^{3}H)	
	1-Bromomorphine	0.04 (^{14}C)	
		0.15 (^{3}H)	
Phalaris tuberosa			
L-[β-^{14}C]Tryptophan	*NN*-Dimethyltryptamine	0.04 (1 day)	117
		0.08	
		(2 day feed)	
	<img of indole with NMe₂ side chain, position 5 labelled>		
	5-Methoxy-*NN*-dimethyltryptamine	0.02 (1 day)	
5-Hydroxy-[*aryl*-^{3}H]tryptophan	5-Methoxy-*NN*-dimethyltryptamine	0.04 (2 days)	
		0.009 (1 day)	
5-Methoxy-[*aryl*-^{3}H] tryptophan	5-Methoxy-*NN*-dimethyltryptamine	0.02 (2 days)	
		0.24 (1 day)	
		0.26 (2 days)	
[β-^{14}C]Tryptamine	*NN*-Dimethyltryptamine	0.32 (1 day)	
		0.33 (2 days)	
	5-Methoxy-*NN*-dimethyltryptamine	0.11 (1 day)	
		0.15 (2 days)	
5-Hydroxy-[3-^{14}C]tryptamine	5-Methoxy-*NN*-dimethyltryptamine	0.48 (2 days)	
5-Methoxy-[*aryl*-^{3}H]tryptamine	5-Methoxy-*NN*-dimethyltryptamine	0.24 (1 day)	
		0.24 (2 days)	

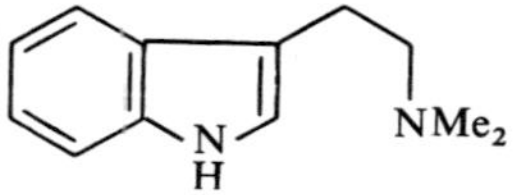

NN-Dimethyl-[2-^{14}C]tryptamine	5-Methoxy-NN-dimethyltryptamine	inactive
5-Hydroxy-[$aryl$-^{3}H]NN-dimethyltryptamine	5-Methoxy-NN-dimethyltryptamine	1.7 (1 day)
		0.36 (2 days)
L-[β-^{14}C]Tryptophan	Activity detected in the potential intermediates:	
(This was fed to the plant 24 h after an inactive potential intermediate, between tryptophan and the alkaloids, had been administered)	Tryptamine 89%	
	5-Methoxytryptamine 42%	
	5-Hydroxytryptamine 33%	
	NN-Dimethyltryptamine 84%	

Rauwolfia serpentia
Sodium [2-^{14}C]acetate   0.00025   86

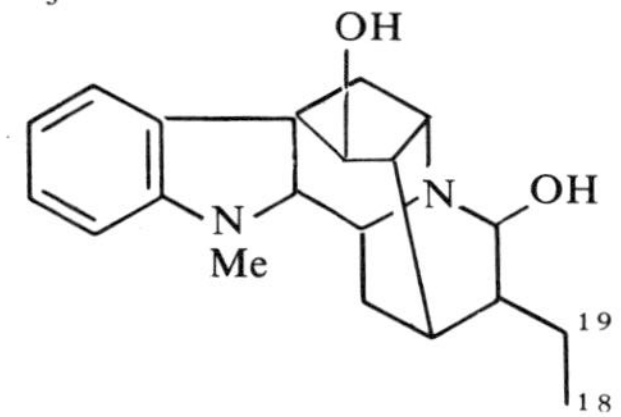

Ajmaline

(C-18 + C-19: 9.5%)
Reserpine +

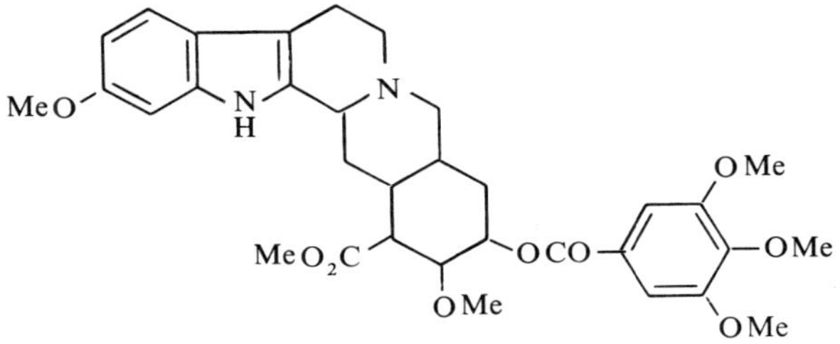

(59.5% of activity in reserpic acid part, 31.8% in 3,4,5-trimethoxybenzoic acid)

Species	Alkaloids labelled	Incorporation/%	Ref.
Compound administered			
Rauwolfia serpentia—contd.			
[2-¹⁴C]Glycine	Ajmaline (C-18: 15.5%, C-19: 0%)	0.0011	
	Reserpine (84% in reserpic acid, 15% in 3,4,5-trimethoxybenzoic acid)	+	
Ricinus communis			
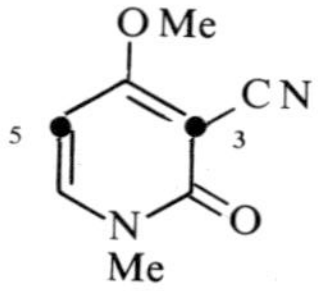 [3,5-¹⁴C]Ricinine	After 28 h: 81% of the ricinine was recovered unchanged, about 20% was found as *N*-demethylricinine		20c

Structure: 4-methoxy-3-cyano-1-methyl-2-pyridone (ricinine), with labels at C-3 and C-5.

(injected with a microsyringe into the petioles of yellow leaves)	After 1 week: 52% of the ricinine was recovered.		118
[7-¹⁴C]Nicotinic acid	Ricinine	Relative activity	
+ No additions		100	
+ Ammonium nitrate (10 mg)		37	
+ Ammonium nitrate (100 mg)		2.5	
+ Sucrose		137	
+ 2-Hydroxypyridine		91	
+ 3-Hydroxypyridine		44	
+ 4-Hydroxypyridine		90	
+ 3-Cyanopyridine		43	
+ Ricininic acid 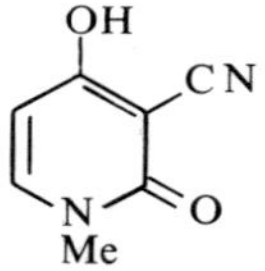		13.5	

Structure: 4-hydroxy-3-cyano-1-methyl-2-pyridone (ricininic acid).

Biosynthesis of Alkaloids

Precursor	Product	Incorporation	Ref.
+ Ricinine			72
+ Asparagine			98
+ β-Cyanoalanine			66
[$^{15}N_2$]Ammonium nitrate	Ricinine (The ricinine contained 30% excess ^{15}N. However, the ratio of numbers of molecules with mol. wts. of 164, 165, and 166 ([$^{15}N_2$]ricinine) was 61:17:22. If there had been uniform dilution of the administered [$^{15}N_2$]ammonium nitrate into the various nitrogen pools one would have expected a distribution of 49:42:9)		
	Ricinine (OMe: 49%, NMe: 45%)	[0.033]	119
[*N-methyl*-^{14}C]Trigonelline			
(structure: pyridinium with CO$_2^-$ and *Me)			
[*N-methyl*-^{14}C,^{3}H]Trigonelline (^{3}H:^{14}C = 31.6)	Ricinine (OMe: 46.4%, ^{3}H:^{14}C = 10.4) (NMe: 43.2%, ^{3}H:^{14}C = 10.3)	^{14}C: [0.105]	
[*N-methyl*-^{14}C]-*N*-Methyl-3-cyanopyridinium salt	Ricinine (OMe: 45.5%, NMe: 48.5%)	[0.01]	
(structure: pyridinium with CN and *Me, X$^-$)			
[*N-methyl*-^{14}C-]-*N*-Methyl-3-cyano-2-pyridone	Ricinine (OMe: 30%, NMe: 70%)	[0.40]	
(structure: 2-pyridone with CN and Me)			

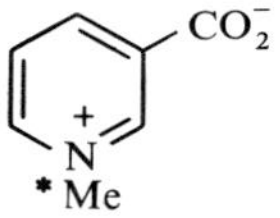

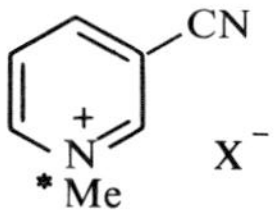

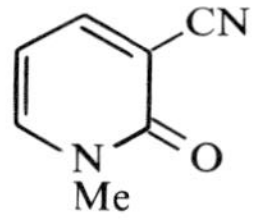

Species / Compound administered	Alkaloids labelled	Incorporation/%	Ref.
Ricinus communis—contd.			
[*methyl*-^{14}C]Methionine	Ricinine (OMe: 43%, NMe: 50%)	[0.83]	120
Santalum album (The cut stems of the plant were placed in the tracer solution which was infiltrated *in vacuo* for 3—4 h, then allowed to grow for 24 h)			
L-[*U*-^{14}C]Arginine	*sym*-Homospermidine NH$_2$(CH$_2$)$_4$NH(CH$_2$)$_4$NH$_2$ Rel. Inc.	5.5	
+ No additions	100		
+ Agmatine	12.7		
+ Putrescine	17.0		
+ Ornithine	14.7		
+ 4-Aminobutanal	100		
DL-[2-^{14}C]Ornithine	*sym*-Homospermidine Rel. incopr.	7.5 (assuming that only the L-isomer is incorporated)	
+ No additions	100		
+ Arginine	20		
+ Putrescine	13.2		
+ Proline	100		
L-[*U*-^{14}C]Glutamic acid	*sym*-Homospermidine 100	0.8	
+ Ornithine	9.1		
L-[*U*-^{14}C]Proline	*sym*-Homospermidine	0.15	
Scopolia lurida (fed to excised roots)			
[7-^{14}C]Hygrine (see under *Datura stramonium*)	Apoatropine	[0.85]	93

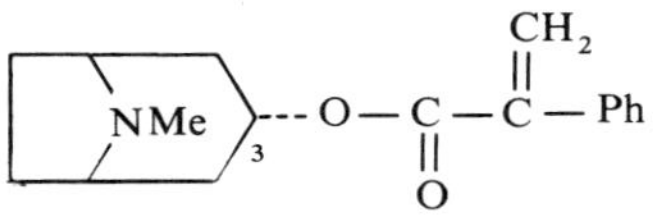

Biosynthesis of Alkaloids

[1.14]

(C-3: >85%)
Cuscohygrine

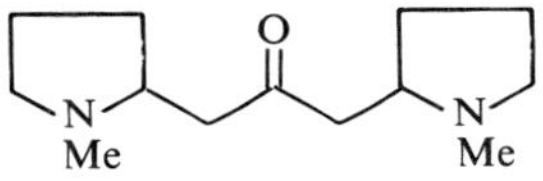

Senecio magnificus

Senecionine

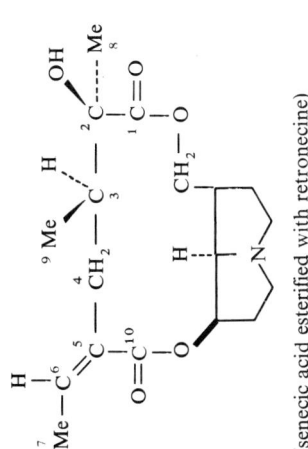

32

(senecic acid esterified with retronecine)

% inc. in senecic acid	% Distribution of Activity						
	C-1	C-2	C-6,7	C-7	C-8	C-9	C-10
98.5	15.8	0.6	18.8	—	0.1	—	12.1 0.25

L-[U-^{14}C]Threonine

Species / Compound administered	Alkaloids labelled					Incorporation/%	Ref.
Senecio magnificus—contd.							
L-[U-^{14}C]Isoleucine	99	9.8				0.44	
DL-[2-^{14}C]Isoleucine*	96	11.6	18.3		8.5	0.48	
	95.2	54.5				0.27*	
	100	50.1				0.44	
DL-[5-^{14}C]Isoleucine	96.6		2.9	49.4	0.35	0.42	
DL-[6-^{14}C]Isoleucine	99.4		49.7		0.51 48.1	0.15	
DL-[2,5-^{14}C$_2$]Isoleucine		15.5			57.5		
(C-5:C-2 = 2.52)		(C-7:C-10 = 2.3 C-9:C-1 = 2.5)					
(*The specifically labelled isoleucines were actually a mixture of DL-isoleucine and DL-alloisoleucine. Incorporations were calculated on the assumption that only the L-isoleucine was an effective precursor)							
L-[*methyl*-^{14}C]Methionine	Senecionine					0.00015	
Skimmia (actual species not given)							
[3-^{14}C]Dictamnine	Skimmianine (no degradations)					1.7–2.1	87
(see under *Choisya ternata*)							
Sophora alopecuroides							
[1,5-^{14}C]Cadaverine	Sophocarpine					0.0014	121

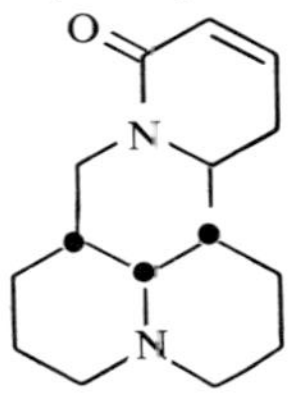

Sophoridine [0.074—0.0063]

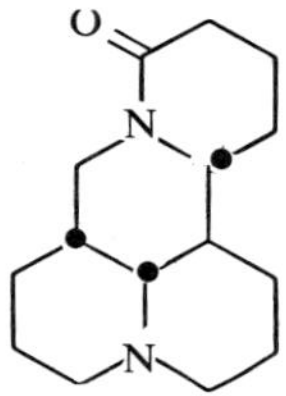

Sophocarpidine [0.0087]

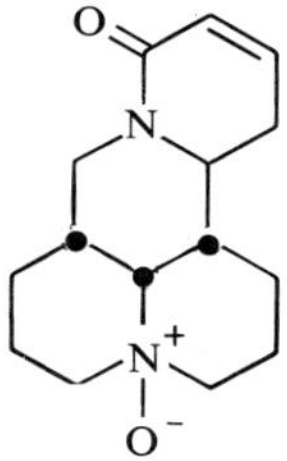

Matrine [0.00064]

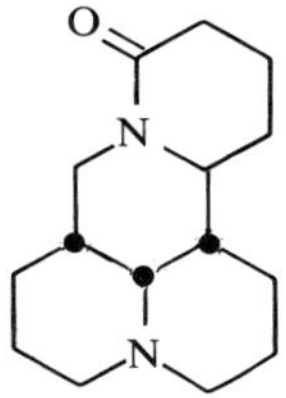

Species	Compound administered	Alkaloids labelled	Incorporation/%	Ref.
Sophora alopercuroides—contd.				
	[*U*-³H]Matrine (made by Wilzbach technique)	Matrine (reisolated)	[4.5]	
		Sophocarpine	[0.23]	
		Matrine-*N*-oxide	[0.19]	
		Sophocarpidine	[0.14]	
		Sophoridine	[0.003]	
	[*U*-³H]Sophocarpine	Sophocarpine (reisolated)	[1.46]	
		Matrine	[0.49]	
		Sophocarpidine	[0.04]	
		Sophoridine	[0.003]	
Thea sinensis (compounds fed to excised tea shoots)	[*methyl*-¹⁴C]Methionine	Caffeine (no degradations) Activity was also detected in *S*-adenosyl-L-methionine isolated from the plants.	+	122
	S-Adenosyl-L-[*methyl*-¹⁴C]methionine	Caffeine (no degradations)	+	
	[2-¹⁴C]Glycine	Caffeine (most of the activity was located on the *N*-methyl groups)	+	123

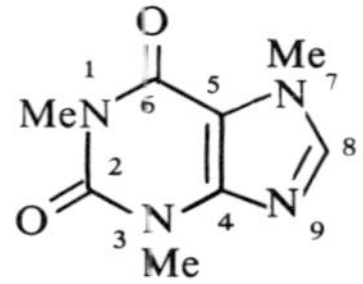

% Distribution of activity
NMe   C-2   C-8   C-4, 5, 6 (by diff.)

Biosynthesis of Alkaloids

Precursor / Plant	Product	Values	Ref.
[N-methyl-^{14}C]-γ-Glutamylmethylamine MeNHCOCH$_2$CH$_2$CHCO$_2$H 　　　　　　　　　　　$|$ 　　　　　　　　　　NH$_2$		25.9　2.6　3.9　67.6	124
[^{14}C]Methylamine		27.9　2.3　6.4　63.4	
Trewia nudiflora [6-^{14}C]Nicotinic acid (injected into the stem of a whole plant)	N-Methyl-5-carboxamide-2-pyridone H$_2$NOC—⟨pyridone ring⟩—N(Me)=O	24 h feed [1] 48 h　　　[1.2] 136 h　　 [7.6]	125
[7-^{14}C]Nicotinamide 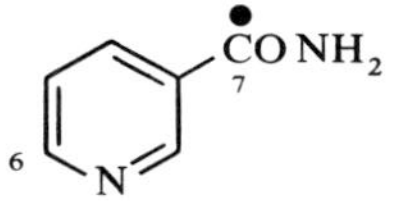 ⟨pyridine-3-carboxamide, C-7 labelled⟩	(no degradations) N-Methyl-5-carboxamide-2-pyridone (no degradations)	136 h feed [8.7]	
Tripterygium wilfordii [6-^{14}C]Nicotinic acid (injected with a micro syringe into soft stems)	Total alkaloids: Wilforine C$_{43}$H$_{49}$O$_{18}$N Wilforgine C$_{41}$H$_{47}$O$_{19}$N Wilfordine C$_{43}$H$_{49}$O$_{19}$N Wilfortrine C$_{41}$H$_{47}$O$_{20}$N	4.40	126

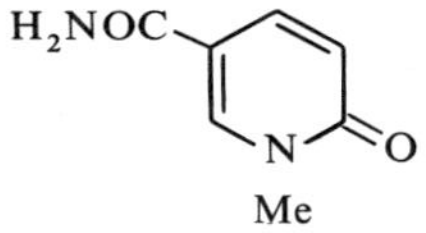

Species	Alkaloids labelled	Incorporation/%	Ref.
Compound administered			
Tripterygium wilfordii—contd.	On hydrolysis these alkaloids yield:		
	Wilfordic acid (no degradations) [structure: pyridine with CO₂H at 3-position and CH₂CH₂CHCO₂H with Me at 2-position]		
	Hydroxywilfordic acid [structure: pyridine with CO₂H at 3-position and CH₂CH₂C(OH)(Me)CO₂H at 2-position]	26.8	
	N'-Methylnicotinamide [structure: N-methylpyridinium with CONH₂, X⁻]		
[*carbonyl*-^{14}C]NAD	Nicotinamide	0.94	
	Nicotinic acid (recovered)	1.63	
	Nicotinamide adenine dinucleotide	0.48	
	Nicotinic acid mono-nucleotide	0.30	
	Deamino NAD	0.38	
	Total alkaloids	2.3	
	N'-Methylnicotinamide	11.25	
	Nicotinamide	0.11	
	Nicotinic acid	13.0	
	Nicotinic acid mono-nucleotide	6.06	
	Deamino NAD	4.38	

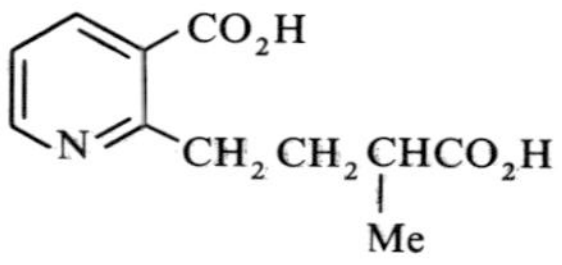

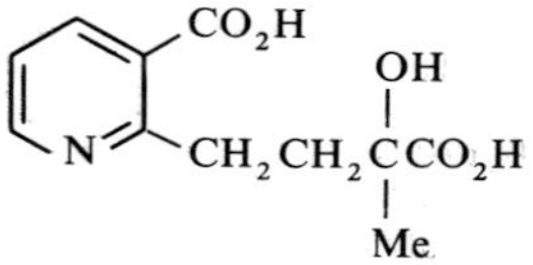

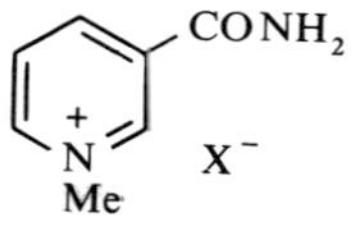

Biosynthesis of Alkaloids

Veratrum grandiflorum
Sodium [1-^{14}C]acetate
(The tracer was fed to the plants which were then cultivated in the dark for 10 or 20 days. The plants were then allowed to grow for various times in the light)

Solanidine

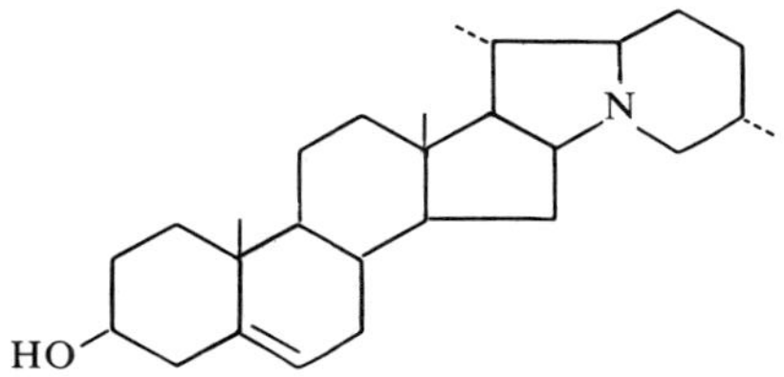

(actually present in the plant as a glycoside)
Veratramine

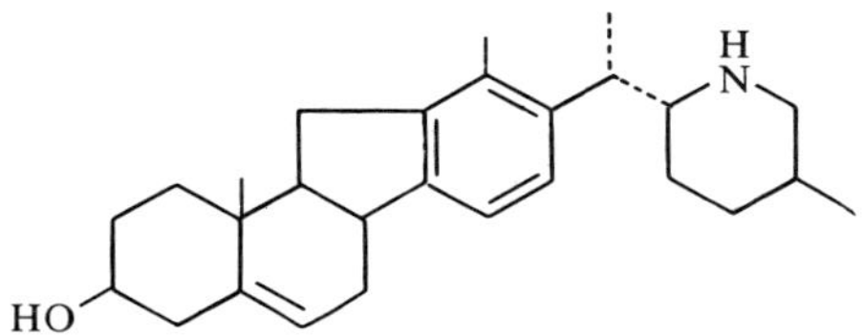

Jervine

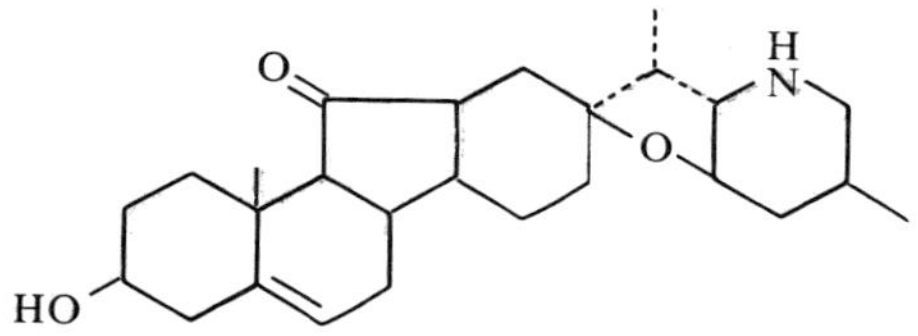

Species / Compound administered	Alkaloids labelled	Incorporation/%	Ref.
Veratrum grandiflorum—contd. Cultivation of plants	Relative Specific Activities		128

Days in:					
Dark	Light	Solanidine	Veratramine	Jervine	
10	0	100	19	14	
10	2	64	43	97	
10	6	23	10	15	
10	10	—	31	57	
20	0	308	23	61	

The above experiments were carried out with mature plants. When etoilated plants at the budding stage were used, the activity of the solanidine also decreased on illumination of the plants (after 5 days growth in the dark), but there was a much slower conversion into veratramine and jervine. Significant activity was detected in a new alkaloid which may be a hydroxylated solanidine.)

Vinca minor
Sodium [2-^{14}C]acetate — Activity found in the total alkaloids.
[2-^{14}C]Tyrosine — No activity in the alkaloids

Vinca rosea — Vindoline — — 129

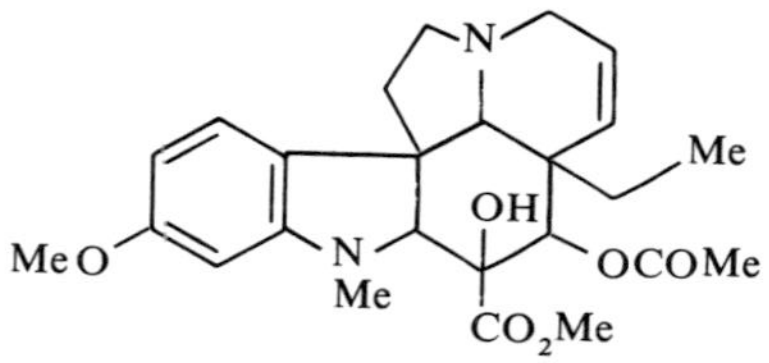

Catharanthine

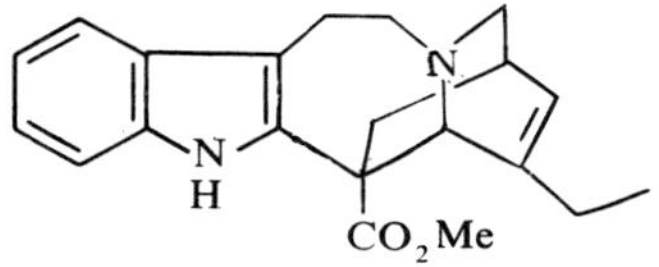

			Incorporation/%	
	Age of Plants	Duration of Feeding	Vindoline	Catharanthine
L-[U-^{14}C]Leucine (fed to cut stems)	6 months	2 days	0.002	0.002
	6 months	7 days	0.007	0.003
	10 months	7 days	0.003	0.002
	10 months	2 weeks	0.006	0.003
	3 years	7 days	0.003	0.002
	3 years	2 weeks	0.005	0.004
	10 months	2 weeks	0.008	0.004
+ puromycin (puromycin fed 24 h before the leucine)	10 months	2 weeks	inactive	inactive

(No degradations carried out on the alkaloids)

[83] L. K. Klyschew, G. K. Kruglychina, N. I. Klimentyeva, and D. Nurakow, ref. 2, p. 213.
[84] N. I. Klimentyeva and L. K. Klyschew, *Trudy Inst. Bot., Akad. Nauk. Kaz. S.S.R.*, 1971, **29**, 164 (*Chem. Abs.*, 1972, **76**, 70243).
[85] E. Brochmann-Hanssen, C.-H. Chen, H.-C. Chiang, and K. McMurtrey, *J.C.S. Chem. Comm.*, 1972, 1269.
[86] A. K. Garg and J. R. Gear, *Phytochemistry*, 1972, **11**, 689.
[87] J. F. Collins, W. J. Donnelly, M. F. Grundon, D. M. Harrison, and C. G. Spyropoulos, *J.C.S. Chem. Comm.*, 1972, 1029.
[88] C. Fuganti and M. Mazza, *J.C.S. Chem. Comm.*, 1972, 936.
[89] H. Keller, H. Wanner, and T. W. Baumann, *Planta*, 1972, **108**, 339.
[90] A. R. Battersby, R. B. Herbert, E. McDonald, R. Ramage, and J. H. Clements, *J.C.S. Perkin I*, 1972, 1741.
[91] A. R. Battersby, T. A. Dobson, D. M. Foulkes, and R. B. Herbert, *J.C.S. Perkin I*, 1972, 1730.
[92] R. H. Wightman, J. Staunton, A. R. Battersby, and K. R. Hanson, *J.C.S. Perkin I*, 1972, 2355.
[93] H. W. Liebisch, K. Peisker, A. S. Radwan, and H. R. Schütte, *Z. Pflanzenphysiol.*, 1972, **67**, 1.
[94] E. Leete, *Phytochemistry*, 1972, **11**, 1713.
[95] R. G. Achari, W. E. Court, and F. Newcombe, *Planta Medica*, 1972, **22**, 38.
[96] M. P. Gupta and M. R. Gibson, *J. Pharm. Sci.*, 1972, **61**, 1257.
[97] Yu. B. Tikonov, *Sb. Nauch. Rab., Vses Nauch.-Issled. Inst. Lek. Rast.*, 1970, No. 1, 126 (*Chem. Abs.*, 1972, **76**, 70248).
[98] E. Leete and A. R. Pinder, *Phytochemistry*, 1972, **11**, 3219.
[99] E. Leete, unpublished work.
[100] H. Rosenberg and A. G. Paul, *Lloydia*, 1971, **34**, 372.
[101] H. Horan and D. G. O'Donovan, *J. Chem. Soc. (C)*, 1971, 2083.
[102] I. J. McFarlane and M. Slaytor, *Phytochemistry*, 1972, **11**, 229.
[103] I. J. McFarlane and M. Slaytor, *Phytochemistry*, 1972, **11**, 235.
[104] A. Rother and A. E. Schwarting, *Phytochemistry*, 1972, **11**, 2475.
[105] H. R. Schütte, ref. 2, p. 103.
[106] I. Murakoshi, S. Ohmiya, and J. Haginawa, *Chem. and Pharm. Bull. (Japan)*, 1971, **19**, 2655.
[107] J. Lundström, *Acta Chem. Scand.*, 1971, **25**, 3489.
[108] J. Lundström, *Acta Pharm. Suecica*, 1971, **8**, 485.
[109] G. J. Kapadia, G. S. Rao, E. Leete, M. B. E. Fayez, Y. N. Vaishnav, and H. M. Fales, *J. Amer. Chem. Soc.*, 1970, **92**, 6943.
[110] J.-C. Braekman, R. N. Gupta, D. B. MacLean, and I. D. Spenser, *Canad. J. Chem.*, 1972, **50**, 2591.
[111] C. Fuganti and M. Mazza, *J.C.S. Chem. Comm.*, 1972, 239.
[112] M. Y. Lovkova, G. S. Iljin, and N. I. Klimentyeva, *Fiziol. Rast.*, 1970, **17**, 409.
[113] S. Sciuto, G. Oriente, and M. Piattelli, *Phytochemistry*, 1972, **11**, 2259.
[114] H. Rönsch, *European J. Biochem.*, 1972, **28**, 123.
[115] H. Bohm, B. Olesch, and Ch. Schultz, *Biochem. Physiol. Pflanz.*, 1972, **163**, 126.
[116] Yu. A. Russkov, *Sb. Nauch. Rab. Vses Nauch.-Issled. Inst. Lek. Rast.*, 1970, No 1, 137 (*Chem. Abs.*, 1972, **76**, 70205).
[117] C. Baxter and M. Slaytor, *Phytochemistry*, 1972, **11**, 2767.
[118] E. Nowacki and G. R. Waller, ref. 2, p. 187.
[119] D. Gross, ref. 2, p. 197.
[120] R. Kuttan and A. N. Radhakrishnan, *Biochem. J.*, 1972, **127**, 61.
[121] J. K. Kuschmuradov, D. Gross, and H. R. Schütte, *Phytochemistry*, 1972, **11**, 3441.
[122] T. Suzuki, *F.E.B.S. Letters*, 1972, **24**, 18.
[123] I. Takao, *Hoshi Yakka Daigaku Kiyo*, 1971, 60 (*Chem. Abs.*, 1972, **77**, 137414).
[124] S. Konishi, T. Inoue, and E. Takahashi, *Plant Cell Physiol.*, 1972, **13**, 695.
[125] S. D. Sastry and G. R. Waller, *Phytochemistry*, 1972, **11**, 2241.
[126] H. J. Lee and G. R. Waller, *Phytochemistry*, 1972, **11**, 2233.
[127] K. Kaneko, M. Watanabe, S. Taira, and H. Mitsuhasi, *Phytochemistry*, 1972, **11**, 3199.
[128] G. Verzar-Petri, *Acta Biol. Acad. Sci. Hung.*, 1971, **22**, 413 (*Chem. Abs.*, 1972, **76**, 110357).
[129] D. C. Wigfield, B. Lem, and V. Srinivasan, *Tetrahedron Letters*, 1972, 2659.
[130] D. G. O'Donovan and M. F. Kenneally, *J. Chem. Soc.*, 1967, 1109.
[131] M. Castillo, R. N. Gupta, D. B. MacLean, and I. D. Spenser, *Canad. J. Chem.*, 1970, **48**, 1893.
[132] E. Brochmann-Hanssen and B. Nielsen, *J. Pharm. Sci.*, 1965, **54**, 1393.

5
Biosynthesis of Polyketides

BY T. MONEY

1 Aromatic Polyketides*

In theory, the acyl–polymalonate biosynthetic route[1–5] (Scheme 1) to naturally occurring phenolic compounds involves condensation of an enzyme-bound carboxylic acid derivative with a variable number of malonate units to produce intermediate enzyme-bound β-polyketo-thiol esters (1). Subsequent intramolecular condensation processes can then lead to phenolic compounds differing widely in structural type. The variety of polyketides produced by this route is made possible by the manipulation of several variables, *viz.* (*a*) the nature of the chain-initiating carboxylic acid unit, RCOSCoA, (*b*) the number of malonyl-coenzyme A units involved,† *i.e.* the chain length of the intermediate β-polyketo-ester (1), (*c*) the nature of the intramolecular condensation-aromatization process, and (*d*) secondary transformations (*e.g.* alkylation, halogenation, reduction, ring-cleavage reactions, oxidation) which can occur before or after formation of the phenolic ring.

Detailed descriptions of the historical development of this biosynthetic route and experimental evidence supporting some of its basic features have been provided in several reviews and books.[2–4] In addition, an excellent new text,[5] which provides a record of achievement in this area, has been published recently. The present Report describes the results of biosynthetic investigations published during the period Jan 1971—Dec 1972. For convenience, the aromatic

---

[1] J. N. Collie, *J. Chem. Soc.*, 1907, **91**, 1806.
[2] A. J. Birch and F. W. Donovan, *Austral. J. Chem.*, 1953, **6**, 360.
[3] R. Robinson, 'Structural Relations of Natural Products', Clarendon, Oxford, 1955; A. J. Birch, *Fortschr. Chem. org. Naturstoffe*, 1957, **14**, 186; *Proc. Chem. Soc.*, 1962, 3; *Science*, 1967, **156**, 202; *Ann. Rev. Plant Physiol.*, 1968, **19**, 321.
[4] (*a*) J. D. Bu'Lock, 'The Biosynthesis of Natural Products', McGraw-Hill, London, 1965; (*b*) J. H. Richards and J. B. Hendrickson, 'The Biosynthesis of Steroids, Terpenes, and Acetogenins', Benjamin, New York, 1964; (*c*) T. A. Geissman and D. H. G. Crout, 'Organic Chemistry of Secondary Plant Metabolism', Freeman–Cooper, San Francisco, 1969.
[5] W. B. Turner, 'Fungal Metabolites', Academic Press, London, 1971.

* The term 'aromatic polyketide' is used to describe phenolic compounds (or derivatives thereof) which are derived, wholly or in part, by the operation of the acyl–polymalonate biosynthetic route. Although fatty acids (and natural products derived therefrom) are biosynthesized by condensation of an acyl-coenzyme A unit with a variable number of malonate or acetate units, the intermediate β-keto-ester is reduced after each condensation step. For this reason the biosynthesis of fatty acids *etc.* will not be considered in this Report.

† In a few isolated cases propionate (methylmalonyl-coenzyme A) and butyrate units may be involved in the chain-propagating process.

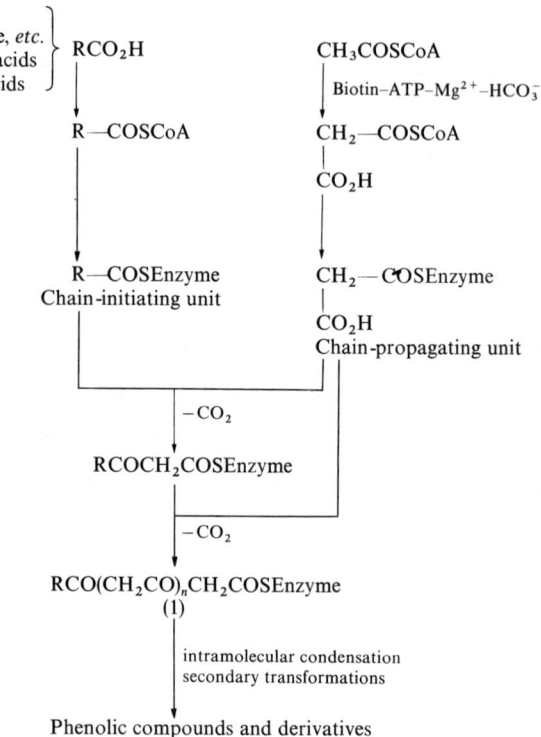

Scheme 1

polyketides will be considered in groups according to the nature of the chain-initiating unit (RCOSCoA) and the number of malonate units involved in their biosynthesis.

Exciting progress[6,7] has been made in the enzymology of systems which elaborate phenolic compounds by the acetate–polymalonate pathway. The purification and properties of 6-MSA (6-methylsalicylic acid) (4) synthetase were reported[7] in 1970 and recent work[8] has provided further information on this enzymic system. Thus it has been reported that an ammonium sulphate protein fraction from *P. patulum* catalyses the synthesis of 6-MSA, TAL (triacetic acid lactone) (2), and fatty acids in the presence of malonyl-coenzyme A and NADPH. The derivation of 6-MSA from one acetate and three malonate

---

[6] Recent reviews: R. J. Light, *J. Agric. Food Chem.*, 1970, **18**, 260; J. W. Corcoran and F. J. Darby, 'Lipid Metabolism', ed. S. J. Wakil, Academic Press, New York, 1970, p. 431.
[7] P. Dimroth, H. Walter, and F. Lynen, *European J. Biochem.*, 1970, **13**, 98; cf. J. Staunton, *Ann. Reports (B)*, 1970, **67**, 553.
[8] A. I. Scott, G. T. Phillips, and U. Kircheis, *Bioorganic Chem.*, 1971, **1**, 380, and references cited therein.

Biosynthesis of Polyketides

units was confirmed and it was shown that both 6-MSA and fatty acid synthetase activity were inhibited by sulphydryl blocking agents.[8] This result is in agreement with previous studies[7] which have shown that the purified 6-MSA synthetase system contains two sulphydryl sites. It has also been shown[8] that the synthesis of 6-MSA is inhibited by 3-hexynoyl-NAC, 3-pentynoyl-NAC, and 2-hexynoyl-NAC* whereas the synthesis of palmitic or stearic acids is

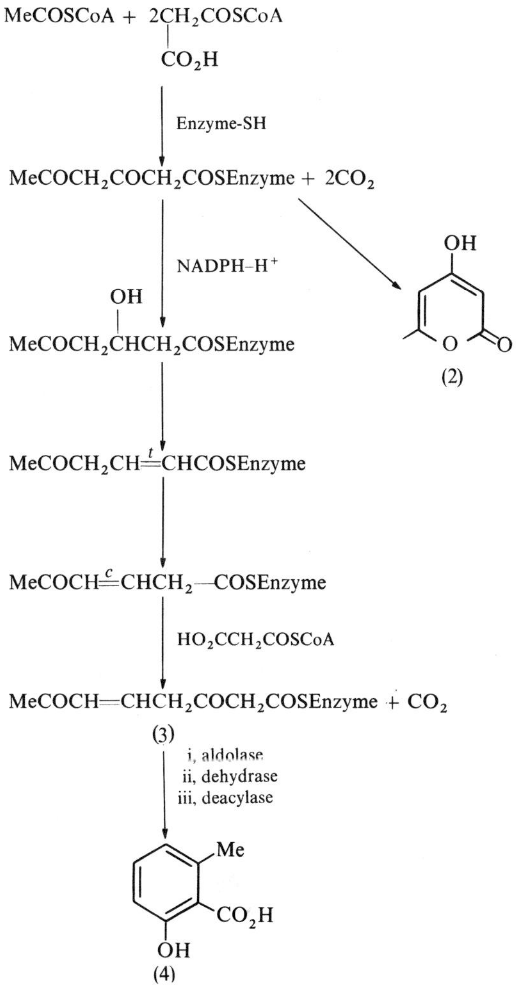

$$\text{MeCOSCoA} + 2\underset{|}{\text{CH}}_2\text{COSCoA}$$
$$\text{CO}_2\text{H}$$

↓ Enzyme-SH

$$\text{MeCOCH}_2\text{COCH}_2\text{COSEnzyme} + 2\text{CO}_2$$

↓ NADPH–H$^+$ → (2)

$$\underset{|}{\text{OH}}$$
$$\text{MeCOCH}_2\text{CHCH}_2\text{COSEnzyme}$$

(2) is a pyranone with OH

↓

$$\text{MeCOCH}_2\text{CH}\overset{t}{=}\text{CHCOSEnzyme}$$

↓

$$\text{MeCOCH}\overset{c}{=}\text{CHCH}_2\text{—COSEnzyme}$$

↓ HO$_2$CCH$_2$COSCoA

$$\text{MeCOCH}=\text{CHCH}_2\text{COCH}_2\text{COSEnzyme} + \text{CO}_2$$
(3)

↓ i, aldolase
ii, dehydrase
iii, deacylase

(4) 2-methyl-6-hydroxybenzoic acid (Me, CO$_2$H, OH on benzene ring)

Scheme 2

* NAC = N-acetylcysteamine.

uñaffected. Related studies[9] have shown that 3-decynoyl-NAC specifically inhibits the biosynthesis of *cis*-unsaturated fatty acids by the fatty acid synthetase of *E. coli*, and this effect has been explained in terms of inhibition of β-hydroxydecanoyl thiol ester dehydrase which is involved in the formation of α,β-*trans*- and β,γ-*cis*-decenol thiol ester. A similar inhibition of 6-MSA synthetase activity by acetylenic thiol ester has been cited[8] as indirect evidence for the presence of an unsaturated enzyme-bound thiol ester intermediate (3) and supports the general biosynthetic scheme[7] outlined in Scheme 2.

Further investigations on the enzymology of conversion of 6-MSA (4) into patulin (10) are described in a recent report[10] which records the isolation and

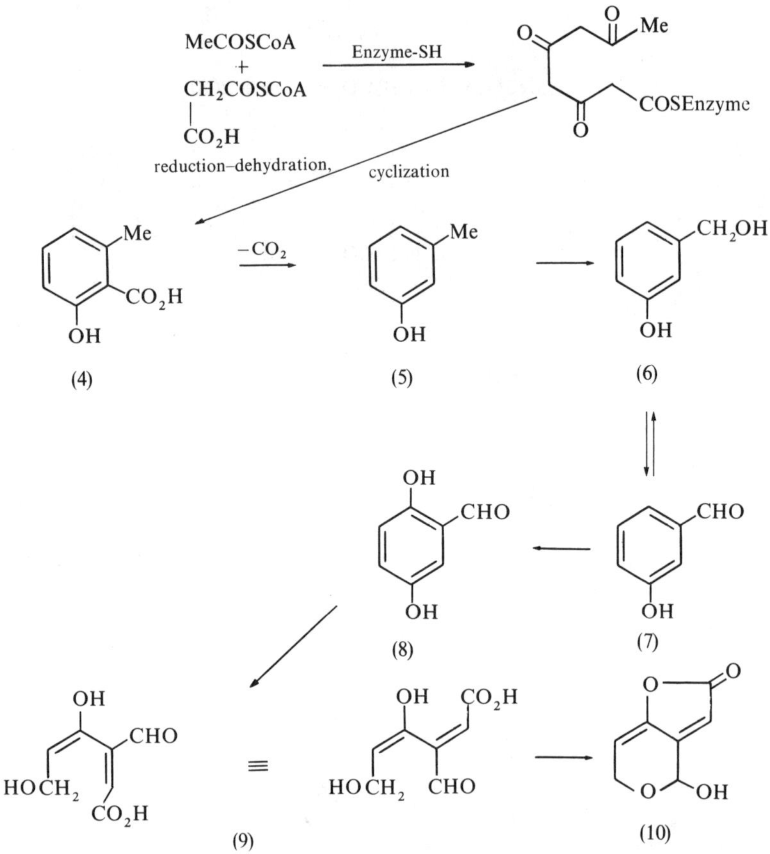

Scheme 3

[9] L. R. Kass and K. Bloch, *Proc. Nat. Acad. Sci. U.S.A.*, 1967, **58**, 1168; K. Endo, G. M. Helmkamp, and K. Bloch, *J. Biol. Chem.*, 1970, **245**, 4293, and references cited therein.

[10] P. I. Forrester and G. M. Gaucher, *Biochemistry*, 1972, **11**, 1108; cf. R. J. Light, *Biochem. Biophys. Acta*, 1969, **191**, 430.

partial purification of *m*-hydroxybenzyl alcohol dehydrogenase from cell-free extracts of *P. urticae*. The enzyme is specific for NADPH and *m*-hydroxy- or *m*-methoxy-benzaldehyde and has a pH optimum of 7.6. The suggestion has been made that *m*-hydroxybenzaldehyde dehydrogenase catalyses the rate-determining step in the overall conversion of 6-MSA into patulin. Related studies,[11] using pulse-labelling experiments and fermenter cultures of *P. urticae* (NRRL 2159A) have shown that the preferred sequence of reactions leading to patulin (10) is represented by Scheme 3. These results support and extend previous results[12] in this area and, in particular, provide experimental support for the intermediate role of *m*-hydroxybenzyl alcohol (6), *m*-hydroxy-benzaldehyde (7), and gentisaldehyde (8) in the biosynthetic pathway (Scheme 3).

The bacterial biosynthesis of 6-methylsalicylic acid (6-MSA) (4) has been studied in detail[13] and the administration of radioactive malonate to *Mycobacterium phlei* has provided 6-MSA with a labelling pattern consistent with its derivation by the acetate–polymalonate biosynthetic route. Some activity from malonate was incorporated into the acetate-derived *C*-methyl group and this partial conversion of malonate into acetate has been recorded previously in fungal studies.[13] Related studies on the biosynthesis of 6-MSA and salicylic acid by *Mycobacterium fortuitum* have shown that acetate was much more efficiently incorporated into the former compound.[14] It has been concluded, therefore, that these structurally similar phenolic acids are derived by different biosynthetic pathways.

Further studies have been reported[15] on the structural elucidation of phenolic metabolites produced by blocked mutants of *Aspergillus rugulosus*. Asperugin C (13) has been isolated from the mycelium of a blocked mutant which no longer produced asperugin (18) and it has been suggested that this compound is a shunt metabolite. The identification of other metabolites (14—17) and shunt metabolites (11—13) of blocked mutants of *A. rugulosus* has previously been recorded by the same group, and the combined results have led to the proposal for the biosynthesis of asperugin (18) which is outlined in Scheme 4.

The sequence of reactions involved in the biosynthesis of mycophenolic acid (23) has been the subject of recent investigations by several research groups.[16–18] Previous studies[4] had established that the basic carbocyclic skeleton of the molecule was acetate-derived and that methionine provided the

---

[11] P. I. Forrester and G. M. Gaucher, *Biochemistry*, 1972, **11**, 1102.
[12] A. I. Scott and M. Yalpani, *Chem. Comm.*, 1967, 945, and references cited therein.
[13] J. G. Dain and R. Bentley, *Bioorganic Chemistry*, 1971, **1**, 374, and references cited therein.
[14] A. T. Hudson, *Phytochemistry*, 1971, **10**, 1555.
[15] J. A. Ballantine, V. Ferrito, and C. H. Hassall, *Phytochemistry*, 1971, **10**, 1309.
[16] L. Canonica, W. Kroszczynski, B. M. Ranzi, B. Rindone, E. Santianello, and C. Scolastico, *J.C.S. Perkin I*, 1972, 2639, and references cited therein ; *cf.* L. Canonica, B. Rindone, E. Santaniello, and C. Scolastico, *Tetrahedron*, 1972, **28**, 4395.
[17] An extensive investigation of the biosynthesis of mycophenolic acid has been directed by N. J. McCorkindale (Glasgow University), W. B. Turner (I.C.I. Ltd.), and their respective co-workers (personal communication).
[18] C. T. Bedford, J. C. Fairlie, P. Knittel, T. Money, and G. T. Phillips, *Chem. Comm.*, 1971, 323; C. T. Bedford, P. Knittel, T. Money, G. T. Phillips, and P. Salisbury, *Canad. J. Chem.*, 1973, **51**, 694.

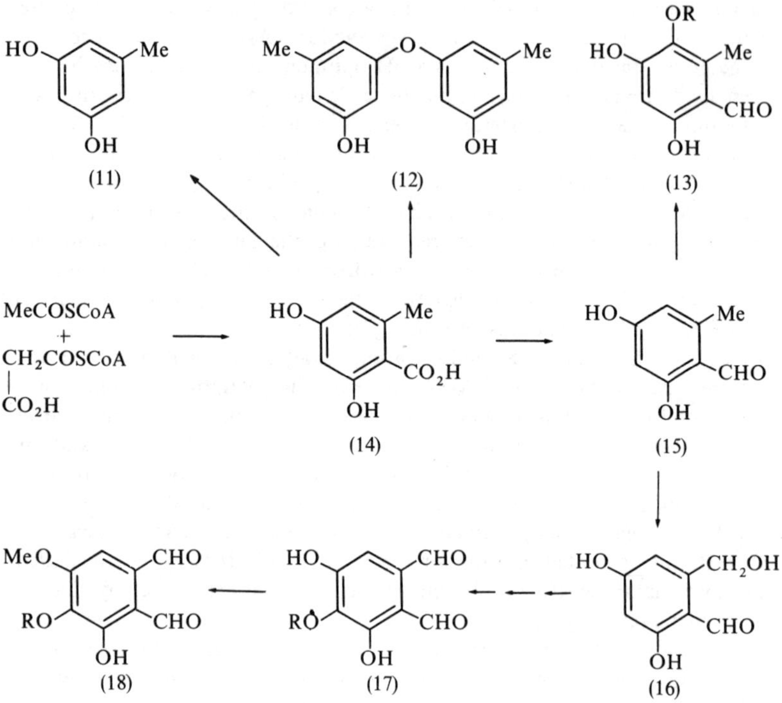

Scheme 4

$R = C_{15}H_{25}$ (farnesyl)

O- and *C*-methyl groups attached to the aromatic ring. In addition, the acidic side-chain has been shown to be derived from two molecules of mevalonic acid. Recent studies of the biosynthesis of mycophenolic acid (23) by *P. brevi-compactum* have shown that 4,6-dihydroxy-2,3-dimethylbenzoic acid (19), 5,7-dihydroxy-4-methylphthalide (20), and 6-farnesyl-5,7-dihydroxy-4-methylphthalide (21) are intermediates in the biosynthetic pathway. It was also demonstrated that orsellinic acid (14) was not a specific precursor and it has been suggested that *C*-methylation is occurring prior to cyclization–aromatization of an enzyme-bound thiol ester of 3,5,7-triketo-octanoic acid. Moreover, the efficient conversion (78.6%) of acid (22) into mycophenolic acid (23) by the culture supports the view that *O*-methylation is the final step in the biosynthetic sequence.* A biosynthetic scheme based on the cumulative efforts of several research groups is shown in Scheme 5.

[19] *Cf.* A. J. Birch, *Ann. Rev. Plant Physiol.*, 1968, **19**, 321.

* It is interesting to note that administration of [1-^{14}C]geraniol, 1,1-ditritiogeraniol, and the corresponding pyrophosphate to *P. brevi-compactum* provided mycophenolic acid which was shown, by degradation studies, to be labelled in a non-specific fashion.[19]

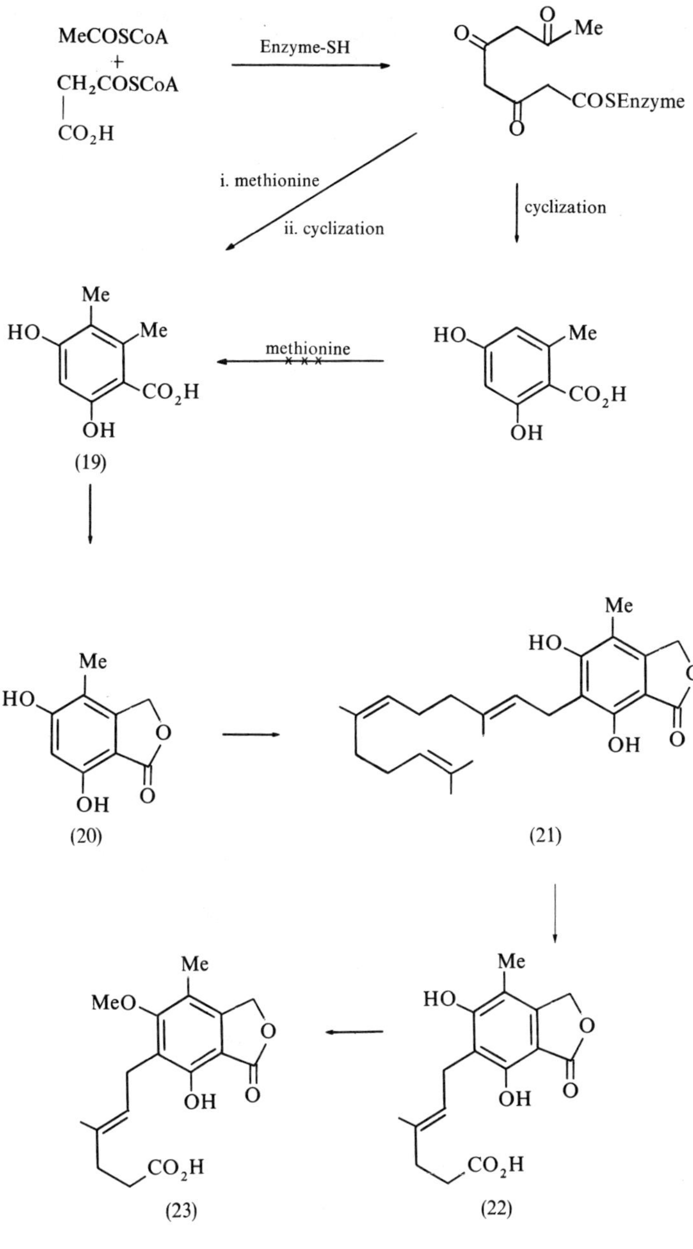

Biosynthesis of Polyketides

Scheme 5

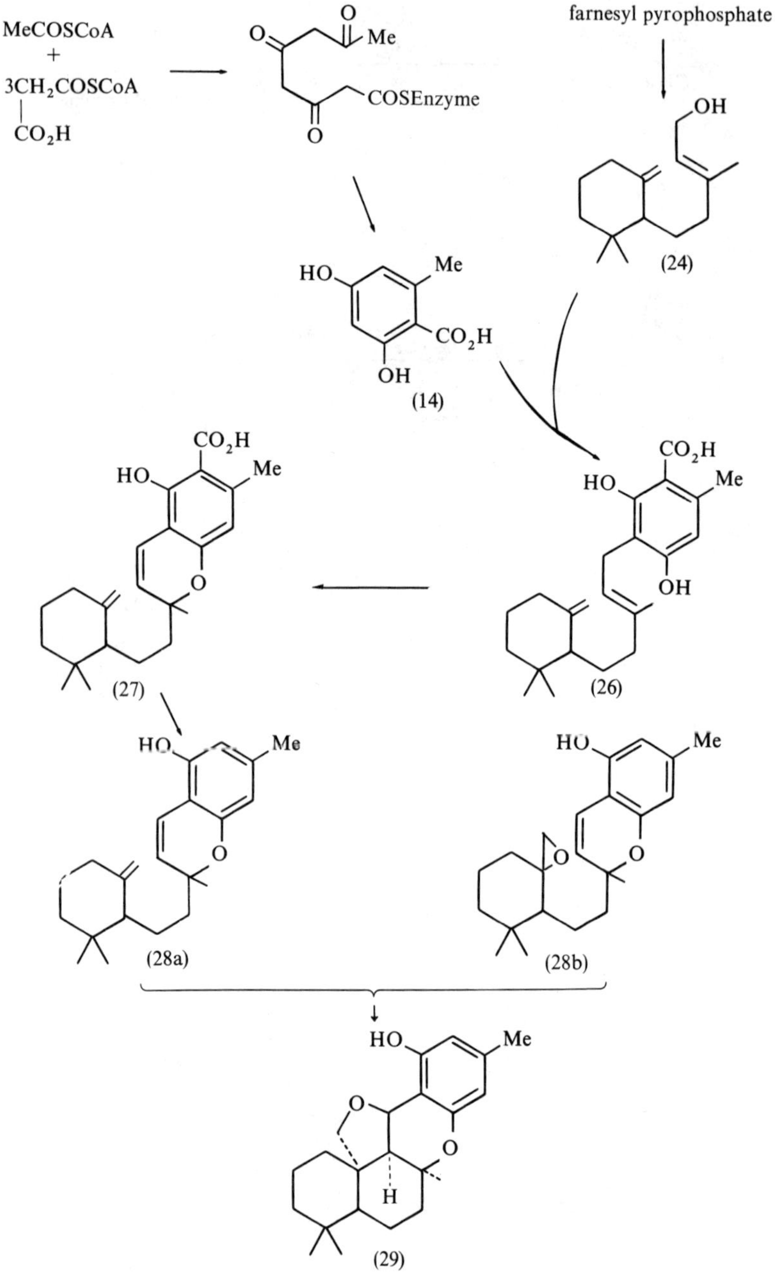
MeCOSCoA
+
$3CH_2COSCoA$
CO_2H
O
O
Me
COSEnzyme
O
farnesyl pyrophosphate
OH
(24)
HO
Me
CO_2H
OH
(14)
CO_2H
HO
Me
O
(27)
CO_2H
HO
Me
OH
(26)
HO
Me
O
(28a)
HO
Me
O
O
(28b)
HO
Me
O
O
H
(29)

Scheme 6

Biosynthesis of Polyketides

Recent investigations[20-22] have provided considerable insight on the biosynthesis of siccanin (29), an antibiotic produced by *Helminthosporium siccans* Drechsler. Initial studies,[20] using intact cell systems of *H. siccans*, had shown that radioactivity from siccanochromene-A (28a) or -B (28b) was specifically incorporated into siccanin (29). Subsequent studies, using cell-free extracts of *H. siccans*, demonstrated[22] that orsellinic acid (14), farnesyl pyrophosphate, pre-siccanochromenic acid (26), and siccanochromenic acid (27) could serve as biosynthetic precursors of siccanochromene-A (28a). Incubation of mevalonic acid with a similar cell-free enzyme preparation resulted in the formation of *trans*-γ-monocyclofarnesol (24), and the detection[21] of this compound as a normal metabolite of *H. siccans* is additional evidence for its role as an intermediate in the biosynthetic pathway (Scheme 6).

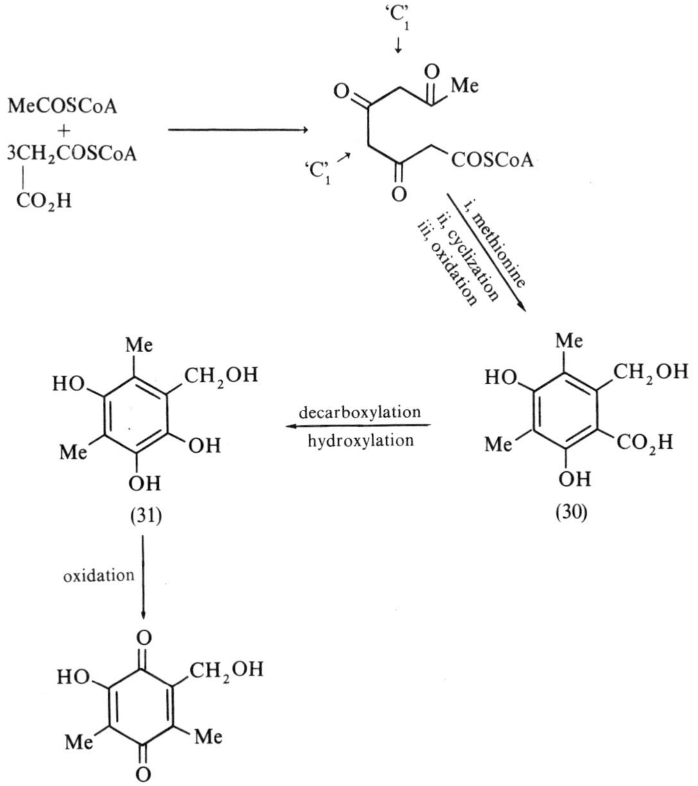

Scheme 7

[20] K. T. Suzuki and S. Nozoe, *Chem. Comm.*, 1971, 527.
[21] K. T. Suzuki, N. Suzuki, and S. Nozoe, *Chem. Comm.*, 1971, 527.
[22] K. T. Suzuki and S. Nozoe, *J.C.S. Chem. Comm.*, 1972, 1166.

The biosynthesis of shanorellin (32), an extracellular product of *Shanorella spirotricha* Benjamin (Ascomycetes), has been shown to involve the acetate–polymalonate pathway, methionine providing the C-methyl groups.[23] Possible precursors such as orsellinic acid (14), 4,6-dihydroxy-2,3-dimethylbenzoic acid, 4,6-dihydroxy-2,3,5-trimethylbenzoic acid, and 4,5-dimethylresorcinol could not be detected in the growth medium and this has led to the suggestion that C-methylation precedes cyclization–aromatization of the postulated β-triketo-thiol ester intermediate (Scheme 7).

Further information on the biosynthetic route to fungal tropolones has been provided.[24] The suggestion[5,25] that 3-methylorsellinic acid (33) or a related compound could act as a biosynthetic intermediate to these compounds has been established by appropriate tracer experiments. Administration of [3-*Me*-^{14}C]3-methylorsellinic acid (33) or 3-methyl[*carboxy*-^{14}C]orsellinic acid (33) to *P. stipitatum* (NRRL 2104) gave radioactive stipitatonic acid (35). Subsequent degradation to stipitatic acid (36) demonstrated that 3-methylorsellinic acid (33) was a specific biosynthetic precursor of stipitatonic acid (35). These results have provided direct support for the presence of an aromatic intermediate in the biosynthetic route to fungal tropolanes and exclude alternative proposals which invoke rearrangement of methylated β-triketo-thiol ester intermediates.[26] The nature of the process which converts methylorsellinic acid (33) into

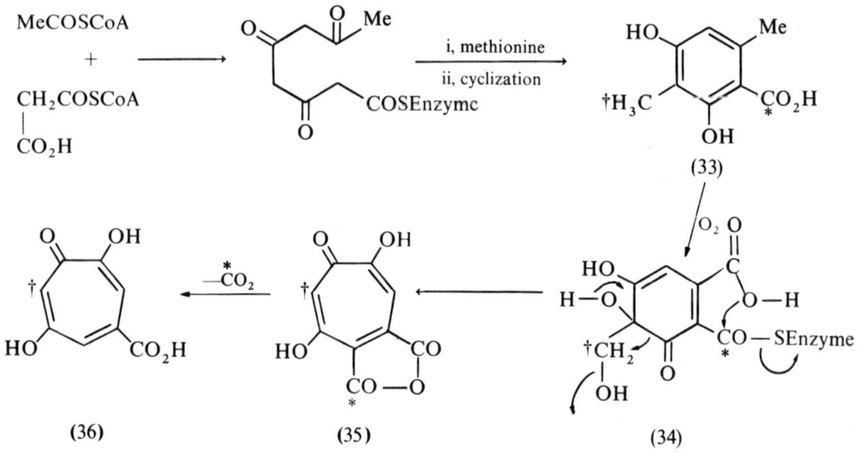

Scheme 8

---

[23] C.-K. Wat and G. H. N. Towers, *Phytochemistry*, 1971, **10**, 103.
[24] A. I. Scott, H. Guilford, and E. Lee, *J. Amer. Chem. Soc.*, 1971, **93**, 3534.
[25] *Cf.* Ref. 4c.
[26] R. Bentley, *J. Biol. Chem.*, 1963, **238**, 1895, and references cited therein.

Biosynthesis of Polyketides

tropolone products is presently uncertain, but recent evidence,[27] obtained by incubation of *P. stipitatum* in the presence of $^{18}O_2$ and sodium [1-^{14}C,^{18}O]-acetate, strongly supports the operation of a mono-oxygenase mechanism [(34), Scheme 8] in the biosynthetic route.

The same research group has also re-investigated the biosynthesis of puberulonic (37) and puberulic acid (38) in *P. aurantio-virens* (NRRL 2138) and have shown that C-9 of (37) is derived from acetate.[28] A common biosynthetic route has been suggested for the fungal tropolones (35)—(38), although attempts to demonstrate the conversion of stipitatonic acid (35) or 3-methylorsellinic acid (33) into puberulonic acid (37) were unsuccessful.[28]

$$Me\overset{*}{C}O_2H \xrightarrow[\textit{aurantio-virens}]{Penicillium}$$

(37)

$$\xrightarrow{-\overset{*}{C}O_2}$$

(38)

Whereas the fungal tropolones described above are derived *via* an enzyme-bound β-triketo-ester, the biosynthesis of sepedonin (41), a metabolite of *Sepedonium chrysospermum*, seems to involve a β-tetraketo-ester intermediate (39). Initial studies, using the more limited ^{13}C–H satellite method,* demonstrated that the methyl group, C-4, and C-5 were labelled from $^{13}CH_3CO_2H$, C-1 was labelled from $Me^{13}CO_2H$, and C-8 from $H^{13}CO_2H$.[30] These conclusions have been confirmed and extended by recent studies utilizing proton-noise-decoupled ^{13}C n.m.r. spectra of sepedonin.[31] Thus separate administration of $^{13}CH_3CO_2H$, $Me^{13}CO_2H$, and $H^{13}CO_2H$ to cultures of *S. chrysospermum* provided sepedonin (41) whose labelling pattern, as determined by ^{13}C n.m.r. spectra, was consistent with a biosynthetic route involving an enzyme-bound β-tetraketo-ester intermediate (39), followed by insertion of a one-carbon unit at an appropriate point in the chain. These conclusions are summarized in Scheme 9, and as a result of related studies with less complex fungal tropolones (*cf.* Scheme 8)[28] it seems reasonable to consider that a C-methylated phenolic compound [*e.g.* (40)] may be involved as an intermediate in the biosynthetic route.

[27] A. I. Scott and K. J. Wiesner, *J.C.S. Chem. Comm.*, 1972, 1075.
[28] A. I. Scott and E. Lee, *J.C.S. Chem. Comm.*, 1972, 655.
[29] J. Staunton, *Ann. Reports (B)*, 1970, **67**, 535; E. McDonald, *ibid.*, 1971, **68**, 395.
[30] J. Wright, D. G. Smith, A. G. McInnes, L. C. Vining, and D. W. S. Westlake, *Canad. J. Biochem.*, 1969, **47**, 945.
[31] A. G. McInnes, D. G. Smith, L. C. Vining, and L. Johnson, *Chem. Comm.*, 1971, 325; *cf.* S. Takenaka and S. Seto, *Agric. and Biol. Chem. (Japan)*, 1971, **35**, 862.

* The application of ^{13}C n.m.r. spectroscopy to biosynthetic problems was first reported in 1966 and an excellent account of the limitations, advantages, and experimental techniques associated with the method has been provided in recent reviews.[29]

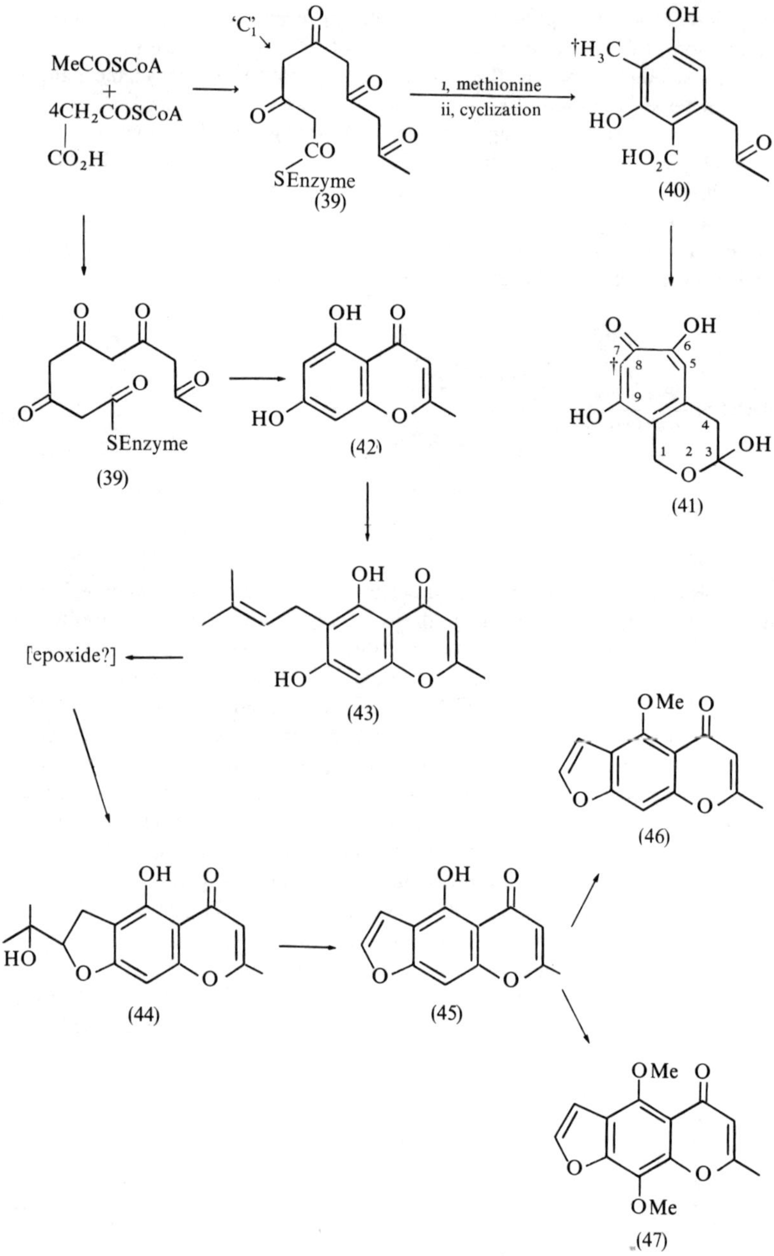
'C'1
MeCOSCoA
+
4CH2COSCoA
CO2H
O
O
O
O
CO
SEnzyme
(39)
i, methionine
ii, cyclization
OH
†H3C
HO
HO2C
O
(40)
O
O
O
O
SEnzyme
(39)
OH
O
HO
O
(42)
O
OH
7
6
†
8
5
9
HO
4
1
2
3
OH
O
(41)
OH
O
HO
O
(43)
[epoxide?]
OMe
O
O
O
(46)
OH
O
HO
O
O
(44)
OH
O
O
O
(45)
OMe
O
O
O
OMe
(47)

Scheme 9

Claisen condensation of an intermediate β-tetraketo-ester (39) could be involved in the biosynthesis of several naturally occurring chromones (Scheme 9). The incorporation of acetate into these compounds supports this viewpoint and a recent study has confirmed that 5,7-dihydroxy-2-methylchromone (42) is a precursor of visamminol (44), visnagin (46), and khellin (47) in *Ammi visnaga* plants.[32] The relative magnitude of activity recorded for these compounds has been cited as evidence in favour of the biosynthetic sequence shown in Scheme 9.

The addition of [1-^{14}C]acetate, [2-^{14}C]acetate, and [1-^{14}C]formate to cultures of *Sclerotinia sclerotiorum* provides radioactive sclerin (48) with labelling patterns which are difficult to reconcile with a straight-chain acetate-derived β-polyketo-ester intermediate methylated at activated methylene positions.[33] It has been suggested,[33] therefore, that the biosynthesis of sclerin (48) involves condensation of two acetate-derived chains (*cf.* Scheme 10), with C-methylation occurring at an unspecified stage in the biosynthetic sequence. In addition, it has been proposed[33] that the same two precursor units could condense in alternative ways to produce the co-metabolites, (50) and (51), of sclerin.

It should be noted, however, that the experimental data do not exclude the intermediacy of a *C*-methylated straight-chain β-pentaketo-ester intermediate (52). Intramolecular cyclization of (52) and appropriate ring cleavage of the bicyclic intermediate (53) could also account for the formation of sclerin and one of its co-metabolites (49) (*cf.* Scheme 11).

Phenalenones produced by *Penicillium herquei* are derived by the acetate–polymalonate pathway, mevalonate and methionine providing the 'extra' carbon atoms. Recent labelling studies with *P. herquei* have provided evidence in favour of a sequential biosynthetic relationship between atrovenetin (54), deoxyherqueinone (55), and herqueinone (57).[34] Although atrovenetin (54) served as a precursor for norherqueinone (56) the latter compound could not be converted into herqueinone (57) by *P. herquei*. It is of interest to note that the plant phenalenones haemocorin (58) and lachnanthoside (59) are derived from phenylalanine, tyrosine, and one acetate (or malonate) unit (Scheme 12).[35]

The second edition of an authoritative text on naturally occurring quinones has been published.[36] The literature is covered up to October 1970 and major changes in the text include a more extensive chapter on biosynthesis and a new chapter on anthracyclinones (naphthoquinones of bacteria).

During the period 1970–1972 additional information on the biosynthesis of acetate-derived anthraquinones was described. Fungal anthraquinones have

---

[32] P. G. Harrison, B. K. Bailey, and W. Steck, *Canad. J. Biochem.*, 1971, **49**, 964.
[33] T. Tokoroyama and T. Kubota, *J. Chem. Soc. (C)*, 1971, 2703.
[34] A. B. Kriegler and R. Thomas, *Chem. Comm.*, 1971, 738.
[35] R. Thomas, *Chem. Comm.*, 1971, 739; J. M. Edwards, R. C. Schmitt, and U. Weiss, *Phytochemistry*, 1972, **11**, 1717, and references cited therein.
[36] R. H. Thomson, 'Naturally Occurring Quinones', Academic Press, New York, 1971.

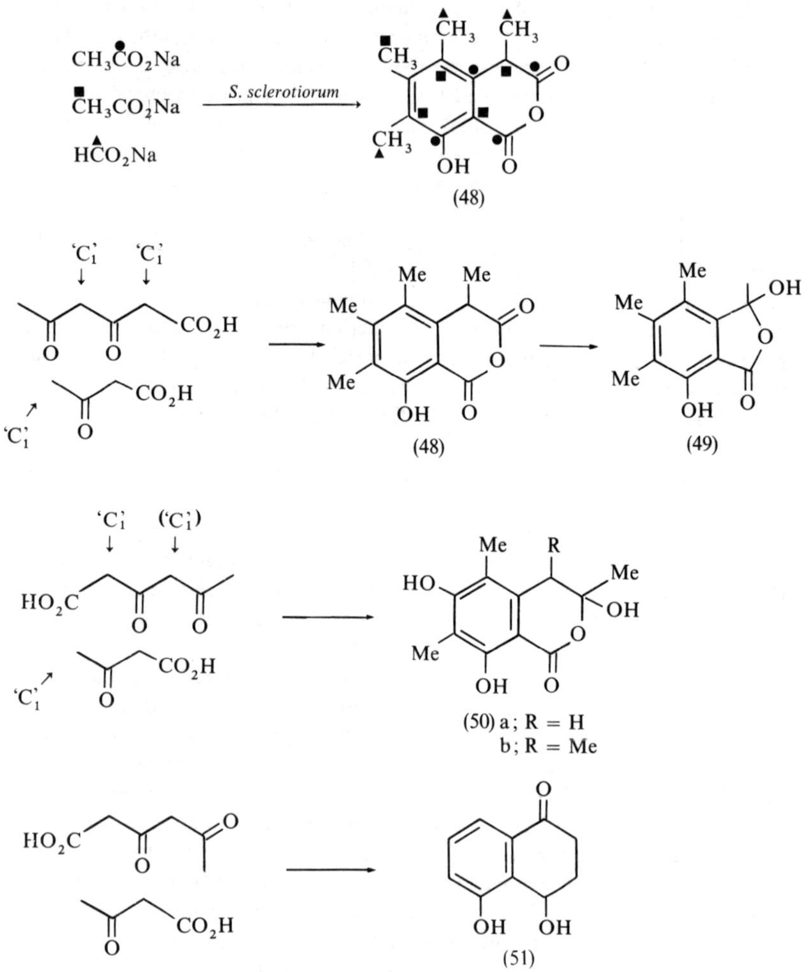

Scheme 10

been shown to be derived by the acetate–polymalonate route[37] whereas plant anthraquinones devoid of hydroxy-groups in one ring, *e.g.* 1,2-dihydroxy-anthraquinone (alizarin), are constructed from shikimic, mevalonic, and, presumably, 2-oxoglutaric acid.[38] Anthraquinones hydroxylated in both

[37] Review: S. Shibata, *Chem. in Britain*, 1967, **3**, 110; *cf.* K. Mosbach, *Angew. Chem. Internat. Edn.*, 1969, **8**, 240.
[38] A. R. Burnett and R. H. Thomson, *Chem. Comm.*, 1967, 1125; E. Leistner and M. H. Zenk, *Tetrahedron Letters*, 1971, 1677, and references cited therein; D. J. Robins and R. Bentley, *J.C.S. Chem. Comm.*, 1972, 232.

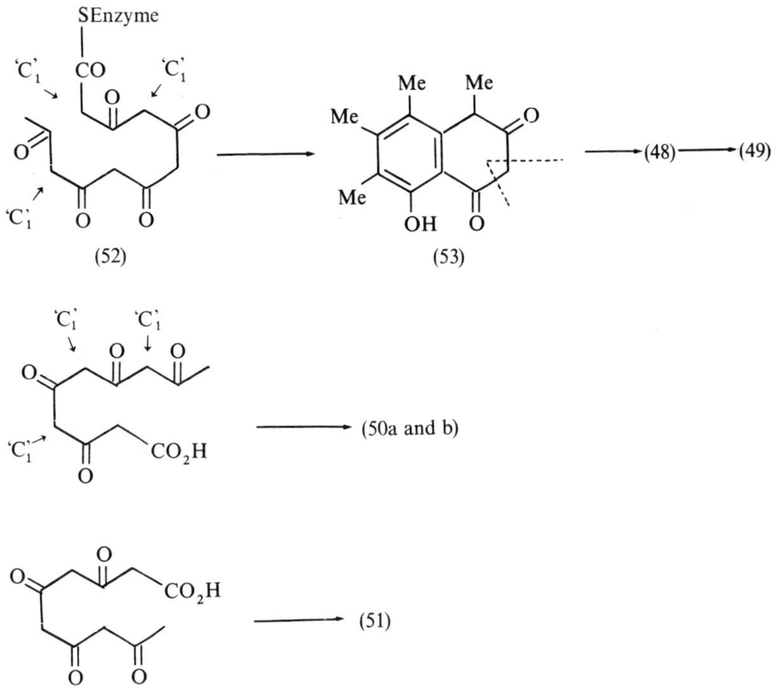

Scheme 11

aromatic rings also occur in plants, and recent investigations[39,40] have shown that, contrary to previous results, the biosynthesis of chrysophanol (62) and emodin (63) in *Rhamnus frangula* and *Rumex alpinus* or in *Rumex obtusifolius*[40] occurs *via* the acetate–polymalonate pathway (Scheme 13).

A useful role for anthraquinones in plants is indicated by the demonstration of rapid use of these compounds in the early stages of fruit development.[40] In a related study[41] of plant anthraquinone biosynthesis it was shown that, in spite of the absence of ring A hydroxy-groups in pachybasin (61), the biosynthetic pathway to this compound in *Phoma foveata* involved acetate and malonate units only. When [1-^{14}C]acetate was used as precursor the C-methyl group in ring C contained greater than average activity. Although this result is in agreement with the utilization of acetyl-coenzyme A as a starter unit in the construction of an enzyme-bound β-heptaketo-ester intermediate (60), it should be noted that a similar result was obtained when [2-^{14}C]malonate was used as precursor.[41]

[39] E. Leistner, *Phytochemistry*, 1971, **10**, 3015.
[40] J. W. Fairbairn and F. J. Muhtadi, *Phytochemistry*, 1972, **11**, 215.
[41] R. F. Curtis, C. H. Hassall, and D. R. Parry, *Chem. Comm.*, 1971, 410.

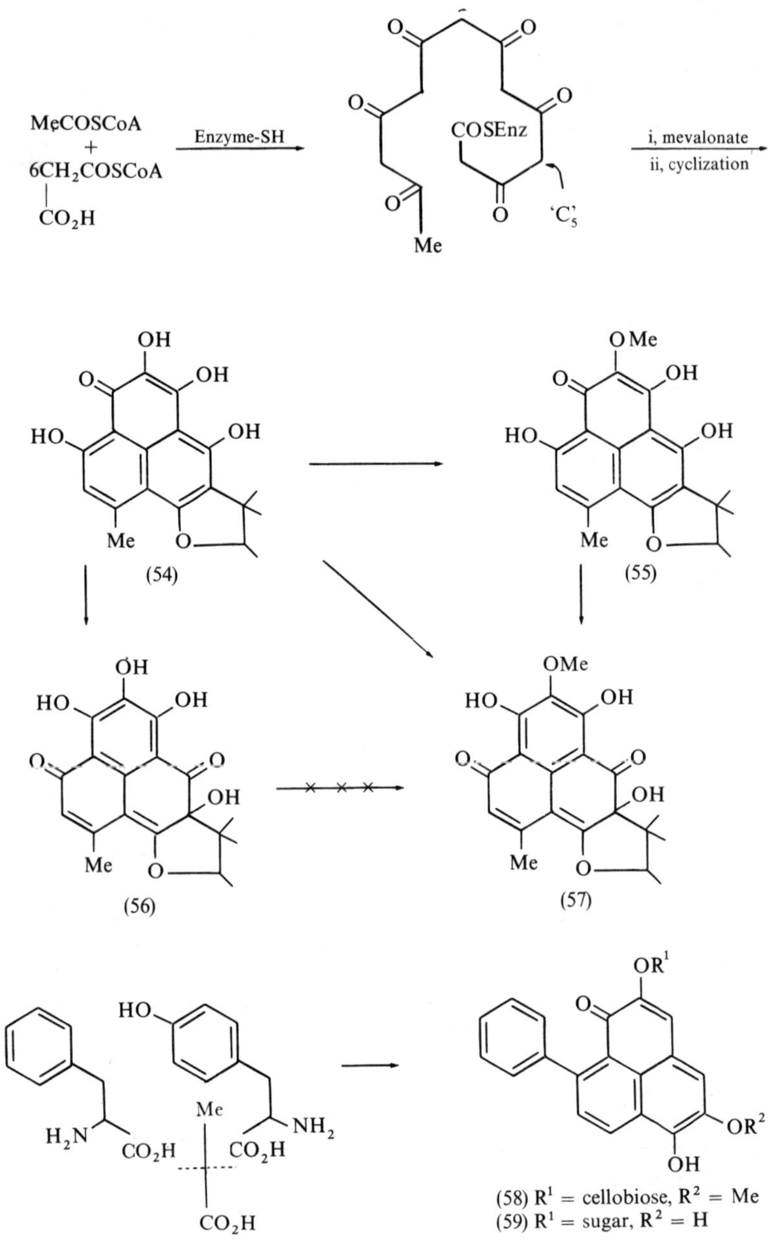
MeCOSCoA
+
6CH2COSCoA
CO2H
Enzyme-SH
COSEnz
Me
'C'5
i, mevalonate
ii, cyclization
(54)
(55)
(56)
(57)
OMe
Me
(58) R1 = cellobiose, R2 = Me
(59) R1 = sugar, R2 = H

Scheme 12

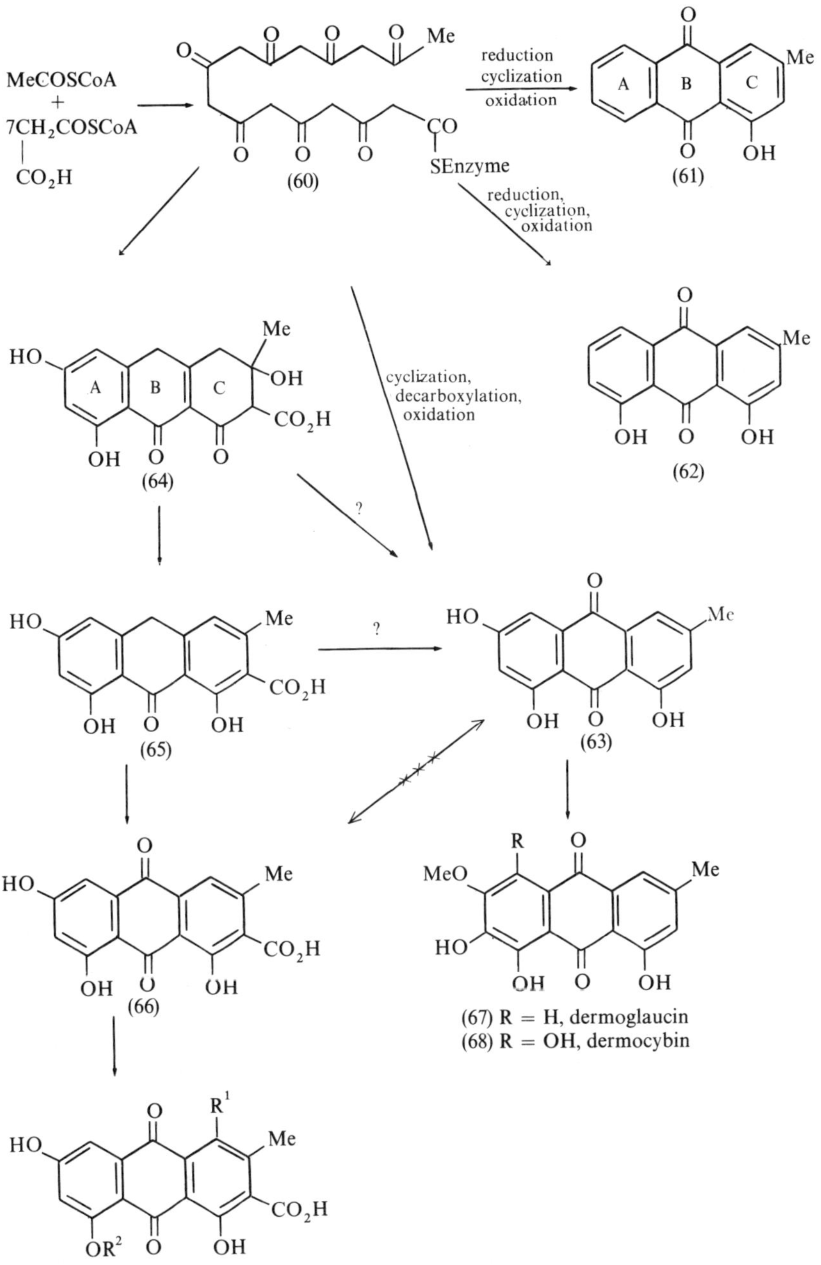

Biosynthesis of Polyketides

MeCOSCoA
+
7CH$_2$COSCoA
|
CO$_2$H

→ (60) polyketide chain with SEnzyme

reduction, cyclization, oxidation → (61)

reduction, cyclization, oxidation → (62)

cyclization, decarboxylation, oxidation → (64)

(64) → (65) → ? → (63)

(65) → (66) → (69)/(70)

(63) → (67)/(68)

(67) R = H, dermoglaucin
(68) R = OH, dermocybin

(69) R^1 = H, R^2 = Me, dermolutein
(70) R^1 = OH, R^2 = Me, dermorubin

Scheme 13

Investigation of the sequence of reactions and intermediates involved in the biosynthesis of anthraquinones has been carried out by several research groups. A recent study[42] has shown that, contrary to popular assumption, endocrocin (66) is *not* a precursor of emodin (63), dermoglaucin (67), or dermocybin (68) in *Dermocybe sanguinea*. Related experiments demonstrated that dermolutein (69) and dermorubin (70) were derived from endocrocin (66) and that in a closely related species (*D. semisanguinea*) emodin (63) was a precursor of dermoglaucin (67) and dermocybin (68). These results do not exclude the attractive possibility that each group of anthraquinones has a common tricyclic precursor which could be endocrocin-9-anthrone (65) or an intermediate (64) in which ring C is non-aromatic.

The hypothesis that anthraquinones or derivatives are metabolically active is supported by a recent study in which the transformation of doubly labelled questin (71; ^{14}C indicated by asterisk) into sulochrin (72) by cultures of *Aspergillus terreus* was demonstrated by appropriate incorporation studies.[43] The ratio of ^{14}C-labelling at the two positions in questin (71) was maintained in the corresponding positions in sulochrin and this result provides support for the proposal that the biosynthesis of sulochrin (72) involves oxidative cleavage of an acetate-derived anthraquinone intermediate. Previous studies, using an enzyme preparation from *Penicillium frequentans*, have also established that questin is a biosynthetic precursor of sulochrin.[44]

The biosynthesis of chartreusin aglycone (74) in *Streptomyces* spp. has been investigated using *trans*-cinnamic acid, sodium diethyl malonate, and methionine as radioactive substrates.[45] Preliminary results indicate that the complete carbon skeleton is derived by the acetate–polymalonate route. Since direct formation of (74) by cyclization of a β-polyketo-ester intermediate is impossible, it has been suggested[45] that a linear tetracyclic compound (73) could be involved as a biosynthetic intermediate (Scheme 14).

The structural diversity exhibited by polyketides is due, in part, to the extensive secondary modifications which can occur at several stages in the biosynthetic route. Many compounds have 'extra' C-methyl groups (or oxidized

---

[42] W. Steglich, R. Arnold, W. Losel, and W. Reininger, *J.C.S. Chem. Comm.*, 1972, 102.
[43] R. F. Curtis, C. H. Hassall, and D. R. Parry, *J.C.S. Perkin 1*, 1972, 240, and references cited therein.
[44] S. Gatenbeck and L. Malmstrom, *Acta Chem. Scand.*, 1969, **23**, 3493.
[45] J. R. Brown, M. S. Spring, and J. R. Stoker, *Phytochemistry*, 1971, **10**, 2059.

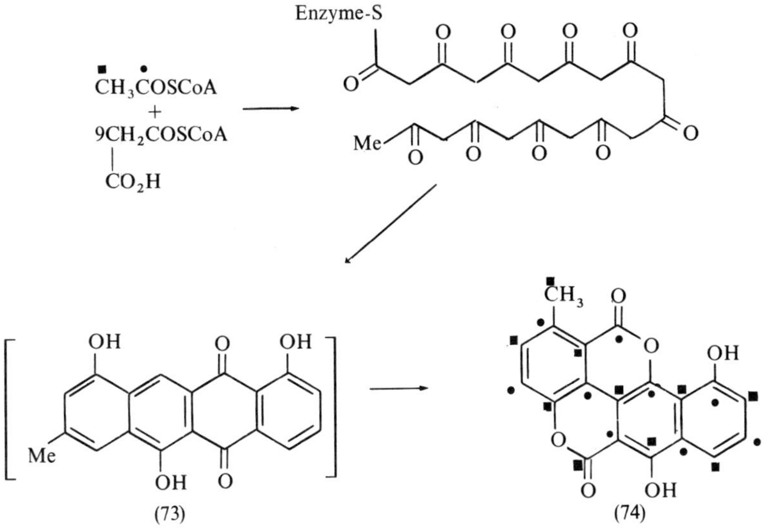

Biosynthesis of Polyketides

Scheme 14

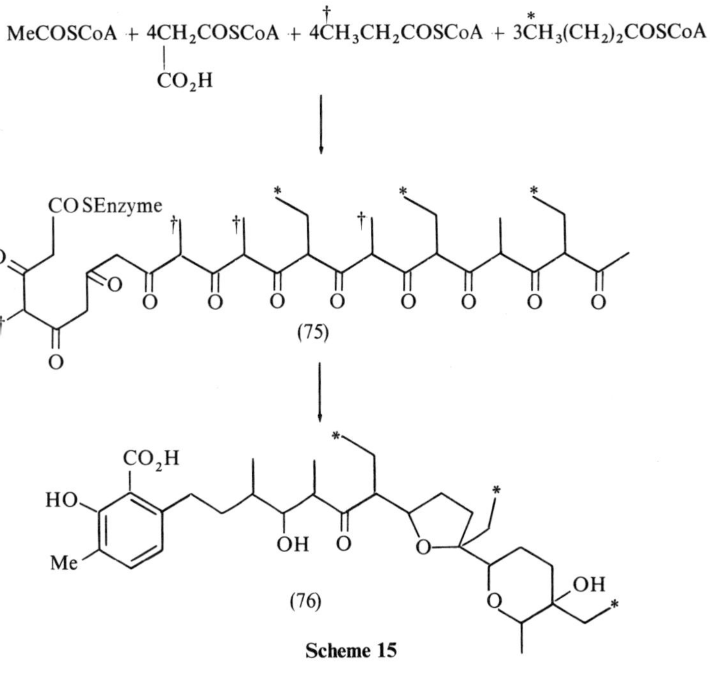

$CH_3COSCoA + 9CH_2COSCoA$ (with CO_2H) → Enzyme-bound polyketide → (73) → (74)

Scheme 15

$MeCOSCoA + 4CH_2COSCoA + 4\overset{\dagger}{C}H_3CH_2COSCoA + 3\overset{*}{C}H_3(CH_2)_2COSCoA$ (with CO_2H)

↓

(75)

↓

(76)

forms thereof) attached to their basic acetate-derived structure, and biosynthetic investigations have shown (a) that the 'extra' carbon atom is derived from methionine and can be inserted at an active methylene position of the β-polyketo-ester intermediate or at a position *ortho* or *para* to the hydroxy-group in a phenolic intermediate or (b) that the 'extra' carbon atom results from the use of propionate (presumably as methylmalonyl-coenzyme A) in the chain-propagating step. Until recently only 'acetate' or 'propionate' units had been implicated in the chain-building process leading to polyketides. However, a recent study of the biosynthesis of antibiotic X-537A (76) has revealed that the C-ethyl groups present in this structure are derived from butyric acid units[46] (Scheme 15). This was elegantly demonstrated by incorporation experiments with sodium [1-^{13}C]butyrate as precursor and is a further demonstration of the value of n.m.r. spectroscopy in biosynthetic investigations. Previous studies, using ^{14}C-labelled acetate, propionate, and butyrate had partially established the biosynthetic pathway but were unable to establish directly whether one, two, or three butyrate units were involved in the process.

According to the acyl–polymalonate hypothesis, chain-starter units other than acetate may be involved in the biosynthetic process (*cf.* Scheme 1). In this connection, the naturally occurring stilbenes and derivatives are produced by condensation of a cinnamate unit with three malonate units followed by aldol condensation of the enzyme-bound β-triketo-ester intermediate (77).[2-5] Secondary transformations (p. 183) may also occur at a pre-aromatic or post-aromatic stage in the biosynthesis. Previous attempts to demonstrate the intermediacy of β-triketo-esters in the biosynthesis of stilbenes resulted in the reported conversion of cinnamoyltriacetic acid (77; R = H, as acid) into pinosylvin (79) by acetone powders prepared from leaves of a mutant of *Eucalyptus sideroxylon*.[48] However, a subsequent report[49] has indicated that this conversion cannot be repeated and this has prompted the explanation that β-polyketo-ester intermediates are bound to the enzyme and do not occur in the free state.* Several different preparations of acetone powder from *E*.

---

[46] J. W. Westley, D. L. Pruess, and R. G. Pitcher, *J.C.S. Chem. Comm.*, 1972, 161.
[47] *Cf.* T. Money, *Chem. Rev.*, 1970, **70**, 553, and references cited therein.
[48] W. E. Hillis and N. Ishikura, *Phytochemistry*, 1969, **8**, 1079.
[49] W. E. Hillis and Y. Yazaki, *Phytochemistry*, 1971, **10**, 1051.
[50] S. Gatenbeck and K. Mosbach, *Acta Chem. Scand.*, 1959, **13**, 1561; *cf.* A. J. Birch, *Science*, 1967, **156**, 202.
[51] M. Hamberg and B. Samuelsson, *J. Biol. Chem.*, 1967, **242**, 5336, and references cited therein; C. Pace-Asiak and L. S. Wolfe, *Chem. Comm.*, 1970, 1234, 1235; J. D. Bu'Lock and G. N. Smith, *J. Chem. Soc.* (*C*), 1967, 332; J. D. Bu'Lock and P. T. Clay, *Chem. Comm.*, 1969, 237; J. Staunton, *Ann. Reports* (*B*), 1969, **66**, 577; E. Leete and J. O. Olson, *J. Amer. Chem. Soc.*, 1972, **94**, 5472, and references cited therein.

* The postulated role of β-polyketo-acid derivatives as intermediates (enzyme-bound or free) in the acyl–polymalonate biosynthetic route to phenolic compounds has not been demonstrated conclusively. Indeed the general acceptance of these compounds as intermediates is based on the demonstration that oxygen atoms from acetate appear as oxygen functionality in the phenolic product.[50] The chemical feasibility of the hypothesis is supported by the conversion, in the laboratory, of β-polyketo-acids and -esters into phenolic compounds of natural type.[47] It is of interest to note that several acetate-derived compounds, whose biosynthesis could readily be explained in terms of partial reduction of a β-polyketo-ester intermediate, are derived from fatty acid precursors.[51]

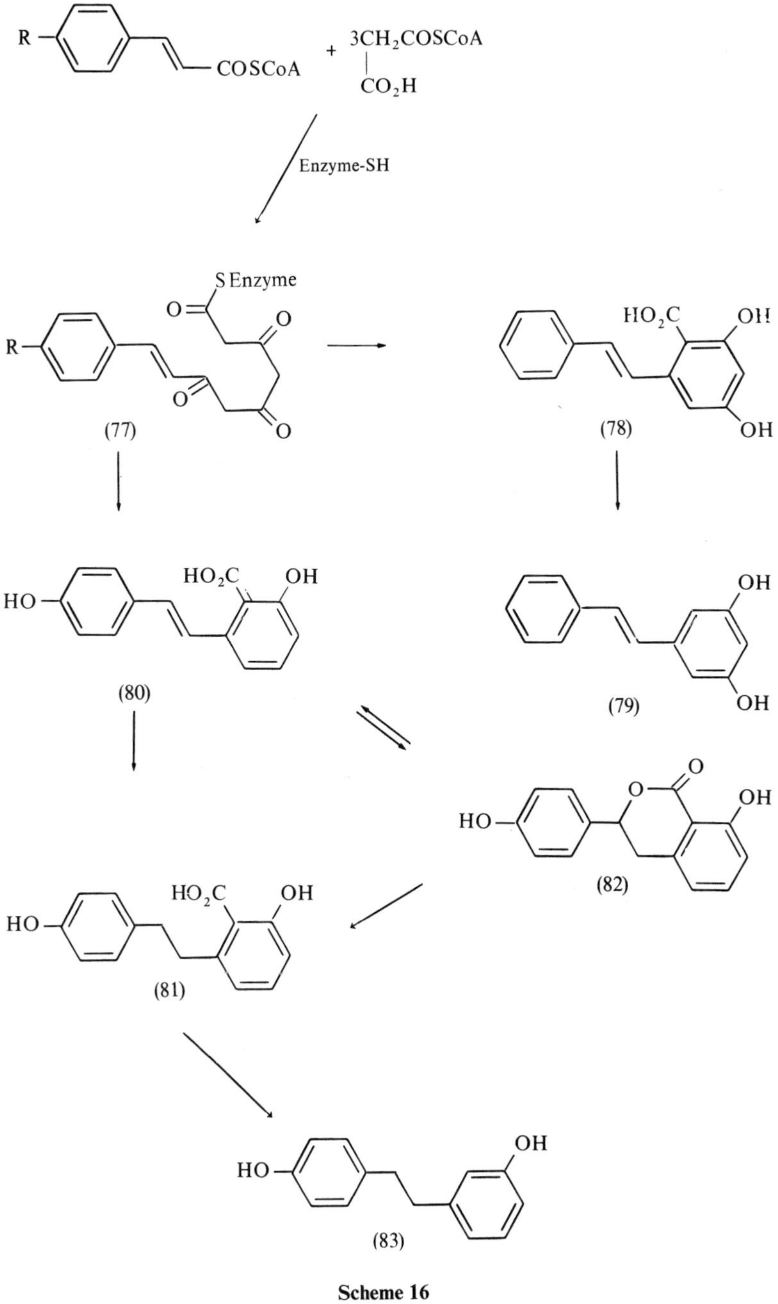

Biosynthesis of Polyketides

Scheme 16

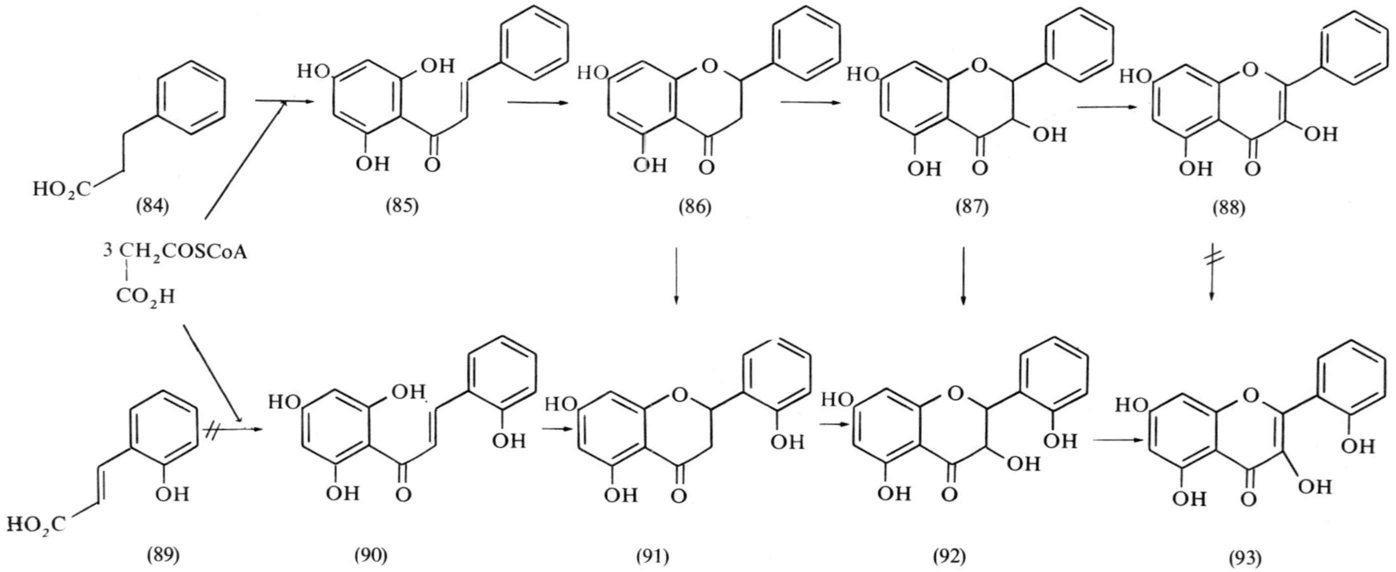

Scheme 17

sideroxylon leaves were also used in unsuccessful attempts to demonstrate a biosynthetic relationship between pinosylvic acid (78) and pinosylvin (79).

Related studies[52,53] have shown that lunularic acid (81), a metabolite of *Lunularia cruciata*, incorporates radioactivity from [U-^{14}C]-L-phenylalanine, [1-^{14}C]acetate, and [^{14}C]hydrangenol (82). Although no degradative studies were reported, it has been suggested that these results support the view that lunularic acid is biosynthesized *via* the acyl–polymalonate pathway (*cf.* Scheme 16) and that hydrangenol (82) and/or hydrangeic acid (80) can serve as intermediates in the process.[52] Furthermore, an investigation of the metabolism of lunularic acid (81) in *L. cruciata* has revealed that one of the primary products is lunularin (83).[53]

The biosynthesis of flavonoids continues to receive considerable attention and contemporary investigations are concerned mainly with establishing the sequence of steps and intermediates involved in the biosynthetic route. Incorporation studies have shown that 2,2′,4′,6′-tetrahydroxychalcone (90), 2′,5,7-trihydroxyflavanone (91), and dihydrogalangin (87) are good precursors of datiscetin (93) in *Datisca cannabina*.[54] In contrast, it was found that *o*-coumaric acid (89) and galangin (88) were not intermediates in the biosynthetic route. A 'metabolic grid' (Scheme 17) has been postulated to account for the biosynthetic relationships established in this plant. The combined results indicate that the 2′-hydroxy-group in datiscetin is introduced into a flavanone or flavanonol intermediate rather than into a flavone precursor, and a similar situation seems to prevail in the biosynthesis of coumestrol (103). The substrate specificity of partially purified chalcone–flavanone isomerase (isolated from young leaves of *D. cannabina*) was also examined. As a result, it was clearly demonstrated[54] that chalcones lacking 6′-hydroxy-groups [*e.g.* (94)] and chalcone glucosides [*e.g.* (95)] are not converted into the corresponding flavanone.

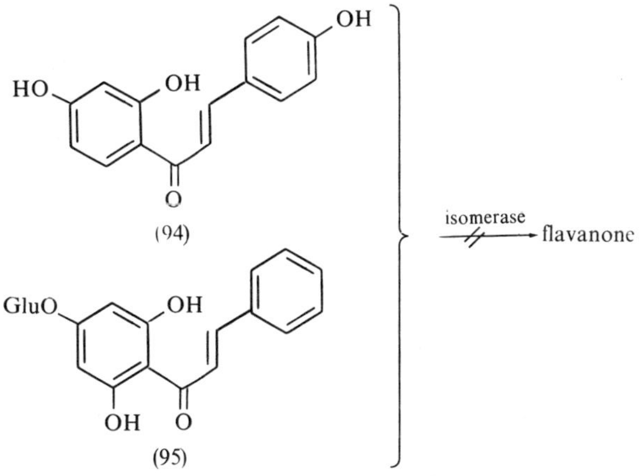

[52] R. J. Pryce, *Phytochemistry*, 1971, **10**, 2679.
[53] R. J. Pryce, *Phytochemistry*, 1972, **11**, 1355.
[54] H. J. Grambow and H. Grisebach, *Phytochemistry*, 1971, **10**, 789, and references cited therein.

(96) a; $R^1 = R^2 = H$
b; $R^1 = OH, R^2 = H$
c; $R^1 = H, R^2 = OH$

In addition, no substrate specificity for the substitution pattern in ring B was observed since chalcones (96a—c) were efficiently cyclized by the isomerase.

Cell suspension cultures of mung bean (*Phaseolus aureus* Roxb.) roots have been used to demonstrate that daidzein (98) and 4′,7-dihydroxyisoflavanone (100) can be converted efficiently into coumestrol (103).[55] In contrast to previous results obtained with intact plants, 2′,4′,7-trihydroxyisoflavone (99) was shown to be a less efficient precursor, and it has been suggested that this compound does not lie on the main biosynthetic route. Earlier proposals for the biosynthesis of coumestrol (103) have been amended to accommodate the new evidence and a revised pathway, in the form of a 'metabolic grid', is shown in Scheme 18. The major pathway (*a*) to coumestrol thus involves 2′-hydroxylation of an isoflavanone (100) (*cf.* datiscetin biosynthesis, Scheme 17), and allylic oxidation of a pterocarpen derivative (102) has been suggested for the final step in the overall process.[55] Cell suspension cultures have also been used to investigate the biosynthesis of cyanidin (106) in *Haplopappus gracilis*.[56] Comparative incorporation experiments using phenylalanine, 4,2′,4′,6′-tetrahydroxychalcone (104), and 3,5,7,4′-tetrahydroxyflavanone (105) have revealed that the latter compound is the most efficient precursor of cyanidin (106). The advantage of using cell suspension cultures rather than seedlings was indicated by higher incorporation rates and shorter feeding times.

The use of sterile germinating seeds of *Amorpha fruticosa* has provided further information on the biosynthesis of amorphigenin (116).[57] Administration of labelled 2′,4,4′-trihydroxychalcone (108), 7-hydroxy-4′-methoxyisoflavone (109; R = Me, formononetin), 7-hydroxy-2′,4′,5′-trimethoxyisoflavone (110), 9-methylmunduserone (111), rotenonic acid (112), or rotenone (115) to the seed system resulted in efficient conversion into amorphigenin (116), and the outline of a biosynthetic scheme for this compound has been proposed (Scheme 19). Of considerable interest was the fact that 7,4′-dihydroxyisoflavone (109; R = H, daidzein) was not an efficient precursor, and this has been cited as

[55] J. Berlin, P. M. Dewick, W. Barz, and H. Grisebach, *Phytochemistry*, 1972, **11**, 1689.
[56] H. Fritsch, K. Hahlbrock, and H. Grisebach, *Z. Naturforsch.*, 1971, **26b**, 581.
[57] L. Crombie, P. M. Dewick, and D. A. Whiting, *Chem. Comm.*, 1971, 1182, 1183, and references cited therein.

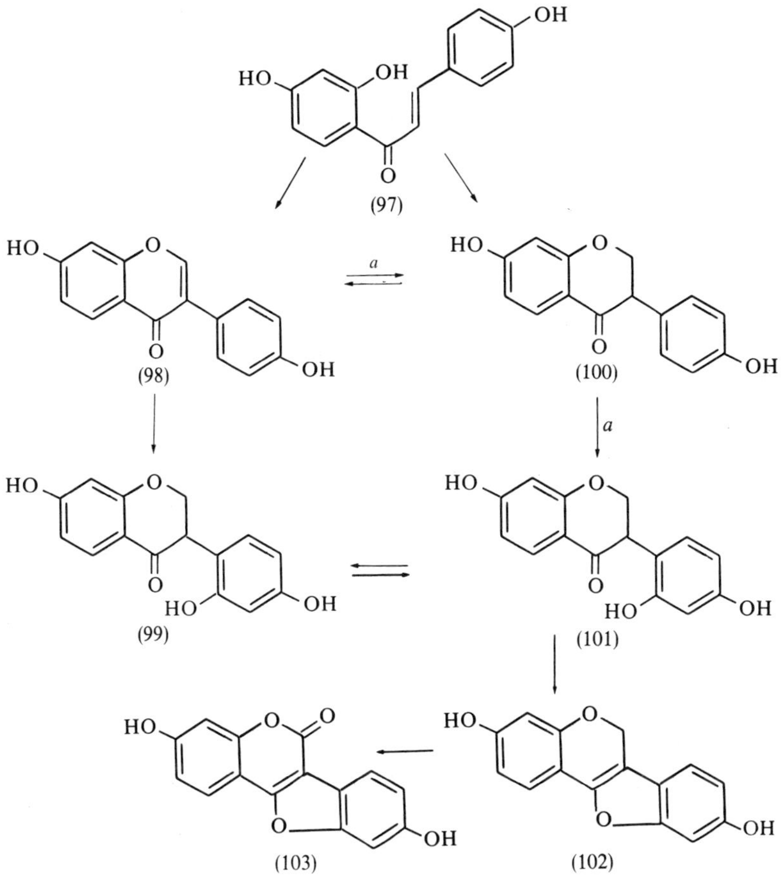

Biosynthesis of Polyketides

Scheme 18

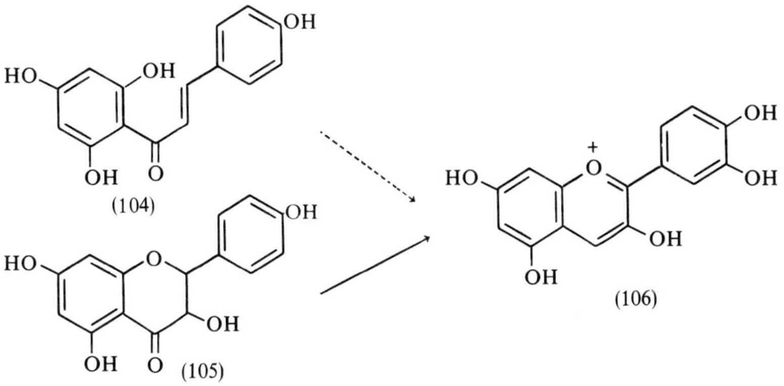

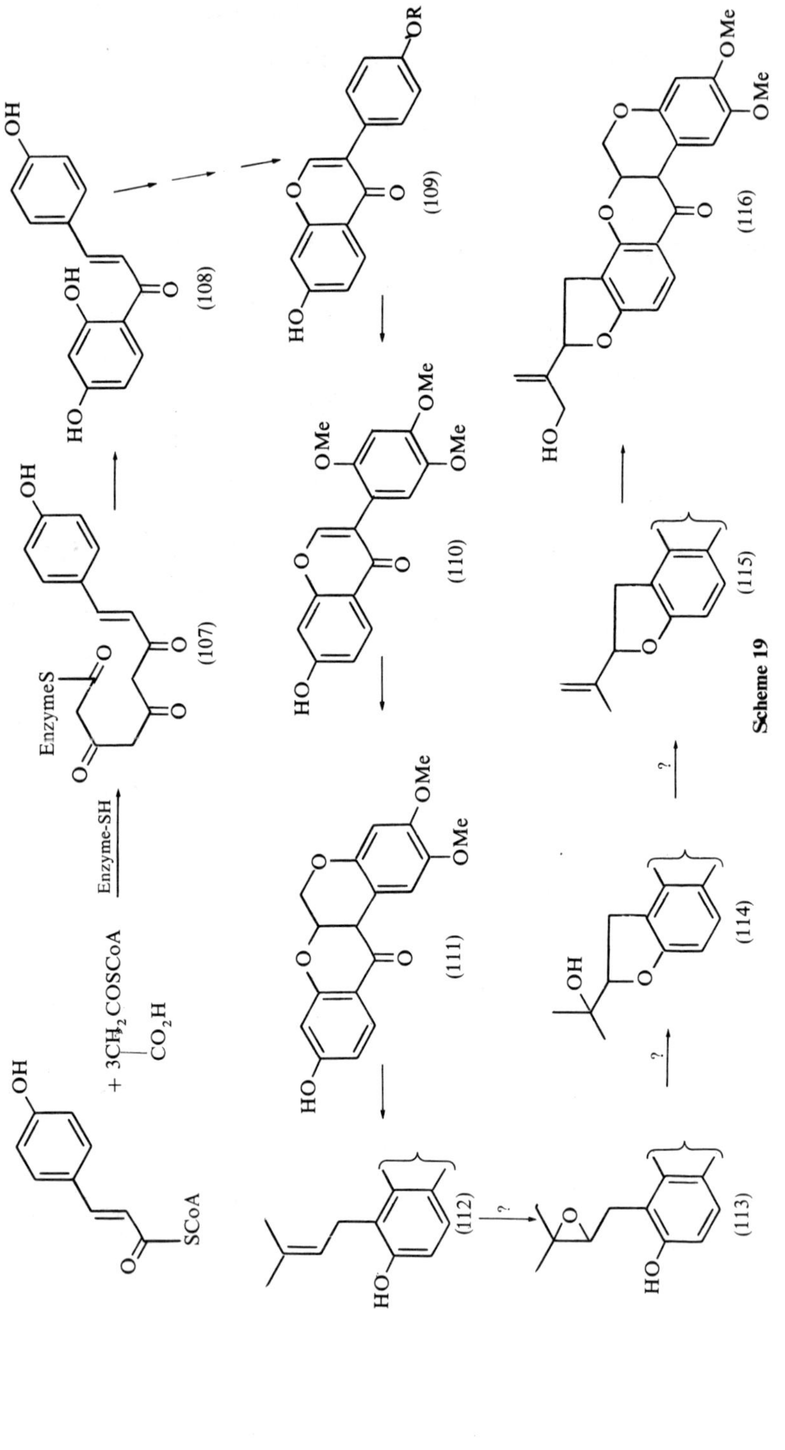

Scheme 19

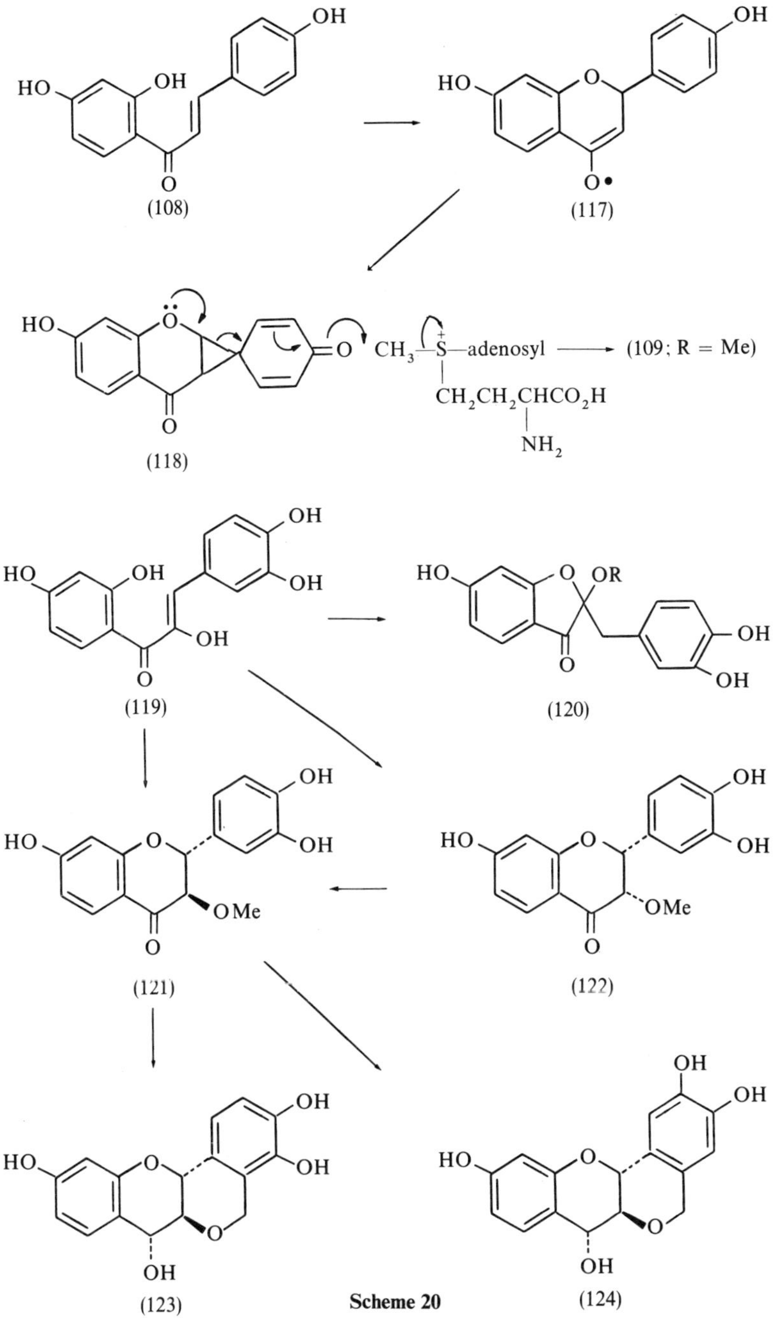

Biosynthesis of Polyketides

Scheme 20

evidence in favour of a spirodienone intermediate (118) (Scheme 20) in the conversion of 2',4,4'-trihydroxychalcone (108) into 7-hydroxy-4'-methoxyisoflavone (109; R = Me, formononetin).[57]

Although appropriate incorporation studies have yet to be reported, the co-occurrence[58] of α,2',3,4,4'-pentahydroxychalcone (119), 2-methoxy-3',4',6-trihydroxy-2-benzylcoumaranone (120; R = Me), 2,3',4',6-tetrahydroxy-2-benzylcoumaranone (120; R = H), 3-O-methyl-2,3-*trans*-fustin (121), and 3-O-methyl-2,3-*cis*-fustin (122) with mopanol (123) and peltogynol (124) in the heartwood of *Trachylobium verrucosum* lends credence to the proposal[59] that the 'extra' skeletal carbon atom in ring D of the latter compounds is derived from a 3-methoxyflavanone intermediate (121).

Further studies on the biosynthesis of plant-derived xanthones have been described.[60] Administration of [3-^{14}C]- or [U-^{14}C]-phenylalanine to intact plants or to tissue cultures of *Gentiana lutea* provided gentisein (127) which was shown to be devoid of radioactivity in ring A. In addition it was established that 2,3',4,6-tetrahydroxybenzophenone (126), which co-occurs in *G. lutea*, was an intermediate in the biosynthetic route (Scheme 21). These results, and those previously reported, support the view that the biosynthesis of gentisein and

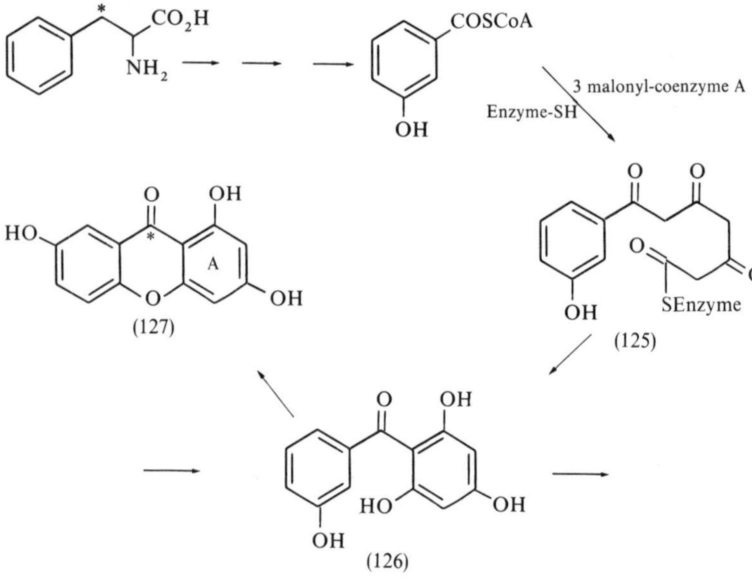

Scheme 21

[58] J. P. van der Merwe, D. Ferreira, E. V. Brandt, and D. G. Roux, *J.C.S. Chem. Comm.*, 1972, 521.
[59] A. C. Waiss and J. Corse, *J. Amer. Chem. Soc.*, 1965, **87**, 2068.
[60] P. Gupta and J. R. Lewis, *J. Chem. Soc. (C)*, 1971, 629.

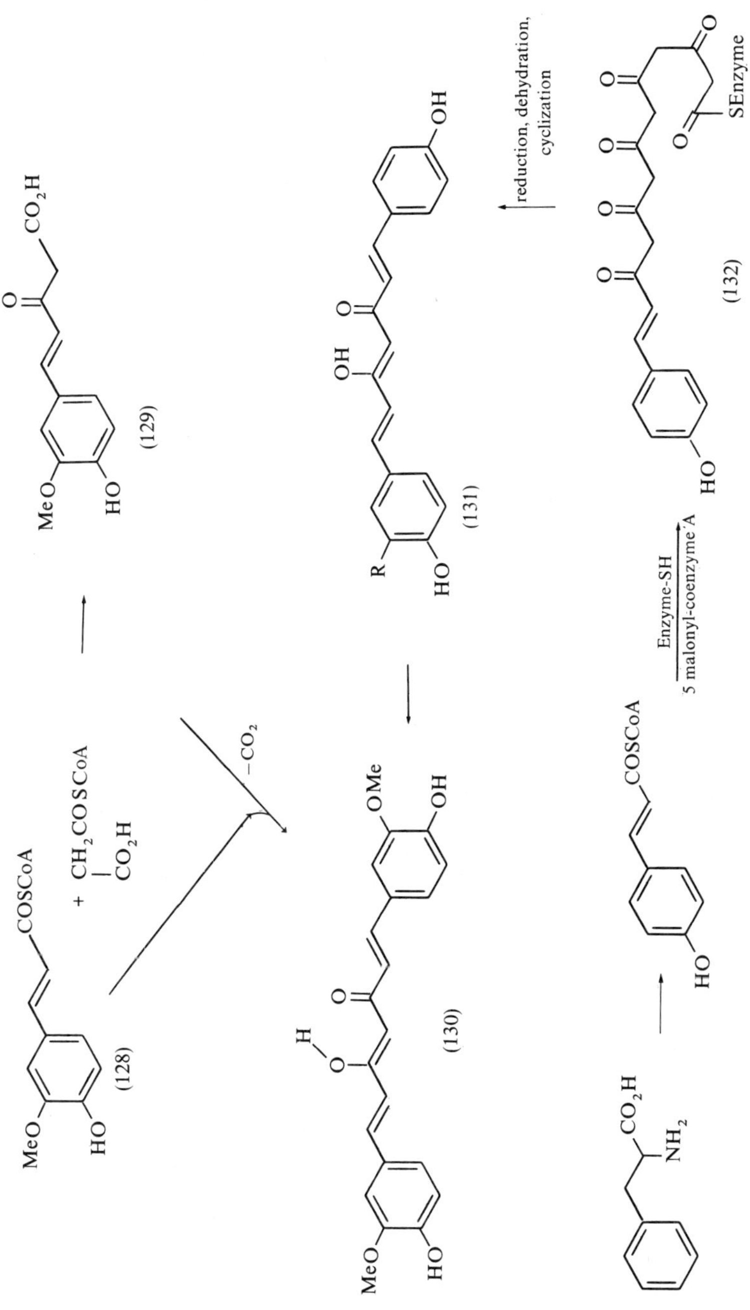

Biosynthesis of Polyketides

Scheme 22

related xanthones involves condensation of a C_6—C_1 unit (*e.g.* *m*-hydroxybenzoic acid derivative derived from phenylalanine) with three malonate units followed by cyclization of the triketo-ester intermediate (125) to benzophenone (126). Subsequent oxidative coupling of (126) could provide the corresponding xanthones.

A study of the biosynthesis of curcumin (130) has produced unexpected results.[61] Administration of labelled acetate, malonate, and phenylalanine to *Curcuma longa* plants revealed that all of these substrates were implicated in the biosynthetic route. Previous biosynthetic proposals suggested that the most plausible biosynthesis of curcumin (130) would involve condensation of two cinnamate or ferulate (128) units (derived from phenylalanine) and one acetate or malonate unit. However, the recent studies[61] have shown that the majority of radioactivity from acetate and malonate was not confined to the central methine carbon atom of curcumin (130). Since [1-^{14}C]phenylalanine was incorporated specifically, the high incorporation of activity from acetate/malonate into the aromatic regions of curcumin has been rationalized in terms of a biosynthetic route (Scheme 22) in which a β-pentaketo-acid derivative (132) is postulated as an intermediate. Further experimental evidence in support of these proposals is anticipated.

2 Non-aromatic Polyketides

It is generally recognized that the biosynthesis of certain non-aromatic acetate-derived natural products may also involve intermediate formation of β-polyketo-acid derivatives. Several compounds formerly included in this structurally diverse group have now been shown to be derived from saturated fatty acids.[51] A distinction between the two biosynthetic processes is often very difficult to demonstrate experimentally since rapid degradation to acetate competes with incorporation of an intact fatty acid precursor. There is a possibility, therefore, that several of the compounds listed in Scheme 22 may be derived from an appropriate saturated fatty acid.

The biosynthesis of asperlin (133), an antibiotic metabolite of *Aspergillus nidulans* (NRRL 3134), has been investigated by growing the culture in the presence of [1-^{13}C]acetate.[62] Examination of the proton-noise-decoupled ^{13}C n.m.r. spectrum (25.15 MHz) provided proof that the biosynthesis of asperlin involves linear condensation of four acetate units. Another example of the use of ^{13}C n.m.r. spectroscopy in biosynthetic investigations has been provided by a recent report on the biosynthesis of prodigiosin (134).[63] Previous studies had shown that radioactivity from acetate, glycine, proline, and methionine could be incorporated into prodigiosin. However, the specificity of the incorporation could not be determined because of the difficulties involved in small-scale

---

[61] P. J. Roughley and D. A. Whiting, *Tetrahedron Letters*, 1971, 3741.
[62] M. Tanabe, T. Hamasaki, D. Thomas, and L. Johnson, *J. Amer. Chem. Soc.*, 1971, **93**, 273.
[63] R. J. Cushley, D. R. Anderson, S. R. Lipsky, R. J. Sykes, and H. H. Wasserman, *J. Amer. Chem. Soc.*, 1971, **93**, 6284, and references cited therein.

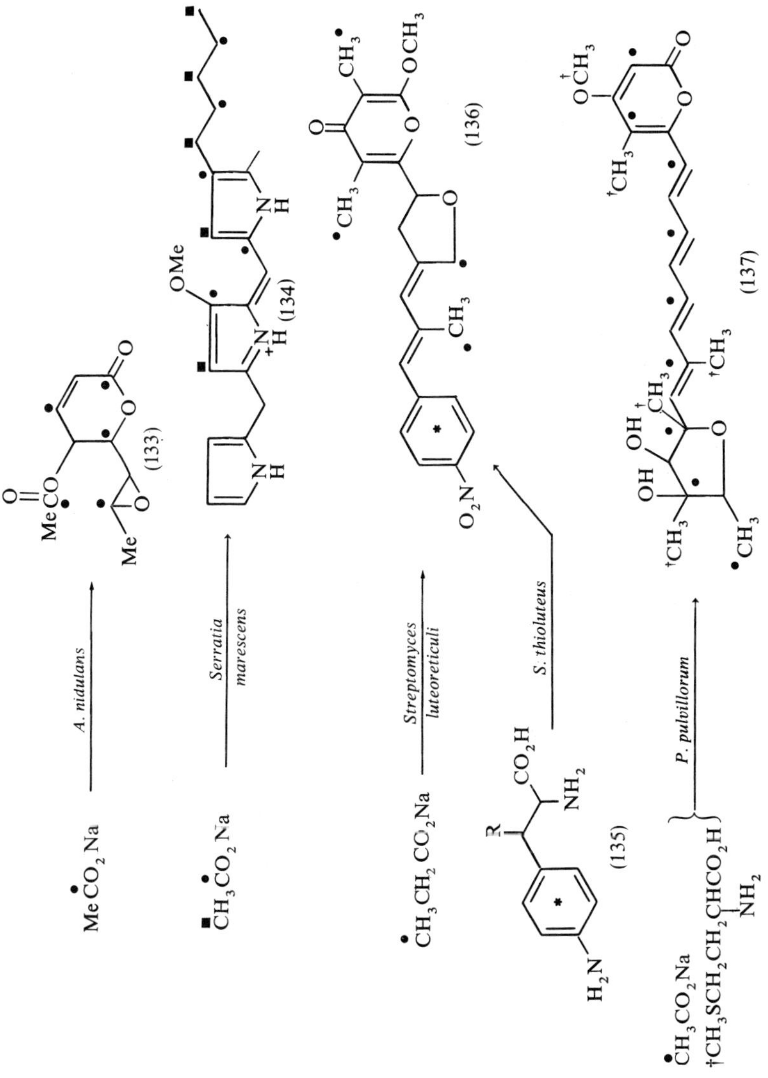

Biosynthesis of Polyketides

Scheme 23

degradation procedures. A solution to this problem was obtained by using [1-^{13}C]- and [2-^{13}C]-acetate as precursors. Subsequent comparison of the proton-noise-decoupled ^{13}C Fourier transform (FT) spectra with the normal ^{13}C FT spectrum of prodigiosin (134) clearly demonstrated the pattern of incorporation shown in Scheme 23. Similar techniques have been used to establish that propionate (presumably as methylmalonyl-coenzyme A) is being used as a chain-propagating unit in the biosynthesis of aureothin (136).[64] The exact nature of the chain-initiating unit, which provides the nitroaromatic ring of aureothin, remains uncertain. However, a recent investigation[65] has shown that *p*-aminophenylalanine (135; R = H) and *erythro*-*p*-aminophenylserine (135; R = OH) are efficient precursors of this portion of the molecule whereas *p*-aminocinnamic acid and *p*-aminobenzoic acid are not. Appropriate incorporation studies using [*Me*-^{14}C]methionine and [2-^{14}C]acetate have shown that the biosynthesis of citreoviridin (137) in *Penicillium pulvillorum* (CSIR 1406) involves successive condensation of nine acetate units (presumably one acetate and eight malonate units), with methionine providing the *C*-methyl and *O*-methyl groups.[66]

---

[64] M. Yamazaki, F. Katoh, J. Ohishi, and Y. Koyama, *Tetrahedron Letters*, 1972, 2701.
[65] R. Cardillo, C. Fuganti, D. Ghiringhelli, D. Giangrasso, and P. Grasselli, *Tetrahedron Letters*, 1972, 4875.
[66] D. W. Nagel, P. S. Steyn, and N. P. Ferreira, *Phytochemistry*, 1972, **11**, 3215.

6
Biosynthesis of Phenolic Compounds Derived from Shikimate

BY J. B. HARBORNE

1 Introduction

1972 has not been a memorable year in the field of phenolic biosynthesis and it is difficult to point to any really major achievements, although much of value has been accomplished in the period under review. Increasing attention has been paid to the enzymology of biosynthesis. It has been demonstrated that ferulic acid can be enzymically reduced, by a preparation from willow bark, to coniferyl alcohol, one of the key precursors of plant lignins. An O-methyltransferase which methylates luteolin to chrysoeriol has been extensively purified from parsley suspension cultures and it has been shown clearly to operate in biosynthesis at the C_{15} and not at the C_9 level. Also, a previously hypothetical 'chalcone synthetase' for converting p-coumaryl coenzyme A and malonyl coenzyme A into a chalcone has been found in suspension cultures of the same plant. The discovery of this enzyme means that all steps in the synthesis of flavones from simple precursors have now been shown to be enzymically mediated.

Much attention has been given to other enzymes of phenolic biosynthesis, and a veritable spate of papers, many of purely physiological interest, has appeared on the topic of phenylalanine ammonia lyase. Full details of the work on the stereochemistry of this enzyme have been published during the year. A paper on the biosynthesis of aromatic amino-acids from shikimate[1] is of general significance, since it indicates that m-carboxy-substituted aromatic acids can be formed without involving prephenic acid as an intermediate. These results are discussed in more detail elsewhere in this volume (see Chapter 3).

Curiously, few purely radioactive tracer studies of phenols have been reported and the emphasis has been on the biosynthesis of the more commonly occurring phenolics. Compounds with exotic structures, particularly of the flavonoid class, continue to be discovered at an ever increasing pace. A whole range of isoprenoid flavones is now known, for example, and some information on their biosynthetic origin and their relationship to terpenoid metabolism would be welcome.

---

[1] P. O. Larsen, D. K. Onderka, and H. G. Floss, *J.C.S. Chem. Comm.*, 1972, 842.

2 Phenols and Phenolic Acids

Salicylic Acid.—It has recently been demonstrated (see last year's Report[2]) that whereas 6-methylsalicylic acid in both micro-organisms and higher plants is acetate-derived, salicylic acid itself is formed in both groups of organisms by the shikimic acid pathway. Further confirmation of this has come from Marshall and Ratledge,[3] who have studied the enzymology of biosynthesis of salicylic acid in *Mycobacterium smegmatis*. These authors isolated the enzyme salicylate synthetase, which catalyses the last step (see Scheme 1) in synthesis of salicylic acid (1) from isochorismic acid (2), formed in turn from chorismic (3) and shikimic acids (4). The enzyme has no cofactor requirements and converts (2) directly into (1). No evidence could be obtained for the presence in bacterial cultures of the possible intermediate 2,3-dihydroxy-2,3-dihydrobenzoic acid.

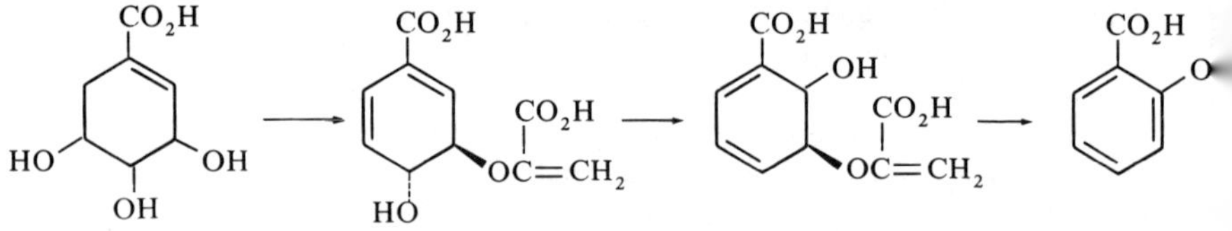

Scheme 1

Salicylate synthetase was also detected in two other *Mycobacterium* species, *M. tuberculosis* and *M. fortuitum*, which contain salicylic acid in their myco-bactins. However, it was definitely absent from *M. phlei*, which lacks salicylic acid but has 6-methylsalicylic acid. Furthermore, in *M. phlei*, salicylic acid failed to operate as a feedback control, as it did in the other three organisms.

3,4-Dihydroxy-2,2-dimethylphenylvaleric acid.—This acid (5) occurs as one of the residues in the cyclic polyester antiobiotic neoantimycin, produced by *Streptomyces orinoci*. Caglioti et al.,[4] in studying its biosynthesis, found it to

$$PhCH_2CH(OH)CH(OH)CMe_2CO_2H$$

(5)

be derived from condensation of a shikimate-derived C_6—C_3 fragment with propionic acid, the first proof that this particular condensation can occur in nature. They also found that one of the methyls of the *gem*-dimethyl group was

---

[2] J. B. Harborne, in 'Biosynthesis', ed. T. A. Geissman (Specialist Periodical Reports), The Chemical Society, London, 1972, vol. 1, p. 119.
[3] B. J. Marshall and C. Ratledge, *Biochim. Biophys. Acta*, 1972, **264**, 106.
[4] L. Caglioti, G. Ciranni, D. Misiti, F. Arcamone, and A. Minghetti, *J.C.S. Perkin I*, 1972, 1235.

derived from methionine, thus accounting for all 13 carbons in (5). [2-^{14}C]-Phenylalanine was fed to *S. orinoci* and the acid produced was degraded to phenylacetaldehyde, where all the activity was located. A similar experiment with [1-^{14}C]phenylalanine gave, as expected, phenylacetaldehyde without radioactivity. [2,3-^{3}H$_2$]Propionate was also fed and was successfully incorporated into (5) in the expected manner.

Lack of Repression of the Shikimic Acid Pathway.—It is well established that a control mechanism on the shikimic acid pathway is by repression of enzymes early in the pathway by the end products, *e.g.* by phenylalanine. Cases where this type of control appear to have broken down have now been reported. Lowe and Westlake,[5] in studying the biosynthesis of the phenolic antibiotic chloramphenicol in *Streptomyces* sp. 3022a, examined several enzymes of the pathway, notably chorismate mutase and anthranilate synthetase, but could find no evidence that end-product control on chloramphenicol synthesis was operating in this organism. Similarly, Chu and Widholm[6] fed phenylalanine and tyrosine to a range of tissue cultures of higher plants, but were not able to observe any evidence of feedback control on chorismate mutase levels.

(6)

In another case with higher plants, Baillie *et al.*[7] deliberately tried to inhibit shikimate dehydrogenase with isonicotinic acid derivatives, with the idea of using them as potential herbicides which would not be toxic to animals. Compounds based on 1,6-dihydroxy-2-oxoisonicotinic acid (6) effectively inhibited the enzyme *in vitro*, but attempts to use these compounds *in vivo* were foiled, probably because the inhibition was reversible and aromatic synthesis was not completely repressed. These inhibitors were anomalous in their behaviour, since they inhibited the enzyme in the back reaction (shikimate → NADP$^+$ + dihydroshikimate), and had only little effect on the forward reaction.[8]

Hydroxybenzoic Acid Turnover.—Hydroxybenzoic acids accumulate, mainly in bound form, in most higher plant tissues and they are undoubtedly subject to active turnover. Their fate, however, has hardly been studied. Recent experiments by Harms *et al.*[9] using plant suspension cultures suggest that

---

[5] D. A. Lowe and D. W. S. Westlake, *Canad. J. Biochem.*, 1972, **50**, 1064.
[6] M. Chu and J. M. Widholm, *Physiol. Plant.*, 1972, **26**, 24.
[7] A. C. Baillie, J. R. Corbett, J. R. Dowsett, and P. McCloskey, *Pesticide Sci.*, 1972, **3**, 113.
[8] J. R. Dowsett, B. Middleton, J. R. Corbett, and P. K. Tubbs, *Biochim. Biophys. Acta*, 1972, **276**, 344.
[9] H. Harms, K. Haider, J. Berlin, P. Kiss, and W. Barz, *Planta*, 1972, **105**, 342.

demethylation and decarboxylation are two steps in their further metabolism. Various benzoic acids, ^{14}C-labelled in the *p*- and *m*-methoxy- and in the carboxy-groups, were tested for turnover in cultures of *Phaseolus aureus* and *Glycine max*. In general, demethylation only occurred in the *p*-position, anisic being 42% converted into *p*-hydroxybenzoic and veratric 47% into vanillic acid. Decarboxylation only took place with acids which had a free *p*-hydroxy-group, *p*-hydroxybenzoic and syringic acids being decarboxylated almost quantitatively. Thus, the enzymes controlling turnover have well-defined specificities and are clearly only accustomed to dealing with the common natural benzoic acids, which all have a free *p*-hydroxy-group.

When hydroxybenzoic acids are artificially fed to intact plants through the petiole, they form glucose esters unless they contain an *o*-hydroxy-group, when *O*-glucosides are formed; hydrogen-bonding between the carboxyl and adjacent hydroxyl presumably prevents ester formation in such cases.[10] Most of the few known naturally occurring derivatives of hydroxybenzoic acids are *O*-glycosides, so that esterification with glucose probably represents a detoxification mechanism which is not part of the normal biosynthetic pathway.

$$\begin{array}{c} \text{CH}=\text{C} \overset{\text{CO}_2\text{H}}{\underset{\text{CH}}{}} \\ \text{HO}_2\text{C} \quad \| \\ \text{HO}_2\text{C}-\text{C} \\ \quad \text{OH} \\ (7) \end{array}$$

Degradation of simple catechol derivatives such as protocatechuic acid by ring cleavage is a well-known process in micro-organisms. A recent paper on the pyrogallol derivatives gallic and syringic acids shows that they too are metabolized by standard routes in *Pseudomonas putida* via muconic acid derivatives [*e.g.* (7)] to oxalacetate and pyruvate.[11] Examples of such ring cleavage occurring in higher plants are very few but, in spite of this, Ettlinger and Eyjolfsson[12] are bold enough to propose that such a degradation has biosynthetic importance in *Triglochin palustris* and *T. maritima*. They suggest that the novel cyanogenic glycoside triglochinin (8) is formed in these plants from a catechol derivative (9) by the route shown in Scheme 2. This Dopa-derived glycoside (9) has yet to be isolated as a natural product, but preliminary feeding experiments with [^{14}C]tyrosine in *T. maritima*[12] indicate incorporation of radioactivity into (8).

3 Phenylpropanoids

Biosynthesis of Dihydroxycoumarins.—From the results of earlier metabolic experiments with unlabelled materials, the most likely route to 6,7-dihydroxy-

---

[10] G. Cooper-Driver, J. J. Corner-Zamodits, and T. Swain, *Z. Naturforsch.*, 1972, **27b**, 943.
[11] B. F. Tack, P. J. Chapman, and S. Dagley, *J. Biol. Chem.*, 1972, **247**, 6438.
[12] M. Ettlinger and R. Eyjolfsson, *J.C.S. Chem. Comm.*, 1972, 572.

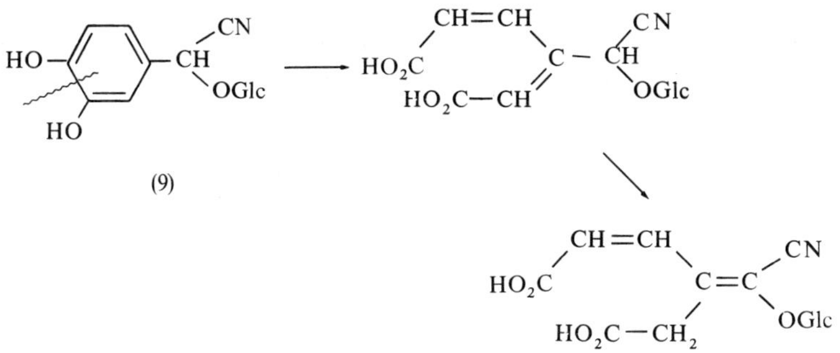

Biosynthesis of Phenolic Compounds Derived from Shikimate

Scheme 2

(8) Triglochinin

Scheme 3

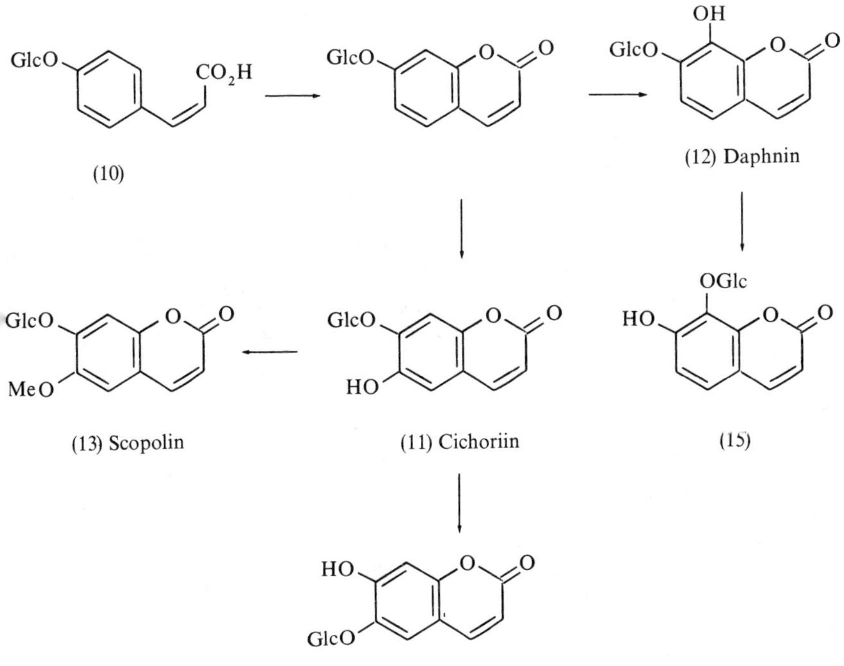

(10)

(11) Cichoriin

(12) Daphnin

(13) Scopolin

(14) Aesculin

(15)

coumarin (aesculetin) seemed to be from caffeic acid, followed by o-hydroxylation and ring closure. As a result of radioactive feeding experiments in *Daphne odora* and *Cichorium intybus*, this route can now be ruled out and it appears that the key intermediate in the synthesis of both aesculetin and its 7,8-isomer daphnetin is p-coumaric acid, probably as the O-glucoside (10).[13] This is o-hydroxylated, ring-closed, and then further hydroxylated in the 6-position to give cichoriin (11) (in *Cichorium*) or in the 8-position to give daphnin (12) (in *Daphne*) (Scheme 3.)

The feeding data[13] in favour of this scheme are very convincing. When labelled precursors were fed to *Daphne* inflorescences, and allowed 20 hours for metabolism, the incorporation of p-coumaric acid into (12) was 27.8%, of cinnamic acid 12.8%, and of caffeic acid only 0.80%. A similar experiment with a 4-day metabolic period gave the results: p-coumaric acid 45%, cinnamic acid 13%, and caffeic acid 2.5%. Experiments with *Cichorium intybus* gave a similar pattern of incorporation into (11). Equal amounts of the three precursor acids were fed and there was similar uptake, but the dilution value for p-coumaric acid was as low as 0.0224, whereas for cinnamic and caffeic acids the values were 1.36 and 1.10, respectively.

The fact that caffeic acid is a poor precursor of aesculetin and daphnetin agrees with the many failures to incorporate it into flavonoid (*e.g.* quercetin) synthesis. These results also raise the question again of how plants synthesize scopolin (13). According to Fritig et al.,[14] it is produced in tobacco tissue culture according to the route: p-coumaric → caffeic → ferulic → scopoletin → scopolin. Such a pathway may not, of course, necessarily operate in intact plants and its synthesis from p-coumaric acid according to Scheme 3, with the final step of O-methylation of cichoriin, is still very plausible. Such a process is supported by the fact that scopolin and cichoriin frequently co-occur in a number of unrelated plant groups (see *e.g.* Harborne[15]). Scopolin is occasionally formed apparently without concomitant production of aesculetin derivatives, *e.g.* in potato tuber tissue in response to blight infection.[16] In such cases, it might be formed by the alternative Fritig route from caffeic and ferulic acids. Further studies of scopolin biosynthesis in several different plant groups would seem to be worthwhile to test these ideas.

The synthesis of aesculin (14) from cichoriin (11) has also been studied[17] and the process is similar to that required for the conversion of daphnin (12) into daphnetin 8-glucoside (15), as predicted in the last Report.[2] The enzyme, isolated from *Cichorium* tissue, catalyses the hydrolysis of (11) to aesculetin and also the transglucosylation from another molecule of (11) to the aglycone, with the formation of (14). The reverse reaction (14) → (11) does not occur. Unlike the enzyme from *Daphne*, this one has two interconvertible com-

---

[13] M. Sato and M. Hasegawa, *Phytochemistry*, 1972, **11**, 657.
[14] B. Fritig, L. Hirth, and G. Ourisson, *Phytochemistry*, 1970, **9**, 1963.
[15] J. B. Harborne, *Biochem. J.*, 1960, **74**, 270.
[16] J. C. Hughes and T. Swain, *Phytopathology*, 1960, **50**, 398.
[17] M. Sato and M. Hasegawa, *Phytochemistry*, 1972, **11**, 3149.

ponents, a high molecular weight protein with predominantly transglucosylase activity and a lower molecular weight form with hydrolytic activity.

The glycosylation of scopoletin has been studied in tobacco seedlings.[18] [4-^{14}C]scopoletin was fed through the roots and its fate determined over the course of several days. After continuous feeding for 4 hours, 91% of the radioactivity in the alcohol-soluble fraction of the plant was scopolin (13), 3% was scopoletin, and 5% was fabiatrin (scopoletin 7-primeveroside). Examination after longer time periods showed that scopoletin was more slowly incorporated into the alcohol-insoluble fraction, into lignin-like material. Compartmentation between a storage vacuolar pool and a turnover metabolic pool appeared to be taking place in this plant.

Biosynthesis of Lignin.—The mechanism of biosynthetic differentiation between the different lignins produced in gymnosperms and angiosperms has been explored by Shimada et al.,[19] employing both enzymic and tracer procedures. The pathway of synthesis of the C_6—C_3 lignin precursors appears to be as shown in Scheme 4. An alternative route to sinapic acid (16), via 3,4,5-

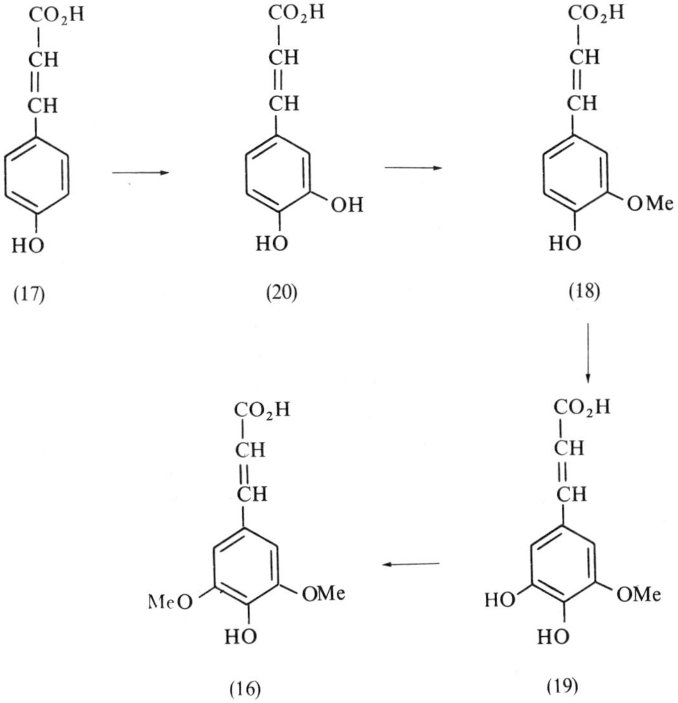

Scheme 4

[18] L. T. Innerarity, E. C. Smith, and S. H. Wender, *Phytochemistry*, 1972, **11**, 1389.
[19] M. Shimada, H. Fushiki, and T. Higuchi, *Phytochemistry*, 1972, **11**, 2657.

trihydroxycinnamic acid followed by O-methylation, has been tested experimentally and shown to be inoperative. The crucial difference between gymnosperm and angiosperm lignins is that the former are derived exclusively from p-hydroxycinnamyl (17) and ferulyl (18) units, whereas the latter contain in addition sinapyl ($\equiv$ syringyl) units. However, the ability to produce 3,4,5-trihydroxycinnamic acid is not absent from the gymnosperms, since lignans, C_6—C_3 dimers, based on 5-hydroxyferulic acid (19) occur in *Thuja plicata*. The basis of differentiation must therefore lie in the step from (19) to sinapic (16), *i.e.* a methylation process.

Experimental evidence that this methylation (19) → (16) does not occur in gymnosperms has now been obtained. An O-methyltransferase catalysing the methylation of caffeic (20) to ferulic acid (18) was extracted from seedlings of Japanese black pine, *Pinus thunbergii*. Caffeic acid was the best substrate of sixteen compounds tested; protocatechualdehyde and 3,4-dihydroxyphenylacetic acid were also methylated to some degree. However, (19), the key intermediate for syringyl lignin, was an extremely poor substrate, the rate of methylation relative to caffeic acid being 5%. In this respect, the pine enzyme differs markedly in substrate specificity from comparable enzymes isolated from angiosperms (bamboo, poplar) which methylate both substrates (19) and (20) at similar rates. Differences in the specificity of the O-methyltransferases, thus, are a controlling factor in determining what type of lignin is produced. The fact that the pine enzyme has slight activity towards 5-hydroxyferulic acid would suggest that *traces* of sinapyl units could be present in gymnosperm lignin, and there is no experimental evidence against this being the case.

The properties of the partly purified O-methyltransferase from pine have also been studied.[19] The enzyme shows a requirement for Mg, is inhibited by Co, Ni, Zn, and Cd, and has a maximal activity at pH 7.5. O-Methyltransferase activity has also been examined in wheat plants, in relationship to growth and lignification.[20] The enzyme activity reached two maxima, one in 10-day-old plants and another at 28—30 days; activity then fell off to a lower level. Enzyme activity was closely correlated with ferulic acid content in the wheat seedlings and the results indicate that there is a plentiful supply of precursor available before lignification begins to take place.

The Japanese workers have also explored the mechanism by which syringyl units are incorporated into bamboo and grass lignins.[21] By feeding [*methoxy*-^{14}C]ferulic acid, they were able to show that (18) was incorporated into syringyl as well as vanillyl units without rearrangement of the label. The ratio of specific activities in syringyl and vanillyl groups was 0.33, a similar ratio being obtained after feeding [U-^{14}C]phenylalanine. In sliced bamboo shoots, some demethoxylation of (18) to (17) occurred but this was only a side reaction.

The three hydroxycinnamic acids, (16)—(18), undergo reduction to the corresponding alcohols before being incorporated into lignin. In spite of the

---

[20] A. D. M. Glass and B. A. Bohm, *Phytochemistry*, 1972, **11**, 2195.
[21] M. Shimada, H. Fushiki, and T. Higuchi, *Phytochemistry*, 1972, **11**, 2247.

importance of this reductive step, it has scarcely been studied, except in the laboratory. The discovery by Mansell et al.[22] of an enzymic system in *Salix alba* tissue which will reduce ferulic acid to coniferyl alcohol is therefore very timely. Cambial tissue was scraped from 7-year-old branches, frozen, powdered, and extracted with 0.1M borate buffer of pH 7.8. This preparation converted ferulic acid into coniferyl alcohol in 40% yield in 2 hours at 27 °C. ATP, coenzyme A, NADH, and NADPH were necessary cofactors. The ferulic acid was tritium-labelled, so that the identity of the product could be established beyond doubt. After one experiment, cold coniferyl alcohol was added to the reaction mixture, the 2,4-dinitrobenzene derivative was prepared, and this was repeatedly recrystallized to constant specific activity. Probable intermediates in the reduction are ferulyl-CoA and coniferyl aldehyde.

Miscellaneous Phenylpropanoids.—New types of C_6—C_3 compound continue to be found in plants, to provide problems for future biosynthetic attack. Gottlieb,[23] for example, has named a new class of lignan in which the two C_6—C_3 units are joined by linkages other than of the common β–β type as neolignans. One such compound is burchellin, from *Aniba burchellii* (Lauraceae),[24] in which the linkage is α–γ [see (21)]. Possible modes of biogenesis of these neolignans are discussed by Gottlieb in a review on natural products of the Lauraceae.[23]

(21)

Another rare type of C_6—C_3 derivative occurs in the Proteaceae, a representative being leucodrin (22) from *Leucodendron* species. The diastereomeric conocarpin (23) has recently been identified in *Leucospermum conocarpodendron* in the same family.[25] The authors[25] suggest that both compounds could be formed from an activated *p*-hydroxycinnamic acid by a Michael-type condensation with L-α-galactonolactone, condensation in one direction giving (22) and in the other (23) (Scheme 5). D-Galactose is abundantly present in the above plants and could provide the source of the lactone. An alternative origin for (22) has been suggested, by condensation of *p*-hydroxycinnamic acid with L-ascorbic acid.[26] The basis of this suggestion is the discovery in several plants of

---

[22] R. L. Mansell, J. Stockingt, and M. H. Zenk, *Z. Pflanzenphysiol.*, 1972, **68**, 286.
[23] O. R. Gottlieb, *Phytochemistry*, 1972, **11**, 1537.
[24] O. A. Lima, O. R. Gottlieb, and M. T. Magalhaes, *Phytochemistry*, 1972, **11**, 2031.
[25] G. W. Perold, A. J. Hodgkinson, A. S. Howard, and P. E. J. Kruger, *J.C.S. Perkin I*, 1972, 2457.
[26] R. Couchman, J. Eagles, M. P. Hegarty, W. M. Laird, R. Self, and R. L. M. Synge, *Phytochemistry*, 1973, **12**, 707.

ascorbalamic acid (24), a condensation product of ascorbic acid and alanine, which shows some structural resemblance to (22).

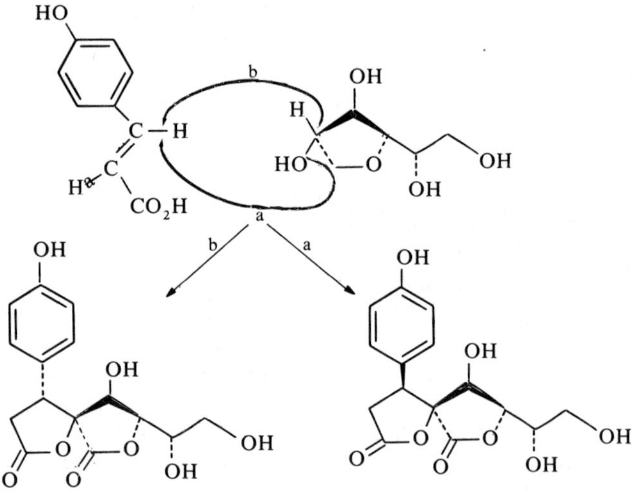

(22) Leucodrin   (23) Conocarpin

Scheme 5

Phenylalanine Ammonia Lyase (PAL).—Work continues at an increasing pace on this key enzyme in phenylpropanoid and flavonoid synthesis and a review of its properties has appeared.[27] Full details of the sterochemistry of the reaction it catalyses in generating *trans*-cinnamic acid from L-phenylalanine by antiperiplanar elimination of NH_3 have been published.[28] The authors describe the synthesis of D- and T-labelled (3S)- and (3R)-phenylalanines and the use of these materials for proving the stereochemistry of the reaction, using PAL from potato. They also show in a second paper[29] that the tyrase activity of PAL isolated from maize follows the same stereochemistry. Using labelled tyrosines, they found that the maize enzyme eliminates the 3-*pro-S* hydrogen of (2S)-tyrosine to give *trans-p*-hydroxycinnamic acid and ammonia. Also, experiments with stereospecifically labelled (2S)-phenylalanine showed that the degree of 3-*pro-S* specificity of the maize enzyme is indistinguishable from that of the potato enzyme. Finally, studies of the reversibility of the two enzyme reactions indicated that the same active site in the maize enzyme was involved in both cases.

[27] K. R. Hanson and E. A. Havir, in 'Recent Advances in Phytochemistry', ed. V. C. Runeckles and J. E. Watkins, Appleton-Century-Crofts, New York, 1972, vol. 4, pp. 48—86.
[28] R. H. Wightman, J. Staunton, A. R. Battersby, and K. R. Hanson, *J.C.S. Perkin I*, 1972, 2355
[29] P. G. Strange, J. Staunton, H. R. Wiltshire, A. R. Battersby, K. R. Hanson, and E. A. Havir, *J.C.S. Perkin I*, 1972, 2364.

```
         HO₂C       NH
             \CH/   \CO
              |       |
             CH₂ ──── C ── OH
                      |
                  ┌── C ── OH
                  |   |
                  O   CH(OH)
                  |   |
                  |   CH(OH)
                  |   |
                  └── CH₂
```

(24)

Other evidence has been obtained that the PAL and tyrase activities of the maize plant reside in one and the same protein.[30] Firstly, both activities are equally enhanced by treating the plant with gibberellic acid. Secondly, the decreases in enzyme activities caused by incubating maize tissue in a moist atmosphere are equally reversed by adding cycloheximide. Thirdly, the ratio of the activities remains the same during a series of enzyme purifications, including chromatography on agarose.

From molecular weight determinations it appears as if the enzyme PAL is very similar in a wide variety of plant sources. Reported values include 306 000 for maize, 300 000 for mustard, 330 000 for potato, and 226 000 for *Streptomyces verticillatus*. Parkhurst and Hodgkins[31] have now determined the molecular weight of PAL from the yeast *Sporobolomyces pararoseus*, using both gel filtration and sedimentation on a sucrose gradient, and again obtained values (275 000—300 000) in the same range. These authors also report that the enzyme is inactivated by carbonyl-, amino-, and sulphhydryl-attacking reagents and that dehydroaniline is present at the active site and is essential for activity. The molecular weight of PAL appears to remain constant throughout the life of the plant, since a study of the effect of ageing on PAL activity in potato tuber tissue[32] showed that there was no change in molecular size, in spite of the fact that the enzyme was continually being turned over and re-synthesized.

The physiology of PAL production in plants has attracted much attention, partly because the control mechanism of PAL synthesis ultimately determines how much phenolic is produced at any one time in plant tissues. Engelsma[33] has found that addition of Mn^{2+} can increase PAL levels in plants, or at least in gherkin hypocotyls. A range of other metal ions was tested and the only other with any effect, a slight one, was Mg^{2+}. The effect of Mn^{2+} is indirect, since it is presumably involved in removing from the site of synthesis hydroxycinnamic acids which normally accumulate and repress and/or inactivate PAL.

[30] P. D. Reid, E. A. Havir, and H. V. Marsh, *Plant Physiol.*, 1972, **50**, 480.
[31] J. R. P. Parkhurst and D. S. Hodgkins, *Arch. Biochem. Biophys.*, 1972, **152**, 597.
[32] J. A. Sacher, G. H. N. Towers, and D. D. Davies, *Phytochemistry*, 1972, **11**, 2383.
[33] G. Engelsma, *Plant Physiol.*, 1972, **50**, 599.

Addition of coumarins to plant tissues can also dramatically effect PAL levels. Pre-soaking pea pods in a solution of the photodynamic coumarin 4,5′,8-trimethylpsoralen, followed by 4 minutes irradiation with 366 nm u.v. light, doubles the PAL activity 3 hours later and increases it twelve-fold 20 hours after.[34] A contrasting effect of added coumarin on PAL activity has been noted in tobacco suspension cultures, where scopoletin addition inhibits PAL synthesis.[35] The growth regulator IAA has the same effect, and, at 1 mmol l^{-1} concentrations, reduces PAL activity by 53%. Although the effect of scopoletin and IAA is probably simple end-product inhibition, that of the photodynamic coumarin must be at a different level, perhaps on plant DNA; psoralen compounds, for example, are known to form stable complexes with the DNA base thymine.

Two isozymes of PAL were reported earlier in *Quercus pedunculata* leaves, one form being sensitive to repression by hydroxycinnamate and the other to repression by hydroxybenzoate (see last year's Report[2]). These two isozymes have now been isolated from the roots of the same plant and a study of their localization indicates that the cinnamate-sensitive isozyme is present in the microsomal fraction, whereas the benzoate-sensitive isozyme is present in the mitochondrial and microbody fraction.[36] The distributions of the two forms of PAL are correlated with those of two other enzymes of phenolic biosynthesis. The enzyme cinnamate 4-hydroxylase accompanies the first isozyme in the microbodies whereas benzoate synthetase is present with the second in the microbodies. Thus phenolic acid biosynthesis is closely compartmented in the oak plant, C_6—C_3 acids being produced exclusively in the microsomes and C_6—C_1 acids in the microbodies.

Phenolases.—The introduction of hydroxy-groups into aromatic nuclei occurs at different stages in phenolic biosynthesis and a range of oxidases with varying specificities is clearly involved. Some progress is now being made in separating these various activities. One of the simplest hydroxylations is of cinnamic to *p*-hydroxycinnamic acid and the enzyme must be widespread since this is the major route to the latter compound, the alternative pathway by deamination of tyrosine being restricted to only a few plant groups. The enzyme, which is not one of the classical phenolases, is apparently so unstable that it has so far defied full characterization. However, its presence can be detected using tissue-slice techniques and it has recently been examined in young grapefruit tissue.[37] Necessary cofactors were found to be NADPH and tetrahydrofolic acid. It only occurred in quantity in very young tissue and its presence was correlated with the stage in fruit development when the flavanone naringenin is being actively synthesized.

Separation of enzyme preparations with monophenolase from those with

---

[34] L. A. Hadwiger, *Plant Physiol.*, 1972, **49**, 779.
[35] L. T. Innerarity, E. C. Smith, and S. H. Wender, *Phytochemistry*, 1972, **11**, 83.
[36] G. Alibert, R. Ranjeva, and A. Boudet, *Biochim. Biophys. Acta*, 1972, **279**, 282; *Physiol. Plant.*, 1972, **27**, 240.
[37] S. Hasegawa and V. P. Maier, *Phytochemistry*, 1972, **11**, 1365.

diphenolase activity have been achieved by Stafford and Dressler,[38] working with shoots and internode tissue of Sorghum vulgare. Separation of a crude phenolase extract on Sephadex G-100 gave three fractions: I, with monophenol and diphenol activity; II, with diphenol activity only; and III, with mainly monophenol activity. Fraction III readily converted p-coumaric into caffeic acid. The diphenol activities of fractions I and II differed in that the ratios of activity with chlorogenic acid and Dopa were in the first case less than 1:1 and in the second greater than 1:1.

An enzyme which specifically oxidizes p-coumaric to caffeic acid has been purified from Streptomyces by Nambudiri et al.[39] It does have diphenol activity, but this activity is completely inhibited by p-coumaric acid. Another unusual feature of this phenolase is that high concentrations of ascorbate inhibit its activity. It is also the smallest tyrosinase so far described, with a molecular weight of only 18 000. The same enzyme in Neurospora, for example, has a molecular weight of 33 000.

A phenolase with only catecholase activity has been purified 17-fold from Prunus avium fruits.[40] Substitution in the 3-position of catechol causes a decrease in affinity of substrate for this enzyme, owing to steric hindrance. Electron-donating substituents in position 4 enhance activity (caffeic and chlorogenic acids are excellent substrates), whereas electron-attracting substituents (as in protocatechuic acid) reduce or prevent oxidation. The specificity displayed by this enzyme suggests it might have a biosynthetic role, but this has yet to be demonstrated.

4 Flavonoid Biosynthesis

Enzymology of Biosynthesis.—The first step in flavonoid synthesis is the condensation of p-coumaryl-CoA and malonyl-CoA to give 2′,4,4′,6′-tetrahydroxychalcone (25), which is then isomerized to the related flavanone naringenin. Evidence that this step is enzyme-mediated has been obtained by Kreuzaler and Hahlbrock,[41] working with illuminated cell suspension cultures of parsley, Petroselinum crispum. A crude enzyme preparation from 11-day-old cells was incubated for 2 hours with [1,3-$^{14}C_2$]malonyl-CoA, p-coumaryl-CoA, and ATP in phosphate buffer of pH 7.5. Cold naringenin was then added and the radioactive naringenin identified by chromatography in five solvents, further characterization being by means of enzymic conversion into radioactive apigenin. Finally, the material was subjected to alkaline cleavage; the phloroglucinol formed was radioactive whereas the p-hydroxybenzoic acid was unlabelled. An enzyme preparation from 21-day-old parsley leaves also catalysed the above reaction.

The second step in flavonoid synthesis is the isomerization of chalcone to flavanone, and the isomerase responsible has been isolated from parsley where

---

[38] H. A. Stafford and S. Dressler, *Plant Physiol.*, 1972, **49**, 590.
[39] H. M. D. Nambudiri, J. V. Bhat, and P. V. S. Rao, *Biochem. J.*, 1972, **130**, 425.
[40] G. Lanzaraini, P. G. Pifferi, and A, Zamorani, *Phytochemistry*, 1972, **11**, 89.
[41] F. Kreuzaler and K. Hahlbrock, *F.E.B.S. Letters*, 1972, **28**, 69.

it has a high specificity for A-ring hydroxylation patterns (see last year's Report).[2] Wiermann,[42] in studying the same enzyme in developing anthers of *Lilium candidum* and a *Tulipa* cultivar, has also found a similar substrate specificity; the enzyme only catalysed the ring closure of 2′,4,4′,6′-tetrahydroxychalcone (25), having no effect on 2′,4,4′-trihydroxychalcone. The main chalcone accumulating early in anther development is different in the two plants studied by Wiermann, although both make the same flavonols, *i.e.* kaempferol, quercetin, and isorhamnetin. In *Lilium*, only naringenin chalcone (25) is present, whereas in *Tulipa* the major chalcone is (26).[43] Probably there is a balance between (25) and (26), depending on how much hydroxylation of the B-ring occurs at the C_{15} level. The amount of intermediate chalcone present is often extremely low, and Endress[44] was only able to detect (26) as a natural intermediate in flavonoid synthesis in *Petunia hybrida* flowers by employing trapping experiments.

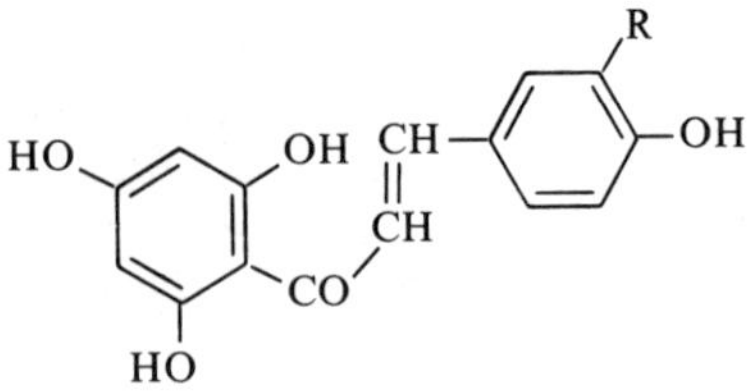

(25) R = H
(26) R = OH

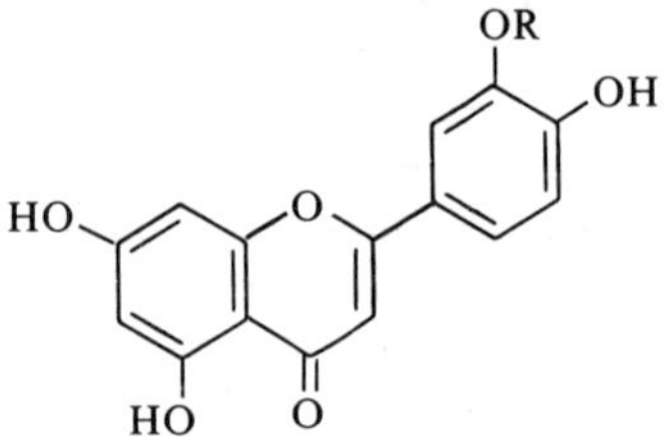

(27) R = H
(28) R = Me

One of the last steps in flavonoid synthesis is methylation and an *O*-methyltransferase for this has been purified 82-fold from cell suspension cultures of parsley.[45] It has a pH optimum of 9.7, requires Mg^{2+}, and the molecular weight is *ca.* 48 000. Unlike catechol methylase from animal tissues, it is not inhibited by *p*-chloromercuribenzoate or by iodoacetamide. Luteolin (27) and its 7-glucoside were the best substrates, giving chrysoeriol (28) and its 7-glucoside respectively; K_m values were 4.6×10^{-5} and 3.1×10^{-5} mol l^{-1}. Eriodictyol and caffeic acid were poor substrates (K_m values 1.2×10^{-3} and 1.6×10^{-3} mol l^{-1}, respectively) and the enzyme was quite specific for catechols and only *O*-methylated a *m*-hydroxy-group.

Naturally Occurring Intermediates.—The possibility that *o*-hydroxydibenzoylmethanes are intermediates in flavone synthesis, particularly of those flavones which lack B-ring substitution, was suggested by the earlier discovery of a compound formulated as 2,4,6-trihydroxydibenzoylmethane 4-glucoside in

[42] R. Wiermann, *Planta*, 1972, **102**, 55.
[43] R. Wiermann, *Z. physiol. Chem.*, 1972, **353**, 129.
[44] R. Endress, *Z. Pflanzenphysiol.*, 1972, **67**, 188.
[45] J. Ebel, K. Hahlbrock, and H. Grisebach, *Biochim. Biophys. Acta*, 1972, **268**, 313.

Malus leaf.[46] That it was an intermediate was indicated by its rapid conversion during chromatographic purification into chrysin 7-glucoside. More recently, a second dibenzoylmethane, namely the 2,6-dihydroxy-4-methoxy-derivative, was reported in *Populus nigra* buds.[47] More careful examination of this latter material,[48] however, showed it to have the isomeric cyclic structure of a 2-hydroxyflavanone, *i.e.* (29). This was proved by i.r., m.s., and n.m.r. studies and also by synthesis. Its co-occurrence in *Populus* buds with the 7-methyl ethers of galangin, chrysin, and pinocembrin suggests it has a biogenetic role. In retrospect, it seems likely that Williams' compound in *Malus* has a similar cyclic structure [*i.e.* (30)] rather than the open-chain form. Re-examination of

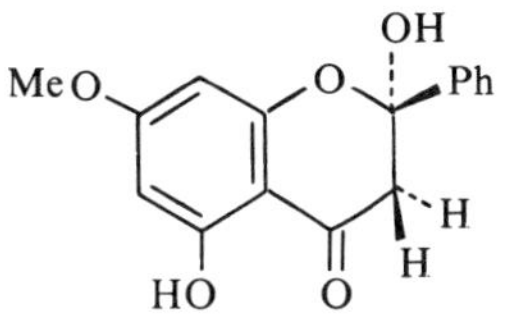

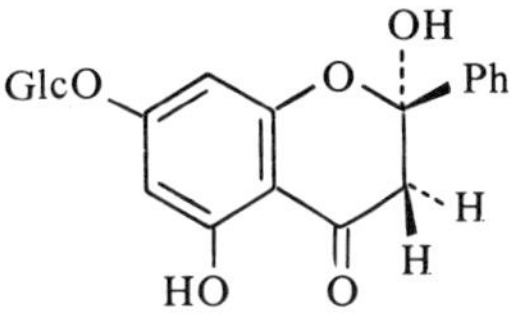

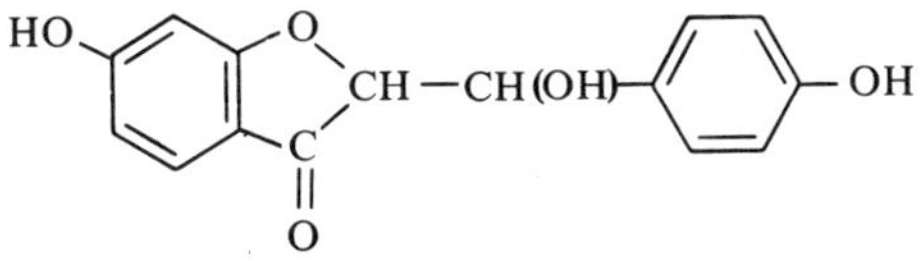

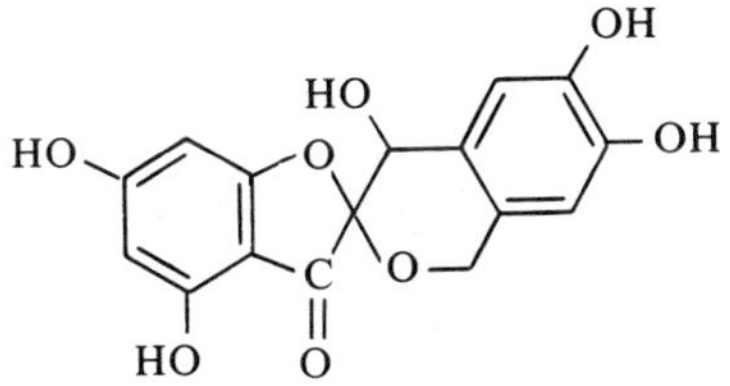

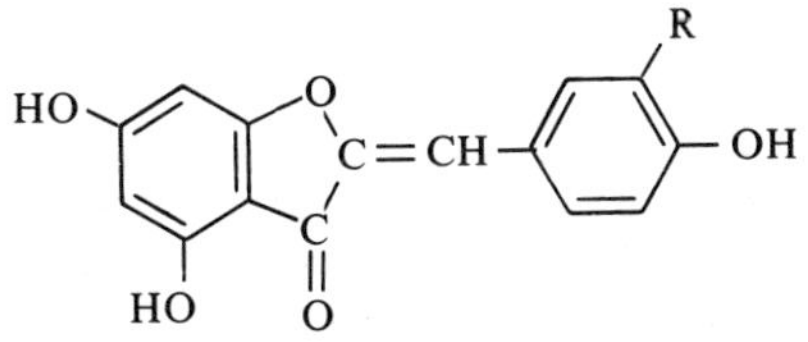

[46] A. H. Williams, *Chem. and Ind.*, 1967, 1526.
[47] M. Chadenson, M. Hauteville, J. Chopin, E. Wollenweber, M. Tissut, and K. Egger, *Compt. rend.*, 1971, **275**, C, 1291.
[48] M. Chadenson, M. Hauteville, and J. Chopin, *J.C.S. Chem. Comm.*, 1972, 107.

his material would certainly be worthwhile, since (29) represents a new class of naturally occurring flavonoid compounds in plants.

Coumaranones such as (31) have been suggested as intermediates in the synthesis of the yellow aurone pigments[49] and it is apposite that a related structure (32) has recently been found in the heartwood of *Acacia crombei*.[50] This represents the first report of a spirocoumaranone in nature. At one time, all known plant aurones had a catechol B-ring [*e.g.* (33)] and their synthesis seemed to be in some way related to this fact. However, an exception to this regularity has now appeared. Asen and Plimmer[51] have discovered an aurone with a single hydroxy-group in its B-ring, namely (34), in yellow flowers of a *Limonium* cultivar, where it occurs in conjunction with (33) and the chalcone (26).

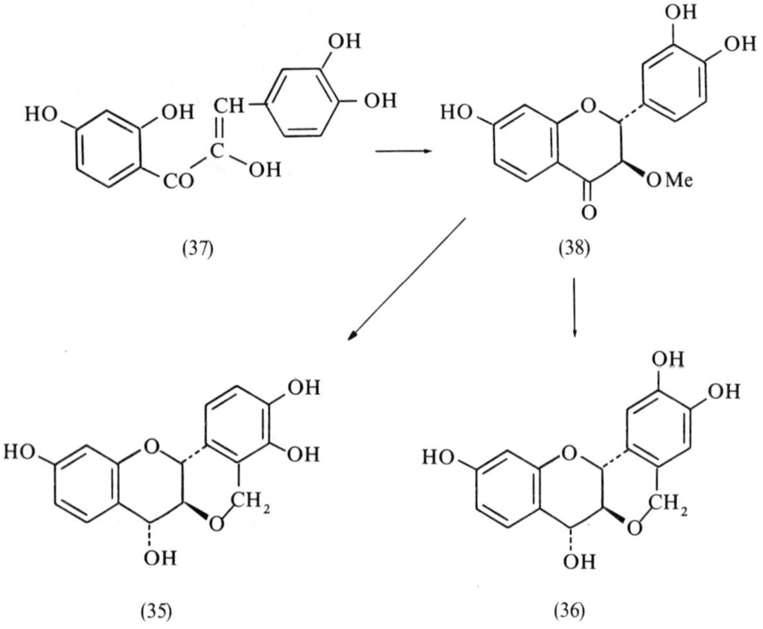

Scheme 6

The biosynthesis of flavonoids in the heartwood of trees is very difficult to study by tracer methods, so that the isolation of possible intermediates, and their conversion *in vitro* into various end-products, is still important. Roux and his co-workers[52] have recently reported in the heartwood of the legume *Trachylobium verrucosum* the presumed precursors of the flavane-3,4-diols,

[49] E. Wong, *Chem. and Ind.*, 1966, 598.
[50] E. V. Brandt, D. Ferreira, and D. G. Roux, *J.C.S. Chem. Comm.*, 1972, 392.
[51] S. Asen and J. R. Plimmer, *Phytochemistry*, 1972, **11**, 2601.
[52] J. P. Merwe, D. Ferreira, E. V. Brandt, and D. G. Roux, *J.C.S. Chem. Comm.*, 1972, 521.

Biosynthesis of Phenolic Compounds Derived from Shikimate

(+)-mopanol (35) and (+)-peltogynol (36). The new 'intermediates' are α,2′,3,4,4′-pentahydroxychalcone (37) and 3-O-methyl-2,3-trans-fustin (38). Biogenesis (Scheme 6) involves α-methoxylation and cyclization of (37) to (38), followed by oxidative cyclization of the 3-O-methyl group and reduction of the 4-position to give (35) and (36).

Glycosylation and Acylation.—These steps in synthesis occur at a late stage and there is much evidence from earlier genetic studies that the enzymes involved have closely controlled substrate specificities. A further example of the effect of genetics on flavonoid synthesis has been published, on pigment inheritance in the garden pea, *Pisum sativum*.[53] There are considerable differences in glycosylation patterns between the anthocyanins, flavones, and flavonols in this plant. There is also a gene, which specifically controls transfer of rhamnose to the anthocyanidins, giving rise to 3-rhamnosides and 3-rhamnoside-5-glucosides and which has no effect on the co-occurring flavones. The contribution of genetic studies to our knowledge of flavonoid production has been reviewed recently.[54]

An enzyme which will catalyse the glycosylation of anthocyanidin has yet to be described, but one which will do this for flavonols has been found in several plants. In an earlier communication,[55] Larson pointed out that maize pollen is a particularly rich source of a glucosyltransferase capable of converting kaempferol and quercetin to their 3-glucosides. The enzyme has now been extracted with distilled water and purified four-fold by $(NH_4)_2SO_4$ fractionation.[56] It has a requirement for UDP-glucose, mercaptoethanol, and Ca^{2+} and possesses an optimum pH of 8.2. The K_m for quercetin was found to be 0.6×10^{-4} mol l^{-1}.

The mechanism of biosynthesis of flavone C-glycosides still remains obscure, but an isomerase which will convert isovitexin (39) into vitexin (40) has recently been detected in the Reporter's laboratory.[57] It occurs in the grass *Briza media*, which contains glycoflavones, and it operates in only the one direction, the pH optimum being about 6.8.

(39) Isovitexin

(40) Vitexin

---

[53] C. M. Statham, R. K. Crowden, and J. B. Harborne, *Phytochemistry*, 1972, **11**, 1083.
[54] A. J. Birch, *Pure Appl. Chem.*, 1972, **33**, 17.
[55] R. L. Larson, *Phytochemistry*, 1971, **10**, 3073.
[56] R. L. Larson and C. M. Lonergan, *Planta*, 1972, **103**, 361.
[57] B. G. Murray and C. A. Williams, *Nature*, 1973, **243**, 87.

Acylation of flavonoids with hydroxycinnamic acids has frequently been recorded; similar acylation with malonic acid is beginning to be recognized as a regular feature of flavonoid metabolism. An isoflavone acylated with malonate was first reported in clover leaves[58] and flavones and flavonols similarly acylated through sugar have more recently been discovered in parsley, *Petroselinum crispum*.[59] A malonyl-CoA:flavone glycoside transferase has also been demonstrated in cell cultures of this latter plant.[60] Incubation of apigenin 7-apiosylglucoside (apiin) with [^{14}C]malonyl-CoA and the enzyme for 1 hour gave radioactive malonylapiin. Identity of the product was confirmed by several procedures, including electrophoretic separation from unchanged apiin. Indirect evidence for the presence of malonylated flavonoids in both *Petroselinum* and *Cicer* was obtained by Davenport and Dupont,[61] who found enzymes in these plants capable of liberating malonate from the flavone glycoside fractions.

Another unusual type of flavonoid conjugation is with potassium bisulphate, the linkage often again being through sugar. Increasing numbers of these conjugates of flavones and flavonols have recently been detected, in such families as the Palmae, Gramineae, and Umbelliferae.[62] Their mode of synthesis and further metabolism remains yet to be determined.

Anthocyanin Synthesis.—Plant tissue culture is favourable for studying anthocyanin synthesis and such pigments have been identified or re-characterized during the year in carrot *Daucus carota*,[63] *Dimorphotheca*,[64] *Happlopappus gracilis*,[65] and *Rosa*.[66] The pigments are identical with those formed in the intact plant. In *D. carota*, the pigment formation is completely inhibited by the presence of gibberellic acid (10^{-4} mol l^{-1}), whereas in *H. gracilis* and *Rosa* anthocyanin is suppressed by 2,4-D (0.1 mg l^{-1}). Light is essential for anthocyanin synthesis in plant cultures and all wavebands except the far-red are effective. In *Happlopappus*, irradiation in blue light apparently suppresses rhamnosyl transferase activity, since the normal 1:1 ratio of the two pigments cyanidin 3-glucoside and 3-rhamnosylglucoside is changed in blue light to 5:1.[65]

Anthocyanin biosynthesis has also been studied in *Impatiens balsamina* in petals cultured on an agar support.[67] Inhibitors of protein and RNA synthesis drastically effect anthocyanin synthesis by preventing extra glycosylation and acylation from occurring. Thus pelargonidin 3-glucoside accumulates at the expense of pelargonidin 3-(p-coumarylglucoside)-5-glucoside (41) usually present. The same thing happened in a white-flowered genotype which, when

[58] A. B. Beck and J. R. Knox, *Austral. J. Chem.*, 1971, **24**, 1509.
[59] F. Kreuzaler and K. Hahlbrock, *Phytochemistry*, 1973, **12**, 1149.
[60] K. Hahlbrock, *F.E.B.S. Letters*, 1972, **28**, 65.
[61] H. E. Davenport and M. S. Dupont, *Biochem. J.*, 1972, **129**, 18.
[62] J. B. Harborne and C. A. Williams, *Z. Naturforsch.*, 1971, **26b**, 490.
[63] M. Schmitz and V. Seitz, *Z. Pflanzenphysiol.*, 1972, **68**, 259.
[64] E. A. Ball, J. B. Harborne, and J. Arditti, *Amer. J. Bot.*, 1972, **59**, 924.
[65] R. G. Stickland and N. Sunderland, *Ann. Bot.*, 1972, **36**, 443, 671.
[66] M. E. Davies, *Planta*, 1972, **104**, 50, 66.
[67] P. C. Bibb and C. W. Hagen, *Amer. J. Bot.*, 1972, **59**, 305.

fed with pelargonidin 3-glucoside normally, synthesized (41); addition of DL-ethionine blocked the conversion.

(41)

The effect of light on anthocyanin and other flavonoids continues to occupy the attention of many plant physiologists and a review has appeared in a symposium volume on phytochrome.[68]

Proanthocyanidin Synthesis.—Although their name suggests a relationship with the anthocyanins, it is now generally accepted that the proanthocyanidins (or leucoanthocyanidins) are biosynthesized by a separate pathway. Data from genetic studies indicate this and a recent reinvestigation of pigments in colour mutants of *Zea mays* leads to a similar conclusion.[69] Anthocyanidins identified in acid-treated tissues were cyanidin, pelargonidin, peonidin, and luteolinidin, but whereas the first three were present as anthocyanins, the latter clearly originated from the proanthocyanidin luteoforol. Significantly, no apigeninidin was detected; also luteolinidin formation was quite unaffected by gene mutations which altered or blocked the pathway of anthocyanin synthesis.

In a major study of the chemistry and distribution of proanthocyanidins, Thompson *et al.*[70] isolated a range of dimeric and trimeric procyanidins with varying configurations from a wide range of plants. The survey showed that the procyanidins were always accompanied by (+)-catechin, (−)-epicatechin, or a mixture of the two, depending on the configuration of the procyanidin(s) present in a particular plant. By contrast, flavane-3,4-diols were absent, and the authors suggest that procyanidin synthesis cannot involve acid-catalysed condensation of these 3,4-diols, as has so frequently been suggested (see *e.g.* ref. 71). Instead, they propose that proanthocyanidins are formed exclusively by oxidative polymerization of flavan-3-ols, and support this idea with the results of a feeding experiment. (−)-[U-^{14}C]Epicatechin (42) was fed to shoots of *Rubus idaeus*; after 7 days, the procyanidin (43) was isolated and was found to have as much as 14% of the original activity of (42).

The biosynthesis of catechin itself has recently been studied and may appropriately be mentioned here. Recent feeding experiments in the tea plant,

---

[68] H. Smith, in 'Phytochrome', ed. K. Mitrakos and W. Shropshire, Academic Press, London, 1972, pp. 433—481.
[69] E. D. Styles and O. Ceska, *Phytochemistry*, 1972, **11**, 3019.
[70] R. S. Thompson, D. Jacques, E. Haslam, and R. J. N. Tanner, *J.C.S. Perkin I*, 1972, 1387.
[71] T. A. Geissman and H. D. G. Crout, 'Organic Chemistry of Secondary Plant Metabolism', Freeman-Cooper, San Francisco, 1969, p.209.

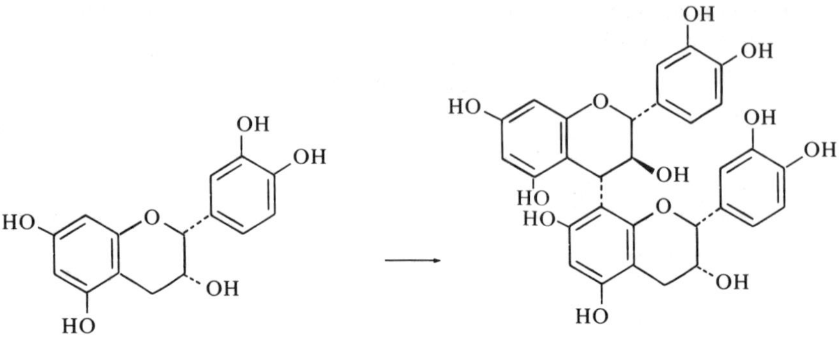

(42) (−)-Epicatechin   (43)

Camellia sinensis,[72] indicate that shikimic acid is a better precursor of the B-ring of catechins such as (42) than are phenylalanine and dihydrokaempferol. Calculations based on label dilution, following the feeding of ^{3}H and ^{14}C precursors, showed shikimic acid to be nearly 100 times more effective a precursor than the other two intermediates.

Indirect support for the views of Thompson *et al.* on procyanidin synthesis comes from a paper on heartwood flavonoids of *Acacia nigrescens*.[73] The compounds present all have the 3′,4′,7,8-tetrahydroxy-substitution pattern. Substitution in the 8-position should prevent proanthocyanidin formation, and indeed no such polymers could be detected. Furthermore, the necessary precursor, 3′,4′,7,8-tetrahydroxyflavan-3-ol was also absent. Present were the chalcone, flavanone, flavonol, dihydroflavonol, and flavane-3,4-diol analogues and also the corresponding benzylcoumaranone (44). It has been suggested[74] that substitution of flavonoids in the 8-position (by hydroxylation, as in *Acacia*, or by glycosylation or C—C coupling) is of special evolutionary significance in plants. Like proanthocyanidin synthesis, it represents a primitive mechanism and is gradually replaced in more advanced taxa by 6-substitution. The mechanism by which functional groups are placed in either the 6- or 8-position of the flavonoid nucleus still remains obscure.

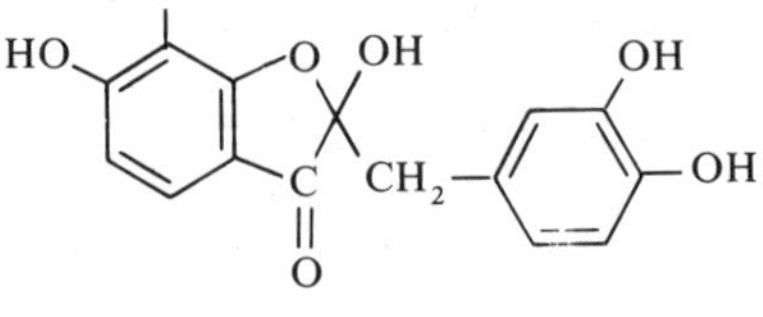

(44) Nigrescin

[72] M. T. Zaprometov and H. Grisebach, *Fiziol. Rast.*, 1972, **19**, 1034.
[73] T. G. Fourie, I. C. Praez, and D. G. Roux, *Phytochemistry*, 1972, **11**, 1763.
[74] J. B. Harborne, *Recent Adv. Phytochem.*, 1972, **4**, 107.

Biosynthesis of Phenolic Compounds Derived from Shikimate 235

Flavonoid Turnover in Higher Plants.—The first step in the degradation of flavonols in higher plants has now been delineated, following the isolation of an enzyme preparation from several plants which converts them into 2,3-dihydroxyflavanones.[75] Kaempferol, quercetin, datiscetin, and morin, but *not* their glycosides, were converted into the corresponding 2,3-dihydroxyflavanones [*e.g.* (45)]. These compounds were identified by spectral studies and their electrophoretic behaviour suggests that the 2,3-dihydroxy-groups have the *cis* configuration. The enzyme was isolated as the acetone powder and then extracted with phosphate buffer of pH 7.5 in the presence of mercaptoethanol.

(45)

The degradative enzyme was found in every higher plant that was tested for it, and in the chick pea, *Cicer arietinum*, it was found in only those organs which are rich in flavonols.[76] The enzyme is produced between the second and sixth day after germination and appears to parallel PAL activity in its time of synthesis. Thus, the regulation of synthesis and degradation of flavonols are interdependent and closely correlated events, and active turnover must occur from the very moment of synthesis.

Flavonoid Degradation by Micro-organisms.—The induced extracellular dioxygenase in *Aspergillus flavus* that oxidizes rutin to carbon monoxide and 2-protocatechuoylphloroglucinolcarboxylic acid has been fully characterized and its specificity determined.[77] It contains two moles of Cu and binds at least two moles of substrate. It is not inhibited by —SH reagents or affected by H_2O_2, but it is inhibited by Cu-chelating agents. The enzyme catalyses the oxidation of flavones that possess a 2,3-double bond and a hydroxy-group at C-3. Hydroxy-groups at other positions affect the relative rates of oxidation and the concentration of substrate required to attain half maximal rate.

The turnover of the flavanone taxifolin (46) by cultures of a *Pseudomonas* species grown on (+)-catechin has been studied by Jeffrey *et al.*[78] A flavoprotein, in the presence of NADPH and molecular oxygen, converts (46) into dihydrogossypetin (47). This intermediate is as yet unknown as a natural product, although the related flavonol is fairly widespread in plants.[79] Compound (47) then undergoes *meta*-cleavage in the A-ring, to form oxalacetic acid

[75] W. Hosel and W. Barz, *Biochim. Biophys. Acta*, 1972, **261**, 294.
[76] W. Hosel, G. Frey, E. Tenfel, and W. Barz, *Planta*, 1972, **103**, 74.
[77] T. Oka, F. J. Simpson, and H. G. Krishnamurty, *Canad. J. Microbiol.*, 1972, **18**, 443.
[78] A. M. Jeffrey, M. Knight, and W. C. Evans, *Biochem. J.*, 1972, **130**, 373, 383.
[79] J. B. Harborne, *Phytochemistry*, 1969, **8**, 177.

and 4-hydroxy-5-(3,4-dihydroxyphenyl)-3-oxovalero-δ-lactone (see Scheme 7). These products are then further metabolized to eventually yield CO_2.

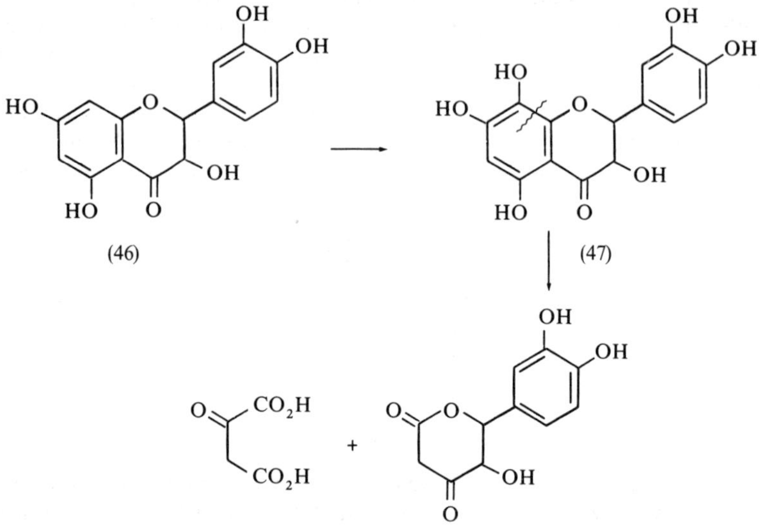

Scheme 7

The degradation of flavonoids in animal tissues is almost certainly mediated by the microbial flora in the intestine, since the pathway is so similar to that detected in micro-organisms in pure culture. This conclusion has been confirmed by Griffiths and Smith,[80] who have studied the fate of both apigenin and myricetin derivatives from *in vivo* feedings and from *in vitro* action of intestinal microflora. Apigenin derivatives undergo ring cleavage to give mainly 4-hydroxyphenylacyl derivatives. Pelargonidin 3,5-diglucoside is broken down to 4-hydroxyphenyl-lactic acid; in contrast, cyanidin, which ought to be more readily cleaved, is not metabolized at all. Myricetin derivatives are largely dehydroxylated at the 4'-position and 3,5-dihydroxy- and 3-hydroxy-phenyl-lactic acids are produced in the urine.

5 Isoflavonoids

Biosynthesis of Hydroxyphaseollin.—The biosynthesis of a newly discovered pterocarpan (48)[81] has been investigated in disease-resistant soya bean hypocotyls.[82] [U-^{14}C]Phenylalanine and [9-^{14}C]isoliquiritigenin were readily incorporated into (48). Label also appeared in daidzein, coumestrol, and sojagol (49). The biosynthetic pathway followed the generally expected route

[80] L. A. Griffiths and G. E. Smith, *Biochem. J.*, 1972, **128**, 901; 1972, **130**, 141.
[81] J. J. Sims, N. T. Keen, and V. K. Honwad, *Phytochemistry*, 1972, **11**, 827.
[82] N. T. Keen, A. I. Zaki, and J. J. Sims, *Phytochemistry*, 1972, **11**, 1031.

via 3,9-dihydroxypterocarpene (50), which was then isoprenylated at C-3' and oxidized in the 6-position. Compound (48) was not present in healthy tissues and also was not detected until 16 hours after inoculation of the plant with *Phytophthora megasperma*. The pterocarpan rapidly accumulated between 16 and 48 hours, whereas the greatest accumulation of daidzein and coumestrol developed 48—72 hours after inoculation. Although anthocyanins and

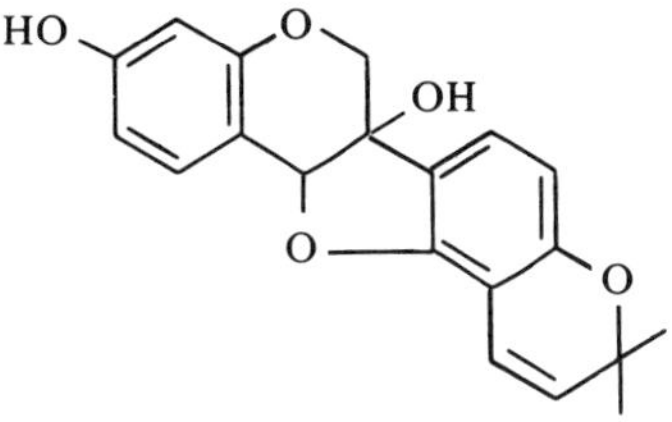

(48)

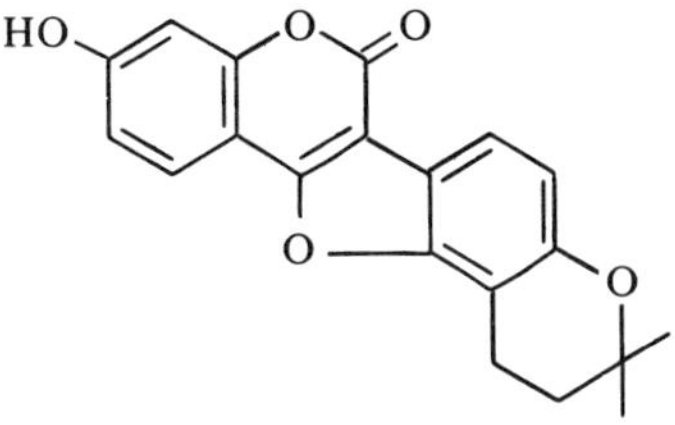

(49)

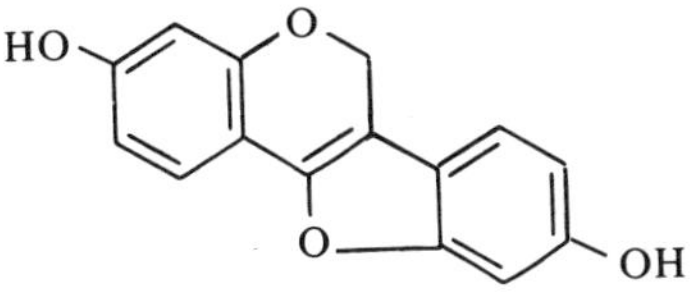

(50)

flavones were present in the soya bean hypocotyls, their levels were quite unaffected by fungal invasion. Thus, isoflavonoid synthesis is a completely specific response to infection and some of the later enzymes in the biosynthetic pathway to (48) must be formed *de novo* for this special purpose.

6 Neoflavonoids

Biosynthesis of Inophyllide.—Evidence that the biosynthesis of neoflavonoids in the Guttiferae occurs *via* phloroglucinolpropionic acid intermediates has

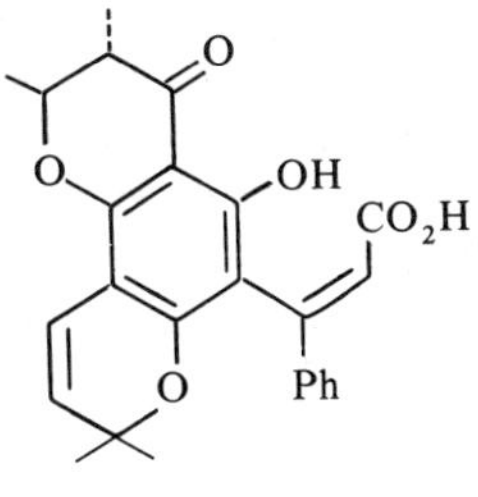

(51)

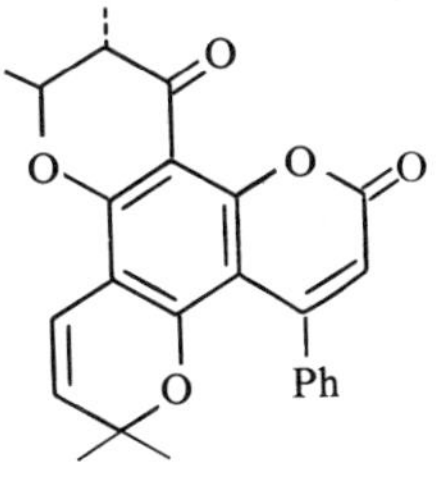

(52)

been provided by Gautier et al.[83] They fed [3-^{14}C]phenylalanine to young shoots of *Calophyllum inophyllum* and obtained significant and sequential incorporation into calophyllic acid (51) and inophyllide (52). A short-term feeding experiment with [U-^{14}C]leucine, a very efficient precursor of the 2,3-dimethylchromanone ring, gave active incorporation into (51) and (52), with five times more activity in the former compound. These results indicate a general pathway to the neoflavonoids of the Guttiferae as indicated in Scheme 8.

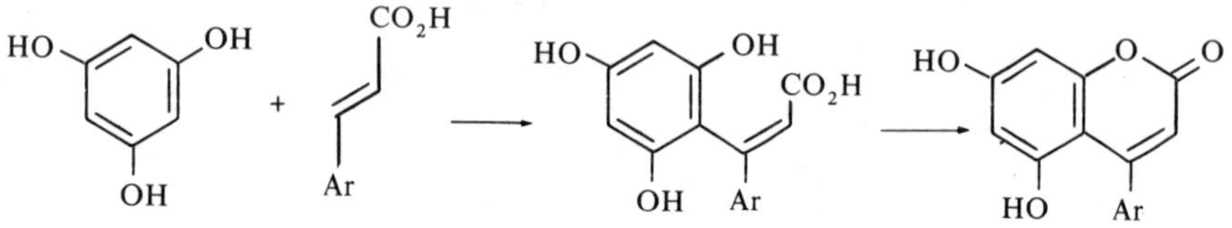

(53)

Scheme 8

The bearing of these results on the biosynthetic pathway to neoflavonoids in the other major family to make them, the Leguminosae, must remain conjectural. The substitution patterns of the legume neoflavanoids are certainly different and intermediates of the type (53) have not been found in the family. The most probable pathway to the legume compounds, following an initial condensation of 1,3,4-trihydroxybenzene with cinnamyl pyrophosphate, is: dalbergiquinol → dalbergione → neoflavene → dalbergin. The latter compounds have the same basic structure as (52). Laboratory experiments supporting the above sequence have been reviewed by Seshadri,[84] but feeding data for confirming this Scheme are still needed.

7 Quinones

Biosynthesis of Lawsone.—The biosynthesis of this naphthaquinone has been investigated for intermediary symmetry by Grotzinger and Campbell.[85] The naphthalene nucleus of lawsone (54) is derived from shikimate with three non-carboxyl carbon atoms of glutamate, and a symmetrical 1,4-naphthaquinone or the acid (55) are both possible intermediates. Feeding experiments in *Impatiens balsamina* now establish that [2-^{14}C]acetate predominantly labels C-2 of (54), so that the unsymmetrical intermediate (55) must be involved (see Scheme 9). An alternative route from (54) to (55) from that illustrated is *via* 1,4-dihydroxy-2-naphthoic acid, but this could not be detected as an intermediate, nor was it incorporated when fed to the plants.

Anthraquinone Biosynthesis.—In the absence of any biosynthetic experiments to report, it is worth mentioning that a biogenetically interesting new dihydro-

---

[83] J. Gautier, A. Cave, G. Kunesch, and J. Polonsky, *Experientia*, 1972, **28**, 759.
[84] T. R. Seshadri, *Phytochemistry*, 1972, **11**, 881.
[85] E. Grotzinger and I. M. Campbell, *Phytochemistry*, 1972, **11**, 675.

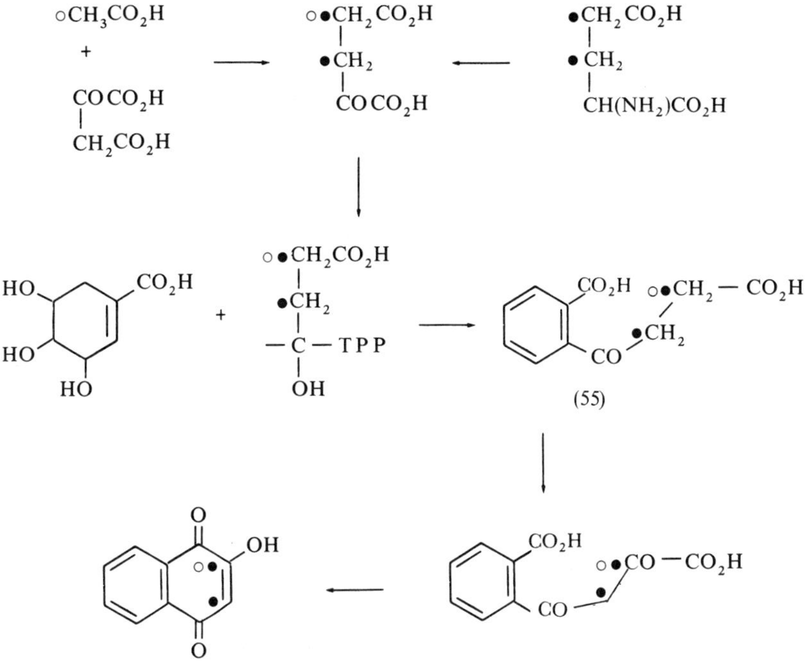

Biosynthesis of Phenolic Compounds Derived from Shikimate

Scheme 9

● = glutamate label
○ = acetate label

anthraquinone has been found in stem of *Morinda lucida* (Rubiaceae). This pigment, oruwal, occurs with a series of shikimate–mevalonate-derived anthraquinones, namely damnacanthal (56), alizarin 1-methyl ether, rubiadin 1-methyl ether, and soranjidol.[86] Oruwal has structure (57) and is intriguing as either an end-product or a possible intermediate in the synthesis of the co-occurring quinones.

Perinaphthenones.—The biosynthesis of haemocorin aglycone, a pigment of *Haemodorum corymbosum*, was mentioned in last year's Report.[2] Edwards *et al.*[87] have shown that a very closely similar pigment, lachnanthoside (58), is formed in *Lachnanthes tinctoria*, by precisely the same general route, *i.e.* from phenylalanine, tyrosine, and acetic acid, the acetate carboxy-group being

---

[86] G. A. Adesida and E. K. Adesogan, *J.C.S. Chem. Comm.*, 1972, 405.
[87] J. M. Edwards, R. C. Schmitt, and U. Weiss, *Phytochemistry*, 1972, **11**, 1717.

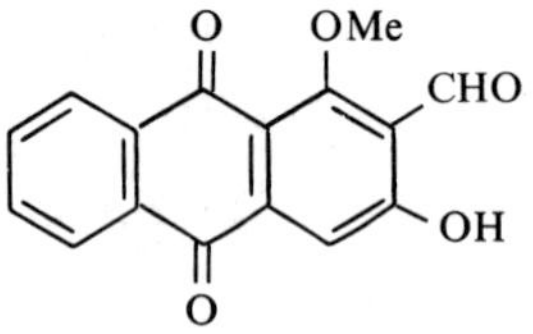

(56)

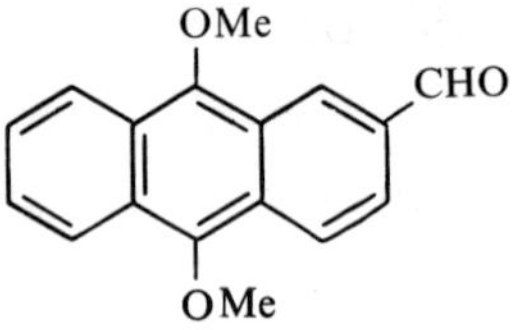

(57)

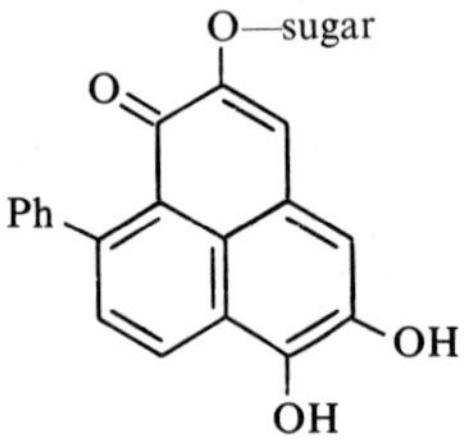

(58)

eliminated in the process. Regrettably, there is no information yet available on possible intermediates formed in this condensation.

7
Stable Isotopes in Biosynthetic Studies

BY M. TANABE

1 Introduction

The use of the stable isotopes ^{2}H, ^{13}C, ^{15}N, and ^{18}O has expanded rapidly with their recent commercial availability[1] in a useful, although limited, array of chemical compositions. The parallel development of more sensitive detection methods for these stable isotopes, which was led by instrumental advances in mass and n.m.r. spectrometry, has also contributed to the increased application of these isotopes to biosynthetic studies. A recent monograph gives general background information and describes the historical development in the biochemical application of these isotopes in tracing intermediary metabolism.[2]

This chapter will be confined to a review of more recent biosynthetic studies that employ ^{2}H, ^{13}C, ^{15}N, and ^{18}O in labelled precursors. Emphasis is placed on the use of ^{13}C-labelled precursors in these studies since the development of n.m.r. detection methods for ^{13}C has resulted in the disclosure of the biosynthetic pathway to an increasing number of complex microbial metabolites. In some cases, comparable results by the conventional ^{14}C technique would be very difficult to obtain.

2 Nuclear Magnetic Resonance Methods

Proton Satellite Method.—In the conventional ^{14}C-labelled method, degradative procedures for the isolation of all labelled carbons in a metabolite may be precluded by formidable structural complexities or the presence of unreactive carbons such as quaternary carbons or carbons in an aromatic framework. The complexity of the chemical degradative procedure is greatly dependent on the molecular structure of the labelled compound. Complex structures require not only complicated, time-consuming procedures, but also tedious procedures for isolation and purification of the molecular fragments. Moreover, since contamination with radioactive impurities influences the experimental results, much care must be taken in the purification steps.

Structures of many new natural products isolated today are determined by advanced instrumentation techniques, without recourse to chemical reactions. Because of this lack of knowledge of chemical reactivity of new metabolites, detailed degradative studies must be initially undertaken, with considerable

---

[1] A. L. Hammond, *Science*, 1972, **176**, 1315.
[2] S. Ratner, in 'Biochemical Applications of Mass Spectrometry', ed. G. R. Walker, Wiley, New York, 1972, p. 1; R. M. Capioli, *ibid.*, p. 735.

expenditure of time and energy, before biosynthetic studies with ^{14}C-labelled precursors can commence.

The recent availability of materials enriched with carbon-13 (a non-radioactive magnetic isotope of carbon) has greatly simplified biosynthetic studies of microbial metabolites. The use of specifically ^{13}C-labelled precursors and ^{1}H and [13]C n.m.r. studies of the labelled products permits detection and identification of the labelled sites without need for chemical degradation.

Spin–spin coupling between a ^{13}C nucleus and a directly bonded proton produces satellite bands on both sides of the main proton signal in the [1]H n.m.r. spectrum. The characteristic satellite coupling constants range from 125 to 250 Hz and depend on the carbon hybridization and substituents. They also show a near linear dependence on the percentage of s-character of the [13]C—H bond.[3]

The degree of enrichment of ^{13}C into the metabolite over the natural abundance of 1.1% is an important consideration that enters into the feasibility of studying a biosynthetic pathway by the ^{13}C-satellite method. The use of a computer for multiscan averaging of the satellite signals aids the examination of metabolites with low enrichment levels ($\sim 2\%$). High enrichment yields often minimize other factors that limit the success of the method, such as (i) yield and purity of the metabolite, since proton peaks from impurities can obscure satellite bands and these impurity peaks accumulate rapidly in multiscan averaging; (ii) low solubility in an organic solvent, which reduces signal intensities; and (iii) complicated spectra with many overlapping peaks, which may also obscure the satellite signals. Spinning side-bands occasionally overlap with satellite signals, but these bands can be shifted by a simple change in spinning rates.

The major advantage of the use of this n.m.r. method for ^{13}C-labelled metabolites is the virtual elimination of chemical manipulation to ascertain labelling patterns. The practical applications of the ^{13}C-satellite method will be reviewed. French and Japanese language articles reviewing biosynthetic applications of ^{13}C have been published.[4]

Griseofulvin (1). Birch[5] established, by the conventional ^{14}C-method with [1-^{14}C]acetate, that griseofulvin (1) is biosynthesized by cyclization of a linear

[3] J. N. Shoolery, *J. Chem. Phys.*, 1959, **31**, 1427; N. Muller and D. E. Pritchard, *ibid.*, pp. 768, 1471.
[4] (a) G. Lukacs, *Bull. Soc. chim. France*, 1972, 351; (b) H. Seto, *Kagaku to Seibutsu*, 1972, **10**, 471, 550.
[5] A. J. Birch, R. A. Massy-Westropp, R. W. Richards, and H. Smith, *J. Chem. Soc.*, 1958, 360.

polyketide chain derived from seven acetate units. The methoxy-groups arise from the one-carbon biochemical pool.[6]

The origin of griseofulvin was the first reported application of the non-degradative ^{13}C-satellite method to a biosynthetic problem.[7] This study, using media fortified with sodium [2-^{13}C]acetate (55%), yielded a ^{13}C-labelled griseofulvin estimated to be of 2.5% enrichment by detection of the satellite bands. The data are presented in Table 1. The enrichment level was obtained by comparison of the satellite peak areas with the satellite peaks of the unlabelled methoxy-groups, which were used as an internal standard.

Table 1 N.m.r. data for griseofulvin

Position	δ/p.p.m.	$J_{^{13}C-H}$/Hz	Enrichment/%
C-6'-CH$_3$	0.97 ($J = 6$ Hz)	120	~ 2.5
C-6'-H, C-5'-CH$_2$	2.3—3.2	—	—
C-2'-, C-4-, C-6-OCH$_3$	3.59, 3.94, 3.89	140	—
C-3'-H	5.51	160	~ 2.5
C-5-H	6.12	160	~ 2.5

Isotope enrichment at three labelled carbons in griseofulvin originating from sodium [2-^{13}C]acetate was established by direct observation of ^{13}C-satellite peaks. The ^{13}C-study confirmed the information obtained earlier from the sodium [1-^{14}C]acetate study by chemical degradation. The complexity of the C-6'-H signal obscured detection of this satellite signal from griseofulvin labelled by [1-^{13}C]acetate.

Fusaric Acid (2). The biosynthetic origin of fusaric acid (2) from two distinct biogenetic units derived from [1-^{13}C]acetate and [2-^{13}C]acetate administered in replacement cultures of *Fusarium oxysporum* was impressively demonstrated by the ^{13}C-satellite method.[8]

In fusaric acid labelled by [1-^{13}C]acetate, satellite peaks from C-4-H, C-6-H, C-8-H, and C-10-H were directly determined. After reduction of the

Table 2 N.m.r. data for fusaric acid

Position	δ/p.p.m.	$J_{^{13}C-H}$/Hz	Enrichment/%	Precursor
C-3=CH	8.27	169	4.0	[2-^{13}C]acetate
C-4=CH	7.88	163	2.0	[1-^{13}C]acetate
C-6=CH	8.87	180	9.0	[1-^{13}C]acetate
C-7-CH$_2$-OH	4.73	143	4.2	[1-^{13}C]acetate
C-B-CH$_2$	2.8	127	9.2	[1-^{13}C]acetate
C-10-CH$_2$	1.1—2.0	128	8.4	[1-^{13}C]acetate
C-11-CH$_3$	0.9	126	5.9	[2-^{13}C]acetate

[6] D. J. P. Hockenhall and W. F. Faulds, *Chem. and Ind.*, 1955, 1390.
[7] M. Tanabe and G. Detre, *J. Amer. Chem. Soc.*, 1966, **88**, 4515.
[8] D. Desaty, A. G. McInnes, D. G. Smith, and L. C. Vining, *Canad. J. Biochem.*, 1968, **46**, 1293.

carboxylic acid to the alcohol, the C-7-H satellite peak appeared to be enriched. One of the satellites from positions C-4, C-6, C-8, and C-7 was visible either on the low-field side of the aromatic resonance or between the aromatic resonance and the aliphatic protons and on the up-field side of the aliphatic resonance. The relative intensities of a common satellite peak appearing in different overlapping regions of the spectrum provided an internal check and served as a standard for the enrichment levels. The n.m.r. data are summarized in Table 2.

This ^{13}C-satellite study demonstrated the biosynthetic origin of fusaric acid from acetate with only a single chemical manipulation (Scheme 1), namely the

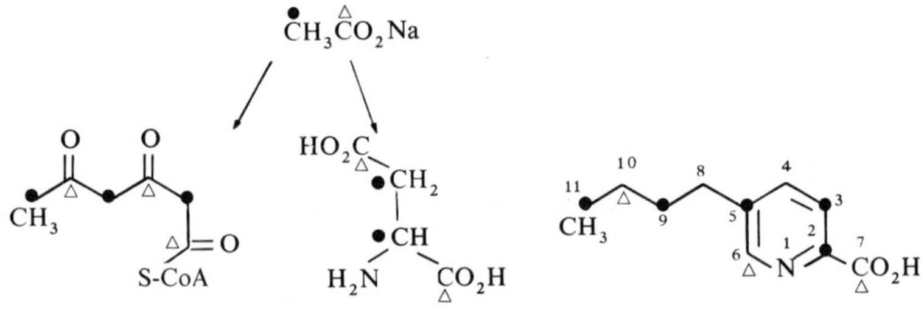

Scheme 1 (2)

conversion of a carboxylic acid into an alcohol. The conventional ^{14}C-method required over 20 different chemical reactions. The incorporation of [4-^{13}C]-aspartic acid into fusaric acid was shown by detection of the C-4-II satellite signal.

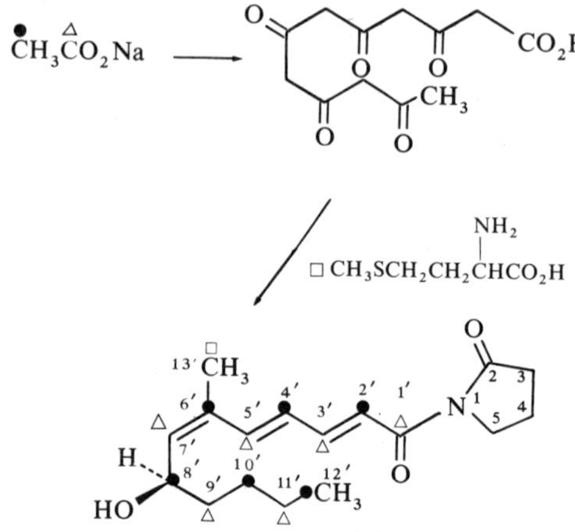

Scheme 2

Stable Isotopes in Biosynthetic Studies

An interesting feature of this study was the preliminary use of ^{14}C-labelled acetates and aspartate to determine optimal growth and culture conditions. These conditions were then applied to growth on the more expensive ^{13}C-labelled substrates, which would result in minimal dilution of isotope.

Variotin (3). The biogenetic origin of the *C*-methyl group of the antifungal agent variotin (3) was established to be methionine by conducting feeding experiments with this [*Me*-^{13}C]-labelled amino-acid.[9] Good incorporation of methionine was evidenced by the very high enrichment value of 34%, which was determined by the ^{13}C-satellite method using the unlabelled *C*-5'-H signal as an internal standard.

Feeding experiments were also conducted with [1-^{13}C]acetate and [2-^{13}C]-acetate, and the biosynthetic origins of six additional carbons of variotin were identified by appropriate detection of the ^{13}C-satellite bands (Scheme 2). The relevant n.m.r. data are summarized in Table 3.

Table 3 *N.m.r. data for variotin*

Position	δ p.p.m.	$J_{^{13}C-H}$/Hz	Enrichment/%	Precursor
C-12'-H$_3$	0.92	128	5.3	[2-^{13}C]acetate
C-10'-H$_2$	1.4	—	2.4	[2-^{13}C]acetate
C-2'-H	7.33	156	3.3	[2-^{13}C]acetate
C-9'-H$_2$/C-11'-H$_2$	1.4	—	4.1	[1-^{13}C]acetate
C-5'-H	6.72	164	—	[1-^{13}C]acetate
C-13'-H$_3$	1.88	128	34.1	[*Me*-^{13}C]methionine

The ^{13}C-satellite enrichment data show the polyketide origin of variotin and a higher enrichment of the starter methyl group of the acetate unit at C-12'. The n.m.r. data also confirm that the pyrrolidine moiety is not derived from acetate.

Piericidin A (4). An earlier ^{14}C-biosynthetic study had shown that piericidin A (4), a naturally occurring insecticide, is formally derived from a linear polyketide formed from four acetates and five propionates (Scheme 3).[10] The use of [3-^{13}C]propionate as precursor and the ^{13}C-satellite method led to the ready identification of the methyl groups C-14, C-15, C-16, C-17, and C-18 as propionate-derived.[11] High enrichment ($\sim$ 10%) was observed in the satellite peaks. The general use of [3-^{13}C]propionate and n.m.r. for establishing the biosynthetic origin of methyl groups derived from propionate in microbial metabolites is advocated, and good enrichment yields can be generally anticipated.

---

[9] M. Tanabe and H. Seto, *Biochim. Biophys. Acta*, 1970, **208**, 151.
[10] N. Takahashi, Y. Kimura, and S. Tamura, *Tetrahedron Letters*, 1968, 4659; Y. Kimura, N. Takahashi, and S. Tamura, *Agric. and Biol. Chem. (Japan)*, 1969, **33**, 1507.
[11] M. Tanabe and H. Seto, *J. Org. Chem.*, 1970, **35**, 2087.

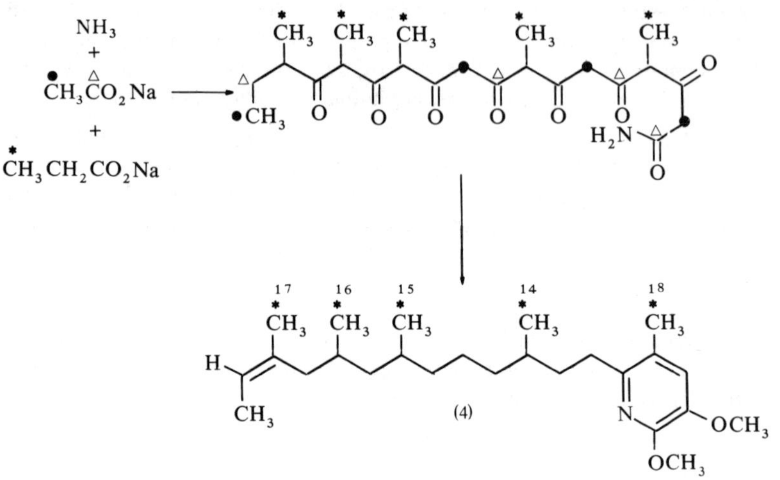

Scheme 3

Mollisin (5). The biosynthesis of this naphthaquinone pigment was initially examined by the ^{14}C-method, using labelled acetates and malonate as precursors.[12] It was concluded from these studies that mollisin (5) is biosynthesized from two separate polyketide chains as shown (Scheme 4). Since the key

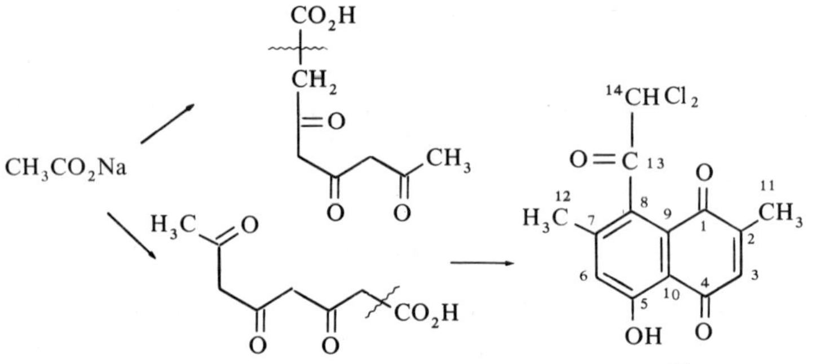

Scheme 4

reaction employed to determine the distribution of radioactivity in the labelled pigment was Kuhn–Roth oxidation to acetic acid, information on the relative specific activities of the two polyketide chains could not be obtained. The ^{13}C-method was particularly amenable to resolution of this problem. Labelling mollisin with [2-^{13}C]acetate yielded the n.m.r. results shown in Table 4.[13]

[12] R. Bentley and S. Gatenbeck, *Biochemistry*, 1965, **4**, 1150.
[13] M. Tanabe and H. Seto, *Biochemistry*, 1970, **9**, 4851.

Stable Isotopes in Biosynthetic Studies

Table 4 N.m.r. data for mollisin

Position	δ/p.p.m.	$J_{^{13}C-H}$/Hz[a]	Enrichment/%
C-11-H$_3$	2.15	128 downfield	2.2
C-12-H$_3$	2.40	130 downfield	1.9
C-3-H	6.83	160 upfield	2.3
		174 downfield	2.1
C-6-H	7.2	160 upfield	1.8
		158 downfield	1.8
C-14-H	6.83	174 upfield	2.3
		188 downfield	2.0

[a] Values found using either the upfield or downfield satellite.

Since an unlabelled proton internal standard was not available in mollisin for ^{13}C-satellite area determinations, these were determined by comparison with an unlabelled sample of mollisin. The enrichment level of the five carbon atoms observed indicates that both polyketide chains – the one commencing from the methyl group C-11 and the other from the methyl group C-12 – are equally labelled.

Further confirmation of the positions of isotopic enrichment of the ^{13}C-labelled mollisin was achieved by means of mass spectrometry. The mass spectrum of mollisin had a base peak at 229 ($M - 83$), which represents the loss of CHCl$_2$. Comparison of this peak of labelled and unlabelled pigment showed a difference of 5.6% or an 0.8% excess at each of seven labelled carbons, assuming equal distribution in the fragment ion. This value corresponds to a total isotopic excess of 1.9% per carbon, including natural abundance. The average value of $\sim 2\%$ found by n.m.r. was in agreement with the mass spectral data.

Sepedonin (6). The ^{13}C-satellite method revealed the origin of all the proton-bearing carbons of the tropolone fungal metabolite sepedonin (6).[14] A detailed study was conducted with [1-^{13}C]acetate, which labels the C-1 position, and with [2-^{13}C]acetate, which labels C-4 and C-5. None of these carbons was

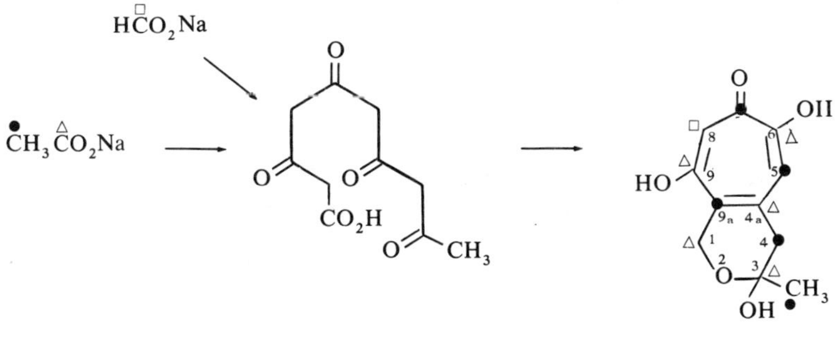

[14] A. G. McInnes, D. G. Smith, L. C. Vining, and J. L. C. Wright, *Chem. Comm.*, 1968, 1669; J. Wright, D. G. Smith, A. G. McInnes, L. C. Vining, and D. W. S. Westlake, *Canad. J. Biochem.*, 1969, **47**, 945.

labelled by [^{13}C]formate. Equal enrichments were observed at C-1, C-4, and C-5, thereby suggesting that a single polyketide chain was involved. Carbon-13-labelled formic acid showed efficient incorporation (> 20%) into C-8; no other carbons were labelled by formate. This result indicates that a one-carbon unit is introduced in the polyketide chain. Carbon-14-labelled substrates were also employed in this thorough study to establish optimal culture conditions for maximum enrichments with ^{13}C-labelled substrates. The n.m.r. data obtained for the ^{13}C-labelled sepedonin are given in Table 5.

Table 5 *N.m.r. data for sepedonin*

Position	δ/p.p.m.	$J_{^{13}C-H}$/Hz	Enrichment/%		
			$^{13}CH_3CO_2Na$	$CH_3{}^{13}CO_2Na$	$H^{13}CO_2Na$
C-3-CH$_3$	1.67	127	4.3	1.1	1.2
C-4-CH$_2$	2.92	128	3.7	1.1	1.0
C-1-CH$_2$	5.09	149	1·0	4.9	1.0
C-5=CH	6.86	153	3.9	1.0	1.1
C-8=CH	7.15	152	1.1	1.1	20.6

The nature of the intermediates involved between the polyketide stage of biosynthesis and sepedonin was not specified, although an alkylated benzene derivative was rejected. In the light of more recent evidence from biosynthetic studies of other naturally occurring tropolones, particularly the splendid work of Battersby on colchicine[15] and the equally elegant work of Scott on stipitatonic acid (7),[16] a benzenoid aromatic precursor is a likely intermediate in sepedonin biosynthesis. The incorporation of radioactive 3-methylorsellinic acid (8) into

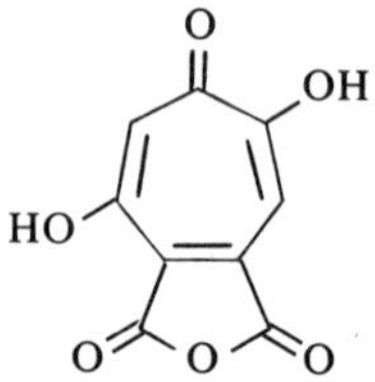

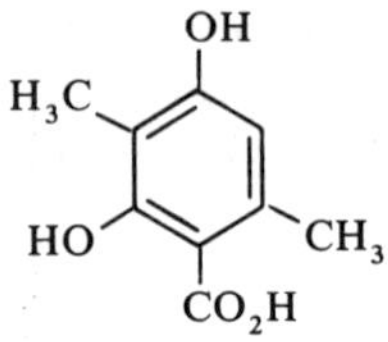

the fungal tropolone metabolite stipitatonic acid (7) indicated that a one-carbon unit is introduced at the lowest oxidation level of the aromatic precursor before ring expansion to the seven-membered ring occurs. By analogy, then, in sepedonin biosynthesis an aromatic precursor is a reasonable intermediate, as outlined in the proposed biogenetic Scheme 5. This scheme is compatible with the ^{13}C-incorporation results.

---

[15] A. R. Battersby, R. B. Herbert, E. McDonald, R. Ramage, and J. H. Clements, *J. C. S. Perkin I*, 1972, 1741.
[16] A. I. Scott, *Pure Appl. Chem.*, 1971, **5**, 34.

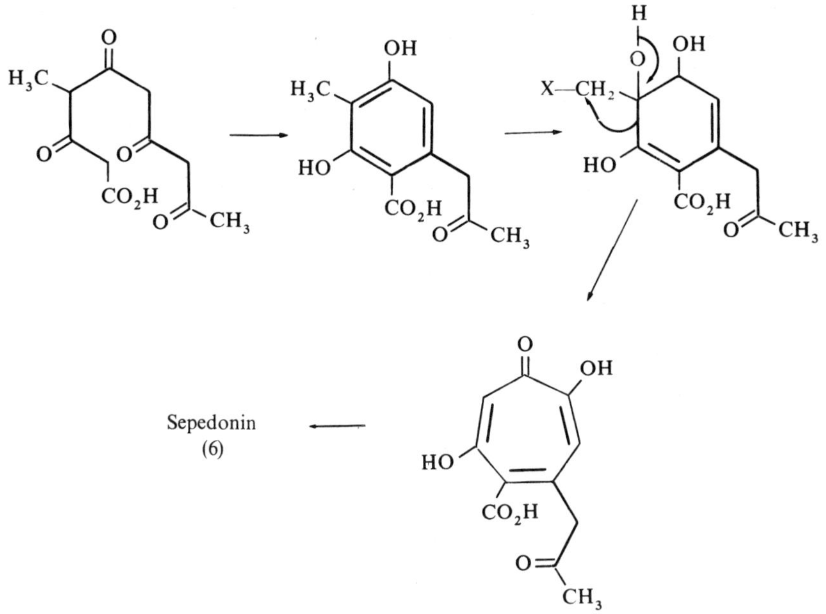

Stable Isotopes in Biosynthetic Studies

Scheme 5

At the time of writing, most of the metabolites examined by the ^{13}C-satellite method also contained labelled carbons not directly bonded to protons, and these labelled nuclei cannot be detected by the satellite method. In some cases, this problem may be circumvented by chemical transformation. For example, in fusaric acid, protons were chemically introduced on to a labelled carbon by reduction to the alcohol (i.e., $CO_2H \rightarrow CH_2OH$). This procedure could, perhaps, be applied to other labelled carbonyl groups and olefinic carbons:

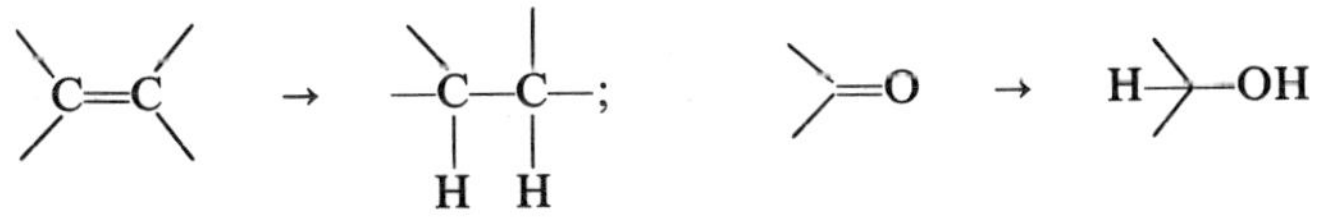

Another limitation of the ^{13}C-satellite method is the obscuring of satellite peaks by complex n.m.r. spectra in which the proton resonances either are not well resolved or exhibit a narrow range of chemical shifts. Investigation of the biosynthesis of steroids and carbohydrates would be particularly difficult because of this limitation.

These disadvantages and limitations of the indirect ^{13}C-satellite method are overcome by the application of direct ^{13}C n.m.r. to ^{13}C-labelled metabolites. All carbons can be recognized by characteristic chemical shifts, and labelled sites can be identified by an increase in signal intensity over that of the natural-abundance peaks. Several timely books and reviews on ^{13}C n.m.r. have been published.[17]

Since the first natural-abundance ^{13}C n.m.r. spectra were reported in 1957 by Lauterbur[18] and Holm,[19] there have been advances in technique and improvements in instrumentation. All these improvements have focused on increasing sensitivity, since the combination of lower intrinsic magnetic moment and natural abundance (1.1 %) makes the carbon signal 1/5700 as sensitive to detection as a proton signal. At an applied field of 23.5 kilogauss, protons absorb at 100 MHz and carbon nuclei at 25.10 MHz.

Proton noise decoupling eliminates multiplicities due to carbon–hydrogen coupling and affords a simplified spectrum in which all carbon signals appear as singlets. Decoupling also results in a positive nuclear Overhauser effect so that the intensity of the decoupled signal is greater than that of the multiplet. A nuclear Overhauser enhancement factor nearly as high as three is possible.[20] The advent of field–frequency locked spectrometers permits multiscan averaging. Sensitivity is therefore increased, since S/N ratios are enhanced. Sensitivity is also increased by the use of larger samples in tubes of 8 or 12 mm outside diameter. Field–frequency stabilization operations in the continuous wave (CW) mode involve homonuclear lock on the carbon signal of a proton-decoupled solvent such as dioxan, benzene, or cyclohexane. More recent spectrometer designs utilize signals from other nuclei, such as deuterium (^{2}H) or fluorine (^{19}F), as the heteronuclear lock signal, provided by either deuteriated solvents or an internal capillary containing a fluorinated material such as hexafluorobenzene.

Although proton noise decoupling simplifies spectra by converting all carbon signals into singlets, it also destroys useful spin-coupling information. This valuable information can still be recovered by an off-resonance decoupling technique while still retaining a substantial nuclear Overhauser effect. The decoupling frequency is offset by several hundred hertz from the optimum value for complete decoupling, and the resulting closely-spaced multiplet signals appear as quartets, triplets, doublets, and singlets, respectively, for methyl, methylene, methine, and quaternary carbons, with residual coupling constants (J_r) small than the normal $J_{^{13}C-H}$ values.[21] Signals from quarternary carbon remain unchanged in the proton-noise-decoupled and off-resonance-decoupled spectra. Methyl and methine carbons exhibit off-resonance-de-coupled spectra and noise-decoupled spectra without coincident lines, and triplet methylene carbons have a coincident centre line in both spectra. These

---

[17] (a) J. B. Stothers, *Appl. Spectroscopy*, 1972, **26**, 1; (b) G. C. Levy and G. L. Nelson, 'Carbon-13 Nuclear Magnetic Resonance for Organic Chemists,' Wiley, New York, 1972; (c) L. F. Johnson and W. C. Jankowski, 'Carbon-13 NMR Spectra,' Wiley, New York, 1972; (d) J. B. Stothers, 'Carbon 13-NMR Spectroscopy,' Academic Press, New York, 1972.
[18] P. C. Lauterbur, *J. Chem. Phys.*, 1957, **26**, 217.
[19] C. H. Holm, *J. Chem. Phys.*, 1957, **26**, 707.
[20] A. J. Jones, D. M. Grant, and K. F. Kuhlmann, *J. Amer. Chem. Soc.*, 1969, **91**, 5013.
[21] H. J. Reich, M. Jautelat, M. T. Messe, F. J. Weigert, and J. D. Roberts, *J. Amer. Chem. Soc.*, 1969, **91**, 7445.

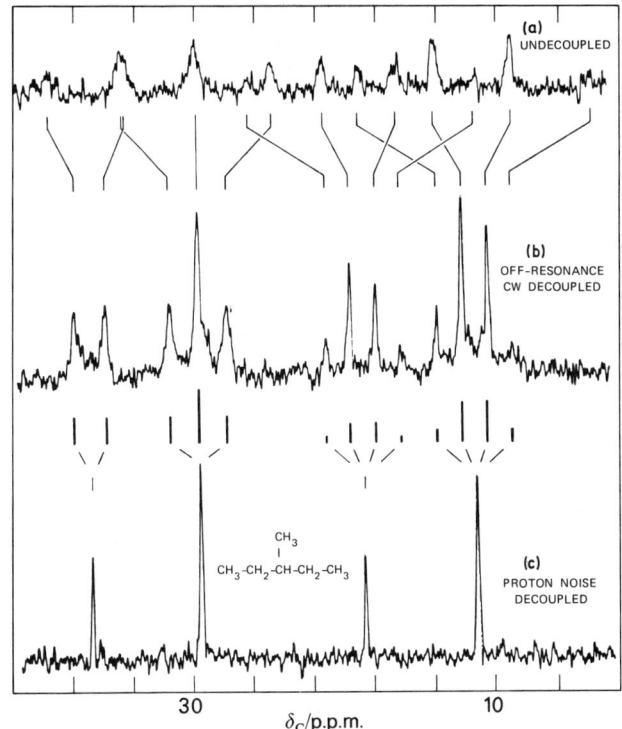

Figure 1 *Spectra of 3-methylpentane (25.15 MHz). Twenty scans were accumulated for the undecoupled spectrum (a) and the off-resonance decoupled spectrum (b). The proton-noise-decoupled spectrum (c) is only a single scan*
(Reproduced by permission of L. F. Johnson).

splitting patterns facilitate identification of carbon multiplicities for proper spectral assignment.

The benefit of spectral simplification derived from proton noise decoupling and off-resonance CW decoupling with its spin-coupling information is shown in the natural-abundance spectra of 3-methylpentane, shown in Figure 1. Figure 1a shows the complexity of the undecoupled spectrum caused by proton splittings. In the proton-noise-decoupled spectrum (Figure 1c), all the carbon signals now appear as singlets with enhanced intensities; however, the individual signals are still difficult to assign. The off-resonance decoupled spectrum (Figure 1b) yielded the closely-spaced, spin-coupled multiplets that confirm the assignment of signals.

The wide range of carbon chemical shifts of approximately 600 p.p.m., compared with proton chemical shifts of 20 p.p.m., is an appealing feature of ^{13}C n.m.r. for obtaining structural information and, hence, for its application in biosynthetic work. For example, the ^{1}H n.m.r. spectrum of cholesterol yields little information other than signals from the two angular methyl groups, the C-3-hydrogen, and the C-6-vinyl proton. In the ^{13}C n.m.r. spectrum, the signals from all 27 carbons of cholesterol are fully resolved.[21]

Carbon chemical shifts follow generally recognized trends and are similar to those of protons.[17] High-field signals are associated with saturated sp^3 carbon atoms; signals from sp^2 carbons are at low field and those from sp carbons are at intermediate field. Electronegative substituents are generally deshielding and produce downfield shifts for the attached carbon atoms. For example, carbonyl groups absorb at the lowest field.

The quickening pace of ^{13}C n.m.r. spectrometry is rapidly adding chemical-shift information; many shift correlations, as well as the development of additive relationships for carbon shifts, have been recorded.[22]

Unfortunately, a standard convention for reporting carbon chemical-shift data has not yet been adopted. Most of the earlier shift data were reported relative to carbon disulphide or benzene as the reference point. More recent data are reported in δ_C tetramethylsilane (TMS) values, with positive shifts downfield from TMS, similar to the convention for proton chemical shifts. The general adoption of this scale is advocated.[23]

The recent application of Fourier-transform techniques to ^{13}C n.m.r. has added an order of magnitude of sensitivity to ^{13}C n.m.r. when compared with the conventional swept-mode time-averaging.[24] A strong broad-band r.f. pulse is employed to excite an entire range of nuclear precession frequencies simultaneously. The resultant free induction signal is measured as a function of time, and the response to repetitive pulses is accumulated. Subsequent Fourier transformation (FT) of the accumulated signals yields the equivalent of the slow, conventionally swept spectrum. The major advantage of FT ^{13}C n.m.r. for biosynthetic studies is the reduction in sample size requirements and spectral observation time. Depending on enrichment levels, conventional CW spectra usually require a sample of 100—200 mg and spectral observation times of 12—72 h, whereas FT operations require 25—100 mg and observation times from minutes to 12 hours.[25]

Many practical applications of ^{13}C n.m.r. to biosynthetic studies of metabolites utilizing the CW and FT modes of operation have been investigated.

Direct Carbon Magnetic Resonance Method.—*Radicinin* (9). The first reported application of ^{13}C n.m.r. to a biosynthetic study was that on the mould metabolite radicinin (9) from *Stenphylium radicinum*.[26] From growing cultures of this organism, ^{13}C-labelled radicinins were prepared by inoculation of the medium with sodium [1-^{13}C]acetate and sodium [2-^{13}C]acetate. The ^{13}C n.m.r. spectra of the enriched radicinins, which are shown in Figure 2, in DMSO solutions were recorded in the CW swept mode and the DMSO

---

[22] Ref. 17(*d*), p. 128.
[23] Ref. 17(*d*), b. 49.
[24] R. R. Ernst and W. A. Anderson, *Rev. Sci. Instr.*, 1965, **37**, 93.
[25] Ref. 17(*b*), p. 19.
[26] M. Tanabe, H. Seto, and L. Johnson, *J. Amer. Chem. Soc.*, 1970, **92**, 2157.

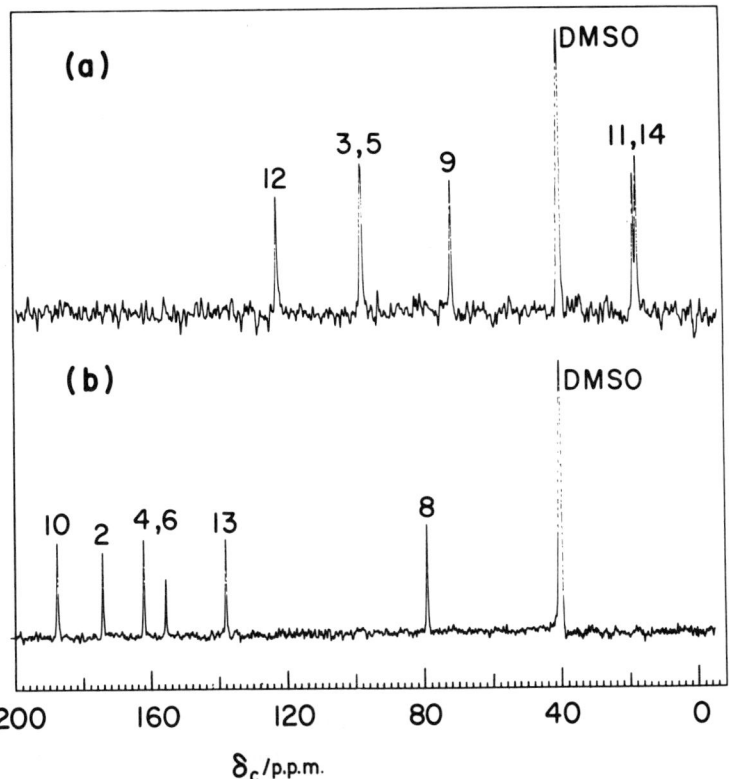

Figure 2 ^{13}C N.m.r. spectra (25.15 MHz) of radicinin from $^{13}CH_3CO_2Na$ (a), and $CH_3{}^{13}CO_2Na$ (b)

solvent peak was the homonuclear lock signal. The $^{13}CH_3CO_2Na$-labelled material (Figure 2a) shows six strong signals at C-3, C-5, C-9, C-11, C-12, and C-14; the six remaining signals are weak and obscured by noise. Although the C-3 and C-5 signals appeared to be a single peak, slow sweeps and horizontal scale expansions showed the presence of two signals. It was, therefore, evident that six carbon atoms of radicinin were derived from the methyl group of acetate. The six strong signals from $CH_3{}^{13}CO_2Na$-labelled material (Figure 2b) that occurred at positions corresponding to C-2, C-4, C-6, C-8, C-10, and C-13 demonstrated that these carbons were derived from the acetate carboxy-group. This labelling of alternate carbons confirms and supports the hypothesis that radicinin is biosynthesized by the combination of two polyketide chains[27] (Scheme 6).

[27] R. Bentley and J. M. Campbell, *Comp. Biochem.*, 1968, **20**, 424.

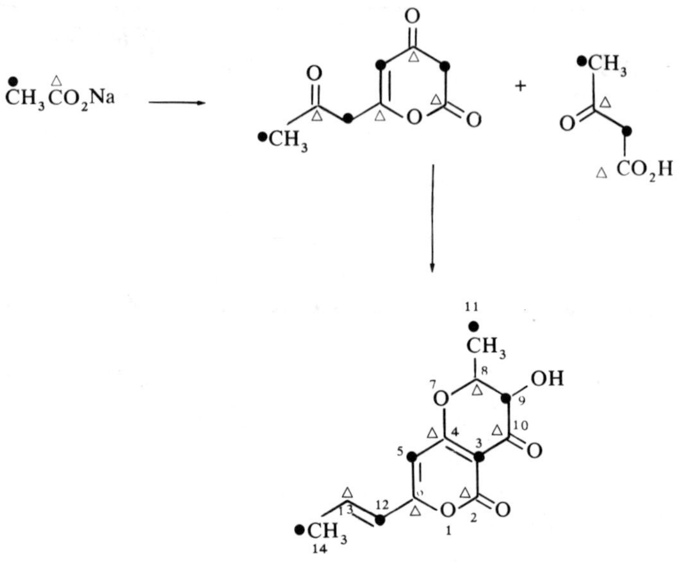

254   *Biosynthesis*

Scheme 6

The assignment of the signals was aided by recognized trends in chemical shifts and by the application of CW off-resonance decoupling where the applied decoupling frequency was set to decouple solvent DMSO protons. The quaternary character of the C-2, C-4, C-6, and C-10 carbons was confirmed, for these signals remained as singlets; the signals of the C-8 and C-13 carbons were split into doublets, showing that these are methine carbons. The decoupled spectrum is shown in Figure 3. A difference in signal intensities between C-4 and C-6 is also apparent. This difference does not directly reflect

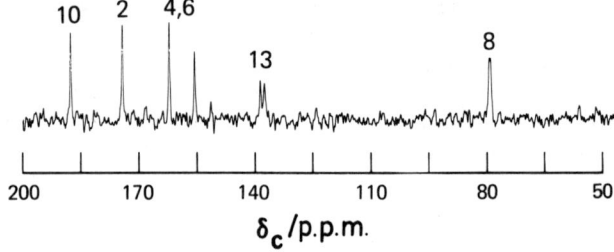

Figure 3 *Continuous-wave decoupled ^{13}C n.m.r. spectrum of radicinin prepared from* $CH_3{}^{13}CO_2Na$

a difference in ^{13}C-enrichments at these sites but, rather, a difference in their respective relaxation times and nuclear Overhauser effects. Carbon-13 spin–lattice relaxation times T_1 are very strong compared with proton relaxation times;[28] therefore, partial saturation effects are very frequently noted in carbon spectra. The accustomed and accepted quantitative information in proton n.m.r. spectra is generally lost in carbon spectra. The one-to-one relationship of carbon signal intensities can only be achieved under special experimental and instrumental conditions. General information on ^{13}C-enrichments can, however, be achieved either by comparison of the signal intensities of similarly substituted carbons that can undergo relaxation by similar mechanisms or by comparison with the ^{13}C n.m.r. spectrum of unlabelled material.

Since natural-abundance peaks in the labelled radicinin were not readily visible in the number of spectral scans employed to make a comparison, the enrichment value of $\sim 17\%$ was determined by integration of the ^{13}C-proton satellite bands of the methyl groups C-11 and C-14, $J_{^{13}C-H} = 126$ Hz, which were evident in the proton 60 MHz spectrum. Combined application with the ^{13}C-satellite method can thus result in a better quantitative approach to carbon enrichments.

By the conventional ^{14}C-method on radicinin, labelled by [1-^{14}C]- and [2-^{14}C]-acetate, 18 chemical steps were required to identify the biosynthetic origin of seven carbons.[29] This ^{14}C study agrees with the ^{13}C study in the necessity for two polyketide chains in radicinin biosynthesis. Neither method could differentiate between a mechanism that involved the condensation of two separate six-carbon polyketide units and one that involved the fusion of an eight-carbon and four-carbon unit, as shown in Scheme 6.

Sepedonin (6). The ability of the ^{13}C n.m.r. method to detect and identify all the labelled carbon atoms in biosynthetic work is illustrated by the tropolone metabolite sepedonin (6).[30] Although the ^{13}C-satellite method provided information on the biosynthetic origin of five (C-1, the C-3-methyl group, C-4, C-5, and C-8) of the eleven carbons of sepedonin by the use of $^{13}CH_3CO_2Na$, $CH_3{}^{13}CO_2Na$, and $H^{13}CO_2Na$ as labelled precursors, the origin of the remaining six carbons, which lack directly attached protons, was not experimentally confirmed.

The proton-noise-decoupled swept ^{13}C n.m.r. spectra of each of the three ^{13}C-labelled sepedonins, dissolved in pyridine solution, showed the carbon signals as singlets, with enhanced intensities for enriched carbons. The carbon assignments were aided by the CW off-resonance decoupling technique. The shift data are summarized in Table 6.

In the $H^{13}CO_2Na$-labelled material, only the C-8 signal at 113.5 p.p.m. is enhanced over natural abundance. This result established the formate origin of C-8 and is in agreement with the satellite data. The C-5 signal appears as a

---

[28] Ref. 17(b), p. 30.
[29] J. F. Grove, *J. Chem. Soc.* (C), 1970, 1860.
[30] A. G. McInnes, D. G. Smith, L. C. Vining, and L. Johnson, *Chem. Comm.*, 1971, 325.

Table 6 ^{13}C N.m.r. data for sepedonin, δ_C/p.p.m.

Carbon	^{13}CH$_3$CO$_2$Na	CH$_3$^{13}CO$_2$Na	H^{13}CO$_2$Na
C-1	—	60.6	—
C-3	—	93.6	—
C-3-CH$_3$	29.0	—	—
C-4	44.0	—	—
C-4a	—	140.6	—
C-5	115.6	—	—
C-6 and C-9	—	161.6 and 166.0	—
C-7	174.7	—	—
C-8	—	—	113.5
C-9a	128.4	—	—

doublet in the CW off-resonance spectrum; since it is the only other ring carbon bearing a proton, it was easily assigned.

The sepedonin from ^{13}CH$_3$CO$_2$Na shows five enhanced signal intensities and its spectrum is similar to that of the CH$_3$^{13}CO$_2$Na-labelled material. These spectra are in accord with the alternate labelling of the carbon chain and with the insertion of C-8 from formate into a linear polyketide chain. They substantiate the ^{13}C-satellite data for the biosynthesis of sepedonin.

Nuclear Overhauser enhancement effects again precluded quantitative enrichment determinations by integration of the peak areas of the carbon signals. The signal intensities of C-5 and C-8 carbons, which probably have similar nuclear Overhauser enhancements, have a 3.9:1 ratio, which is identical with that obtained by the ^{13}C-satellite method.

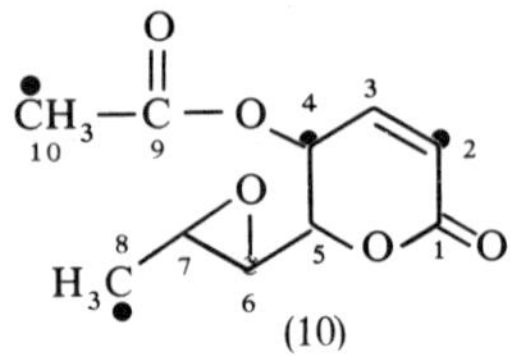

Asperlin (10). Labelling studies with ^{13}CH$_3$CO$_2$Na confirmed the predictable biosynthetic origin of the antibiotic lactone asperlin (10) from a linear tetra-acetyl polyketide.[31] The swept ^{13}C n.m.r. spectrum of the labelled antibiotic in dioxan solution showed signals from C-2, C-4, C-6, C-8, and C-10 of intensity greater than those of the natural-abundance peaks at C-1, C-3, C-5, C-7, and C-9 (Figure 4).

An interesting feature of this biosynthetic study was the determination of the magnitude of the residual ^{13}C–proton coupling constants (J_r) when the decoupler was changed from a random-noise to a single-frequency offset

---

[31] M. Tanabe, T. Hamasaki, D. Thomas, and L. Johnson, *J. Amer. Chem. Soc.*, 1971, **93**, 273.

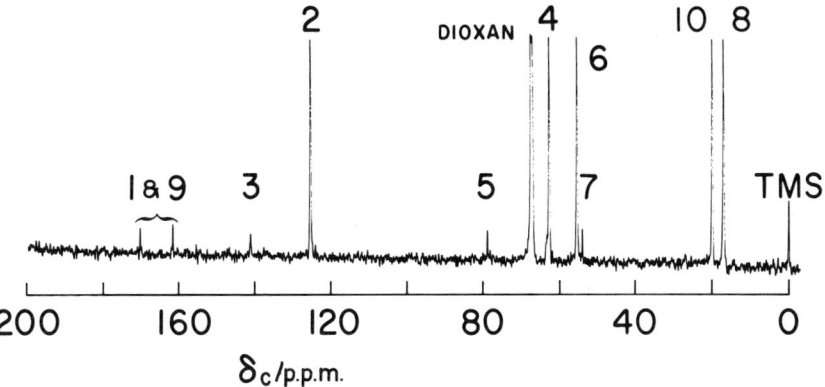

Figure 4 ^{13}C *N.m.r. spectrum* (25.15 MHz) *of asperlin from* $^{13}CH_3CO_2Na$

position. The magnitude of these single-frequency residual couplings was then related to the proton chemical shifts of asperlin by the relationship[32]

$$J_r = \frac{\Delta f J}{\gamma H_2/2\pi}$$

where J_r is the residual coupling, Δf is the separation (in hertz) of the proton chemical shift and the applied decoupling frequency, J is the ^{13}C–H coupling constant, and $\gamma H_2/2\pi$ is the applied decoupling field strength. This relationship between proton chemical shifts and carbon chemical shifts can be of considerable value in the correct assignment of carbon signals in the ^{13}C n.m.r. spectrum.

In asperlin, the pairs of carbons at C-4 and C-5 and at C-6 and C-7 are in similar chemical environments. All are deshielded by electronegative oxygen and are attached to a proton, so they all appear as doublets in the off-resonance-decoupled spectrum. Since an unambiguous assignment of the ^{1}H n.m.r.

Table 7 *Residual couplings of asperlin*[a]

Carbon	J_r(obs)	δ_H/p.p.m.	Δf	J	J_r (calc)
C-4	15	5.22	215	152	15.5
C-5	24	3.90	347	145	24
C-6	32	3.05	432	165	31.9
C-7	36	2.93	444	—	—
C-8	13.5[b]	1.31[b]	239	126	14.3
C-10	10.0[b]	2.10[b]	160	130	9.9

[a] $\gamma H_2/2\pi$ set at 4200 Hz, all values given in Hz.
[b] Measured in dioxan ($\delta = 3.70$).

[32] R. R. Ernst, *J. Chem. Phys.*, 1966, **45**, 3845.

signals at C-4, C-5, C-6, and C-7 can be made, the magnitude of the J_r values therefore defines the assignment of their carbon shifts in the off-resonance ^{13}C n.m.r. spectrum.

The J_r values of asperlin were determined in benzene solution, which was used to obtain a larger offset frequency (Δf). The applied decoupling frequency

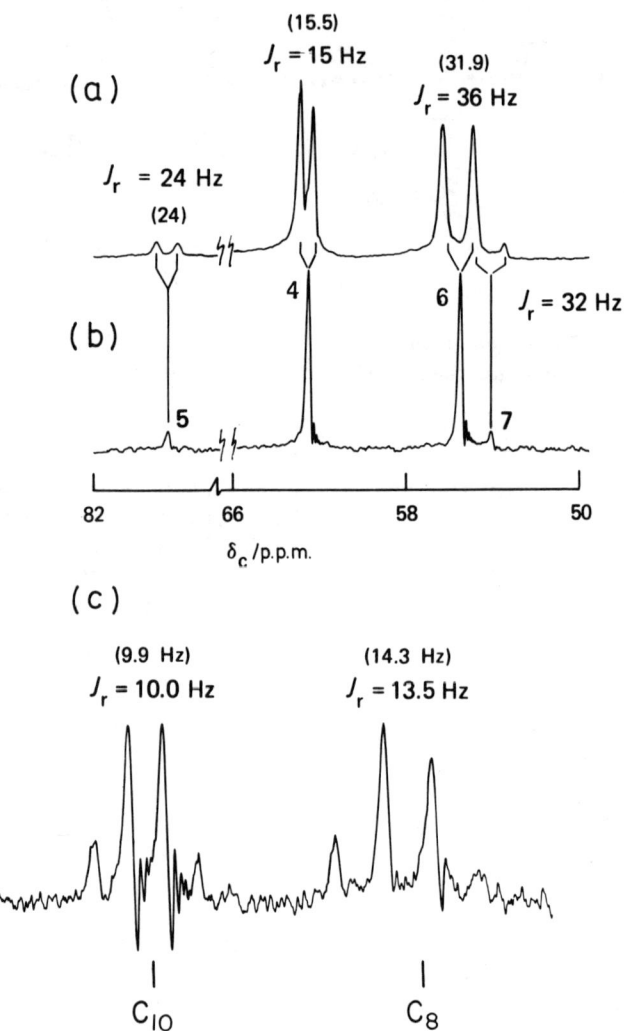

Figure 5 *Portions of the ^{13}C n.m.r. spectra of asperlin from ^{13}CH$_3$CO$_2$Na: (a) continuous-wave decoupled spectrum (in benzene); (b) proton-noise-decoupled spectrum (in benzene); (c) continuous-wave decoupled spectrum (in dioxan)*

was set to decouple benzene protons ($\delta = 7.37$). These J_r values are compiled in Table 7 and are in good agreement with calculated values. The residual couplings in the off-resonance decoupled spectrum of asperlin in dioxan solution ($\delta = 3.70$) helped to determine the carbon assignments of the two methyl groups C-8 and C-10. Both sets of decoupled spectra are shown in Figure 5.

Because of the availability of ^{13}C n.m.r. spectrometers with heteronuclear lock capabilities, the applied decoupling frequency is no longer restricted to decouple only at the solvent proton frequency in off-resonance decoupling experiments. The improved instrumentation should simplify the determination of residual coupling constants.

Table 8 ^{13}C *Enrichment levels of labelled asperlin by* $^{13}C–H$ *satellite area*

Carbon	$J_{^{13}C-H}$/Hz	Enrichment/%
C-8-H$_3$	126	11.5
C-10-H$_3$	130	10.9
C-4-H	152	10.0
C-2-H	174	10.8
C-6-H	165	—^{a}

^{a} Complexity of satellite band prevented estimation of this value.

The ^{13}C-enrichment levels of the labelled asperlin were determined from the $^{13}C–H$ satellite areas and are summarized in Table 8.

Cephalosporin C (11). The biogenesis of this biologically prominent antibiotic was examined by ^{13}C n.m.r. spectroscopy.[33] Administration of sodium [1-^{13}C]acetate and sodium [2-^{13}C]acetate to submerged cultures of *Cephalosporium acremonium* yielded the ^{13}C-labelled antibiotics, which were purified

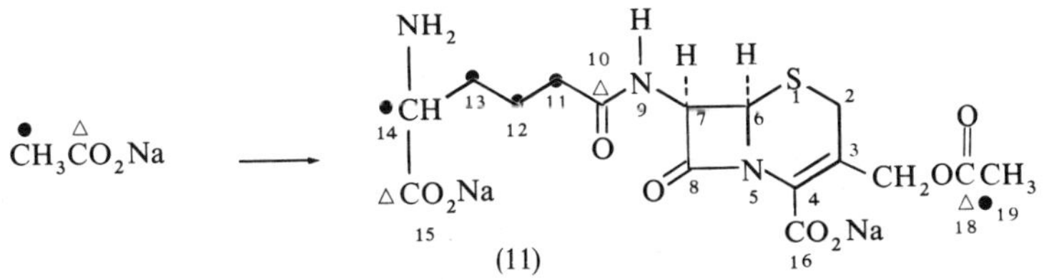

and crystallized as the sodium salts (11). The swept 25.2 MHz ^{13}C n.m.r. spectra of a dioxan solution of the labelled salt showed 16 clearly resolved carbon signals (Table 9).

[33] N. Neuss, C. H. Nash, P. A. Lemke, and J. B. Grutzner, *J. Amer. Chem. Soc.*, 1971, **93**, 2337.

Table 9 ^{13}C N.m.r. data for cephalosporin C

		Relative intensities		
Position	δ_C/p.p.m.[a]	Natural abundance	$^{13}CH_3CO_2Na$	$CH_3{}^{13}CO_2Na$
C-2	25.0	1.0	1.0	1.0
C-3	134.3	1.0	1.2	1.0
C-4	118.3	1.2	1.2	1.2
C-6	57.6	1.0	0.8	0.8
C-7	59.6	1.0	1.0	1.0
C-8	168.5	1.0	1.0	1.4
C-10	180.8	1.0	1.2	2.2
C-11	34.7	1.0	2.0	1.0
C-12	20.2	1.0	1.8	1.2
C-13	29.7	1.0	1.8	1.2
C-14	55.6	1.0	4.6	1.0
C-15	178.7	1.0	1.0	5.0
C-16	172.7	0.8	0.8	1.0
C-17	65.0	1.0	1.0	0.8
C-18	178.2	1.0	1.0	3.6
C-19	19.6	0.8	3.6	0.8

[a] Values recalculated from literature values by using $\delta_C(TMS) = 192.8 - \delta_C(CS_2)$.

Since [1-^{13}C]acetate was only incorporated into the carbonyl groups at C-10, C-15, and C-18, the C-8 β-lactam carbonyl and the C-16 carboxy-group of the cephem moiety of cephalosporin were not derived from acetate. The ^{13}C n.m.r. spectrum of the [2-^{13}C]acetate-labelled material showed incorporation into C-11, C-12, C-13, and C-14 of the α-aminoadipic acid fragment of the antibiotic as well as the expected enrichment at C-19 of the acetate group. There were no enriched carbons in the cephem nucleus. The label in the α-aminoadipate at C-11, C-12, and C-13 had about one-half the enrichment of that in the C-14 methine group. This amount corresponds to the label dispersion expected from acetate cycling through the Krebs cycle during α-aminoadipic acid synthesis.[34] The relative signal intensities of these carbons were obtained from the ^{13}C n.m.r. spectrum of unlabelled cephalosporin C, which was recorded under the same instrumental conditions. This technique offers a means of overcoming different nuclear Overhauser enhancements of carbons when signal intensity comparisons are made.

The thoroughness of this study is reflected in the complete assignment of all 16 carbon resonances by off-resonance decoupling and single-frequency proton decoupling experiments and by the use of model compounds. The pH dependence of the chemical shift of the carboxy-group C-16 aided in its differentiation from the β-lactam C-8 carbonyl group.[35]

Prodigiosin (12). The first reported biosynthetic application of Fourier-transform (FT) ^{13}C n.m.r. techniques was to the study of prodigiosin (12).[36] Earlier

[34] H. R. Mahler and E. H. Cordes, 'Biological Chemistry,' Harper and Row, New York, 1966, p. 525.
[35] A. Allerhand, D. W. Cochran, and D. Doddrell, *Proc. Nat. Acad. Sci. U.S.A.*, 1970, **67**, 1093.
[36] R. J. Cushley, D. R. Anderson, S. R. Lipsky, R. J. Sykes, and H. A. Wasserman, *J. Amer. Chem. Soc.*, 1971, **93**, 6285.

Stable Isotopes in Biosynthetic Studies

261

[Structure of prodigiosin (12) with labelled positions, derived from CH$_3$CO$_2$Na]

^{14}C-biosynthetic incorporation work established that acetate, glycine, proline, and methionine were prime precursors of prodigiosin. However, the structural complexity, sensitivity, and lability of this molecule made the isolation of fragments difficult. Studies on the mechanism of this biosynthesis were thus ideally suited to and amenable to the ^{13}C method. The mechanism of the pyrrole ring formation in prodigiosin biosynthesis is also unrelated to porphyrin biosynthesis and is thus of special interest.

Labelled prodigiosins were prepared by feeding 1-^{13}C- and 2-^{13}C-enriched sodium acetate (63%) to the bacterium *Serratia marcescens*. The enhanced sensitivity available by the ^{13}C FT mode of operation is underscored in this study since the ^{13}C n.m.r. spectra were recorded on 360 mg ml^{-1} of natural-abundance prodigiosin in 18 minutes, on 180 mg ml^{-1} of the material from CH$_3$^{13}CO$_2$Na labelling in 20 minutes, and on 17.9 mg ml^{-1} of ^{13}CH$_3$CO$_2$Na-labelled pigment in 6.7 hours.

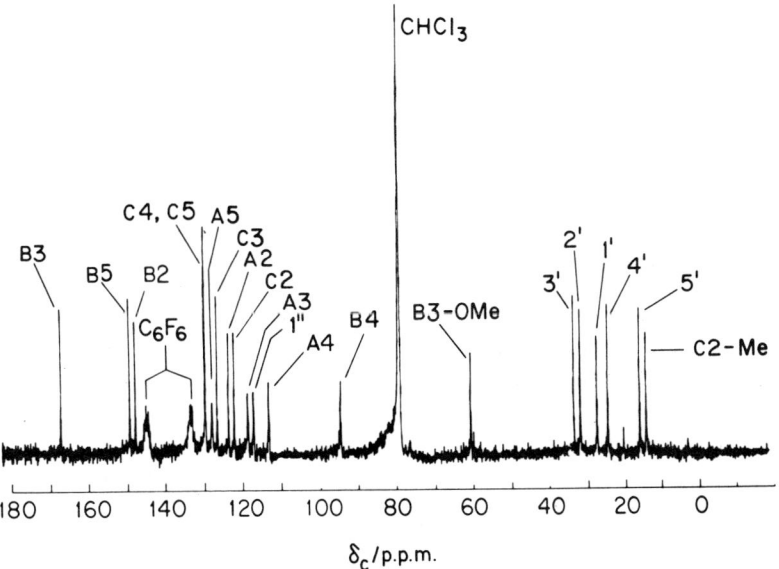

Figure 6 *Proton-noise-decoupled ^{13}C n.m.r. spectrum of prodigiosin hydrochloride* (Reproduced by permission of R. J. Cushley).

The ^{13}C n.m.r. spectrum (Figure 6) of the 20-carbon natural-abundance material showed 19 resolved resonances and a pair of overlapping signals. A very notable accomplishment was the complete assignment of all of these carbon resonances by applying most of the available techniques for carbon shift assignments. These shift values relative to TMS are presented in Table 10. In the original paper, values were reported relative to chloroform as the standard.

Table 10 ^{13}C N.m.r. chemical shifts of prodigiosin hydrochloride in $CHCl_3$

Position of ^{13}C	Chemical shift δ_C/p.p.m.[a]		
	Natural abundance	$CH_3{}^{13}CO_2Na$	$^{13}CH_3CO_2Na$
A2	121.92	—	—
A3	116.95	—	—
A4	111.45	—	—
A5	126.23	—	—
B2	145.98	—	—
B3	165 50	165.60	—
B4	92.73	—	92.84
B5	147.42	—	—
B3-Methoxy	58.48	—	—
Bridgehead (1″)	115.44	—	—
C2	120.51	—	—
C3	124.83	124.98	—
C4 C5	127.96	128.11	128.33
C2-methyl	11.99	—	—
1′	24.98	—	25.40
2′	29.41	29.52	—
3′	31.19	—	31.35
4′	22.18	22.23	—
5′	13.71	—	13.88

[a] Relative to ^{13}C resonance of TMS, recalculated from literature values by using δ_C(TMS) = δ_C(CHCl$_3$) + 77.02 p.p.m.

Single-frequency decoupling was employed to assign the methine carbon shifts at A5, A4, A3, B4, C4, and 1″. The known $J_{^{13}C-H}$ satellite coupling constants obtained from the ^{1}H n.m.r. spectrum were used in the assignment. The signals from methyl groups attached to B3 (OCH$_3$) and C2 (CH$_3$) and from 5′ were assigned in the same way.

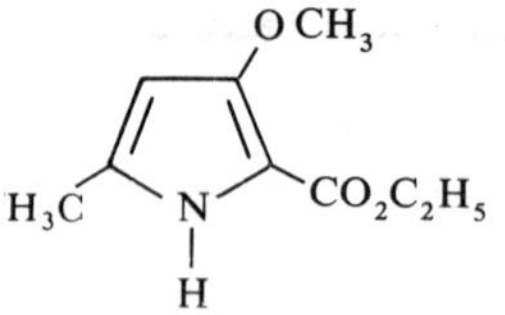

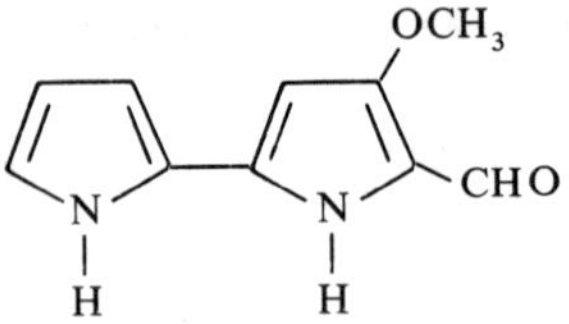

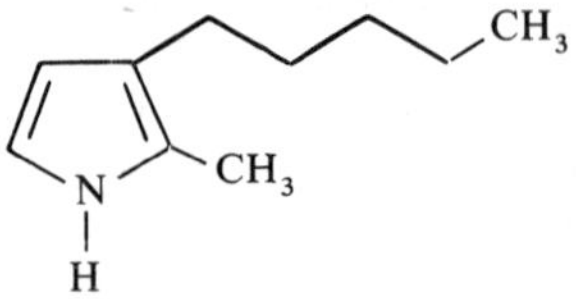

The assignment of the seven quaternary carbon signals was aided by comparison with the shift values of model compounds (13), (14), and (15). Further confirmation of the quaternary A2 and B2 shift positions was secured with an A3,A5,B4,C4-tetradeuterio-derivative, in which the B2 carbon remained coupled to the 1″ portion in the deuteriated material. Progressive saturation experiments also confirmed the presence of six quaternary carbons, at A2, B2, B3, C2, C3, and C5.

The methylene carbon resonances were assigned by strong CW irradiation in the methylene ^{1}H n.m.r. absorption region of the n-amyl side-chain. This resulted in residual coupling (J_r) values consistent with the ^{1}H n.m.r. shifts of the side-chain, which have been unequivocally assigned. Comparison with known carbon shift values of n-alkanes also aided the assignments.

An interesting result that emerged from the carbon shift assignments of prodigiosin is that the alkyl groups directly attached to the pyrrole ring at C2 and C3 show a marked upfield shift relative to the shift of other saturated carbons in the n-amyl side-chain.

Each of the labelled prodigiosin hydrochlorides showed ^{13}C incorporation at five positions, as shown in Table 10. This distribution is consistent with the biosynthetic origin of the amylpyrrole moiety of ring C from an eight-carbon polyketide. The acetate origin of B3 and B4 was also established from these ^{13}C feeding experiments. The notable absence of acetate incorporation into ring A indicates that there is a novel but undisclosed route to this pyrrole ring.

The enhanced carbon signal intensities at enriched carbons were clearly differentiated from nuclear Overhauser enhanced signals, and the level of ^{13}C content was estimated to be $\sim 8\%$ as a result of $CH_3{}^{13}CO_2Na$ labelling and $\sim 11\%$ from using $^{13}CH_3CO_2Na$. Signal intensity differences are particularly accentuated in the FT mode because of the rapid pulse repetition rate (~ 0.1 to 1 second), and saturation effects become more apparent.[37] The signals from carbons with long relaxation times are then less intense than those with the shorter relaxation times. The variable signal-intensity differences noted in the natural-abundance prodigiosin FT spectrum are illustrative of this effect.

Sterigmatocystin (16). The biosynthetic origin of all the carbons of the fungal metabolite sterigmatocystin (16) was determined by feeding experiments conducted with sodium [1-^{13}C]acetate (56%) and sodium [2-^{13}C]acetate

(16)

[37] Ref. 17(b), p. 30.

(61%).[38] Location and identification of the ^{13}C-labelled sites of the metabolite in dioxan were accomplished by swept spectra at 25.15 MHz with simultaneous proton noise decoupling.

The spectrum (Figure 7a) for the sterigmatocystin sample derived from [1-^{13}C]acetate shows nine carbon resonances of enhanced intensity between 100 and 182 p.p.m. downfield from TMS. Similarly, the spectrum shown in

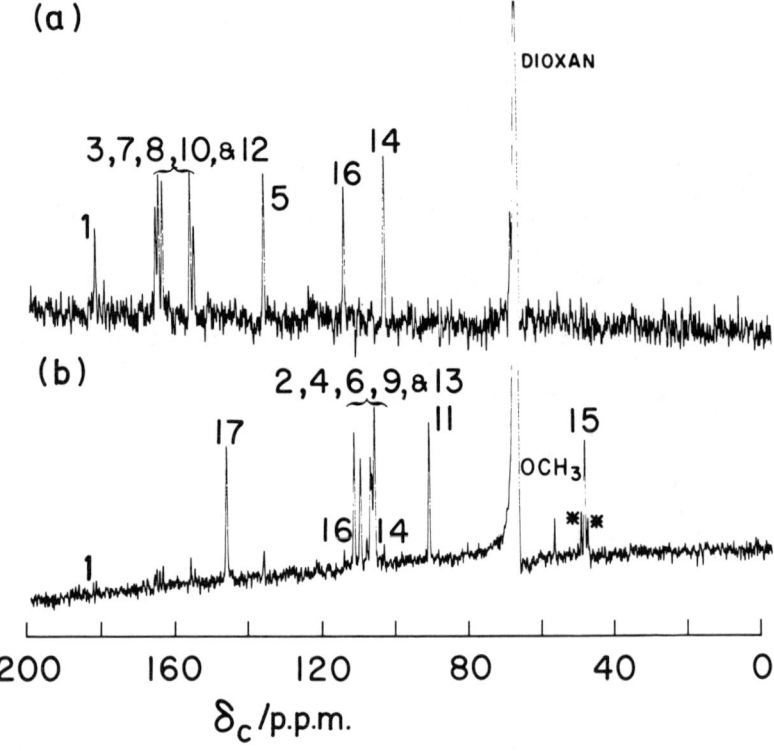

Figure 7 ^{13}C N.m.r. spectra (25.15 MHz) of sterigmatocystin in dioxan: (a) from CH$_3$^{13}CO$_2$Na; (b) from ^{13}CH$_3$CO$_2$Na. * Indicates C–C coupling between C-9 and C-15.

Figure 7b shows signals of enhanced intensity for eight carbons of sterigmatocystin derived from [2-^{13}C]acetate. The assignment of resonances in Figures 7a and 7b was aided by comparison with the known shift values of model compounds. The data are given in Table 11. The cluster of aromatic resonances between 154 and 164 p.p.m., which are from the C-3, C-7, C-8, C-10, and C-12 aromatic carbons that are directly attached to oxygen, was not individually

[38] M. Tanabe, T. Hamasaki, H. Seto, and L. Johnson, *Chem. Comm.*, 1970, 1539.

Table 11 ^{13}C N.m.r. data for sterigmatocystin

Sterigmatocystin carbon	Source[a]	Shift, δ_C/p.p.m.[b]	Reference compound	Carbon	Shift, δ_C/p.p.m.[c]
C-1	A	181	Aromatic ketones	Carbonyl	190—200
C-5	A	135	1,3-Dimethoxybenzene	C-5	131
C-3, C-7, C-8, C-10, and C-12	A	154—164	1,3-Dimethoxybenzene	C-1 and C-3	162
C-2, C-4, C-6, C-9, and C-13	B	104—112	1,3-Dimethoxybenzene	C-4 and C-6	107
C-11	B	91	1,3-Dimethoxybenzene	C-2	102
C-14	A	103	Glucose, carbohydrates	C-1	94—104
C-15	B	49	Isopropylbenzene	α	33
C-16	A	115	Furan	C-3 and C-4	110
C-17	B	146	Furan	C-2 and C-5	144

[a] A: from sodium [1-^{13}C]acetate [spectrum 7(a)]; B: from sodium [2-^{13}C]acetate [spectrum 7(b)].
[b] δ_C values are downfield from dissolved TMS, calculated from the lock signal dioxan using the relationship δ_C(TMS) = δ_C(dioxan) + 67 p.p.m.
[c] δ_C values are p.p.m. downfield from TMS calculated from reference data.

assigned, nor was differentiation made for the resonances of C-2, C-4, C-6, C-9, and C-13, which are aromatic carbons situated *ortho* or *para* to oxygen functions.

The labelling pattern shown in Scheme 7 supports the novel biogenetic hypothesis that sterigmatocystin, as an early precursor of the aflatoxins,

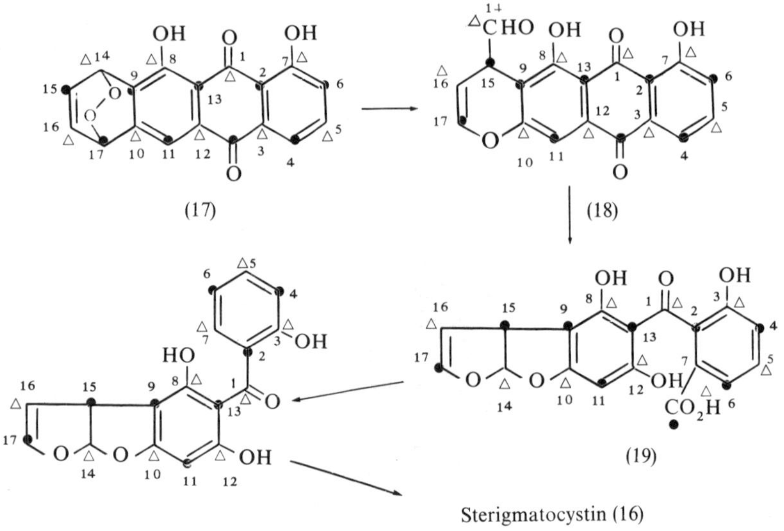

Scheme 7

arises from a naphthacene *endo*-peroxide (17) that originates by cyclization of an 18-carbon polyketide.[39] Rearrangement of (17) generates the elements of the difuran ring in (18) with the adjacent carbons C-9 and C-15 derived from the methyl of acetate. The biological equivalent of a Baeyer–Villiger cleavage at the quinone carbonyl derived from the methyl of acetate yields (19). Compound (19), in turn, undergoes decarboxylation, which accounts for the loss of the acetate-methyl label before cyclization to the xanthone, sterigmatocystin.

The observation of carbon–carbon coupling between C-9 and C-15 is firm evidence for the adjacent labelling of these sites, which was proposed in the biosynthetic scheme. Since it appeared difficult to resolve the downfield C-9 and C-15 splitting in the aromatic carbon resonance region at 25.15 MHz, the FT ^{13}C n.m.r. spectrum of a [^{2}H$_5$]pyridine solution of sterigmatocystin derived from ^{13}CH$_3$CO$_2$Na was recorded at 55.33 MHz in a superconducting solenoid at 53.7 kg.[40]

[39] M. Biollaz, G. Büchi, and G. Milne, *J. Amer. Chem. Soc.*, 1970, **92**, 1035.
[40] M. Tanabe and L. F. Johnson, unpublished results.

This FT spectrum, Figure 8, shows the downfield C-9 signal well resolved, with $J_{^{13}C-9 - ^{13}C-15}$ of 48 Hz. This value was also observed for the upfield splitting. Another benefit derived from the FT spectrum is the ability to distinguish between the aromatic carbon resonances at C-4 and C-6, which have attached protons and higher signal intensities, and the quaternary C-2 and C-13 resonances of lower intensity. The differences are due to differences in nuclear Overhauser effects and T_1 relaxation times.

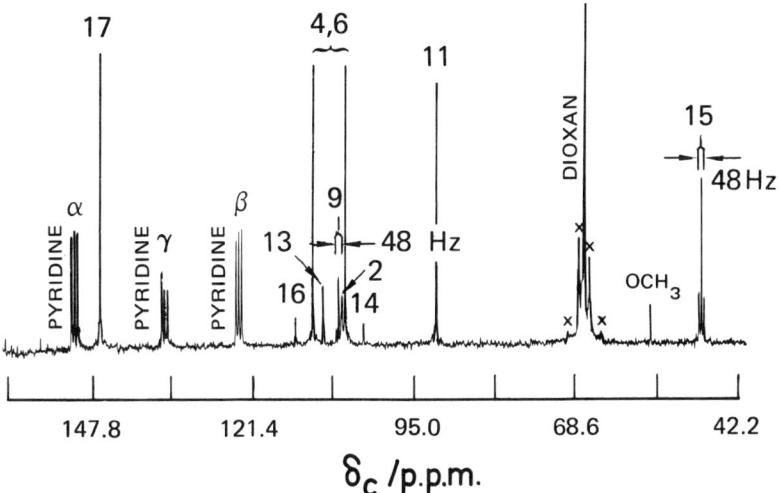

Figure 8 *FT* ^{13}C *n.m.r. spectrum* (55.33 MHz) *of sterigmatocystin prepared from* $^{13}CH_3CO_2Na$

The unexpectedly high proportion of ^{13}C at the adjacent carbon atoms C-9 and C-15 indicates that under the experimental conditions for production of sterigmatocystin, normal isotopic acetate has been nearly depleted from the carbon pool and biosynthesis of the metabolite proceeds in the presence of enriched sodium [2-^{13}C]acetate.

In the spectrum shown in Figure 7b, most of the carbons at natural abundance are visible. Comparison of peak heights shows a ^{13}C-enrichment of about 5%, assuming a constant nuclear Overhauser enhancement factor.

Avenaciolide (20). Several schemes for the biosynthetic origin of the bislactone avenaciolide (20), a metabolite of *Aspergillus avenaceus*, have been proposed.[41] Evidence from ^{13}C-labelling studies conclusively established that avenaciolide is biosynthesized by the pathway outlined in Scheme 8.[42]

---

[41] C. Mentzer, in, 'Comparative Phytochemistry,' ed. T. Swann, Academic Press, New York, 1966, p. 26; W. B. Turner, 'Fungal Metabolites,' Academic Press, New York, 1971, p. 292.
[42] M. Tanabe, T. Hamasaki, Y. Suzuki, and L. F. Johnson, *J.C.S. Chem. Comm.*, 1973, 212.

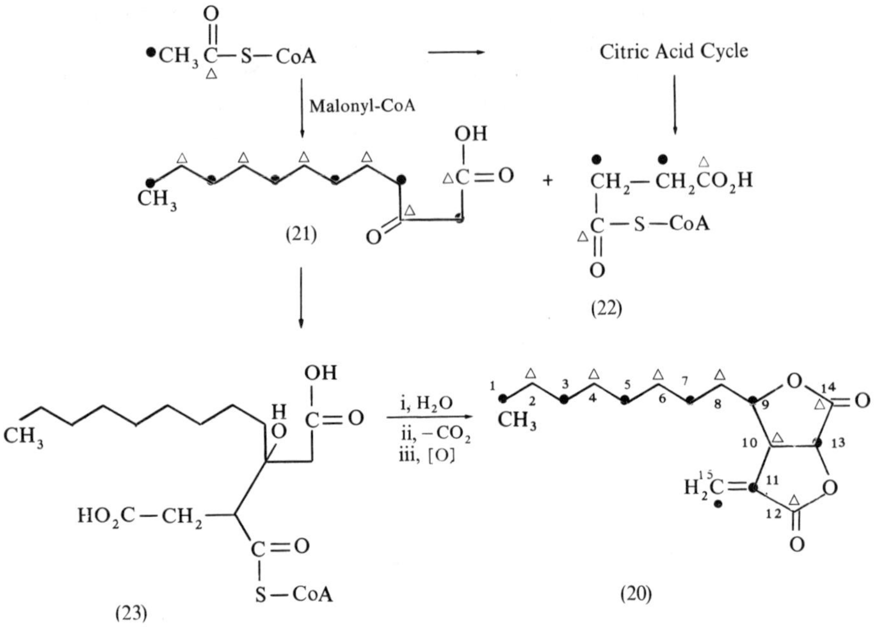

Scheme 8

^{13}C-Enriched avenaciolides were prepared from the producing organism in media fortified with sodium [1-^{13}C]acetate (90%) and sodium [2-^{13}C]acetate (60%). The FT ^{13}C n.m.r. spectra of the isolated enriched bislactones are shown in Figure 9.

From sodium [2-^{13}C]acetate-labelled material (Figure 9a), it is evident that in the upfield region of the spectrum, the alternating carbons (C-1, C-3, C-5, and C-7) of the n-octyl side-chain are enriched sites, and their chemical-shift assignments are in accord with the reported shift values for octan-2-ol. The two oxygenated carbons at C-9 and C-13 are also enriched sites and in the expected shift region. The two unsaturated carbons C-11 and C-15 were also labelled and had signals at 134.9 p.p.m. and 126.2 p.p.m., with a carbon–carbon coupling ($J_{^{13}C-11-^{13}C-15}$) value of 75 Hz, which is in agreement with a reported coupling constant of 67.6 Hz for ethylene. In the absence of off-resonance-decoupling data, the lower-field and lower-intensity 134.9 p.p.m. signal was assigned to the quaternary carbon C-11, since its reduced intensity relative to the C-15 signal at 126.2 p.p.m. is in accord with a diminished nuclear Overhauser effect and increased T_1 relaxation time, which are caused by the absence of directly bonded protons. The succinic acid origin of C-15, C-11, and C-12 is nicely confirmed by the carbon–carbon coupling observed for C-11 and C-15,

Stable Isotopes in Biosynthetic Studies

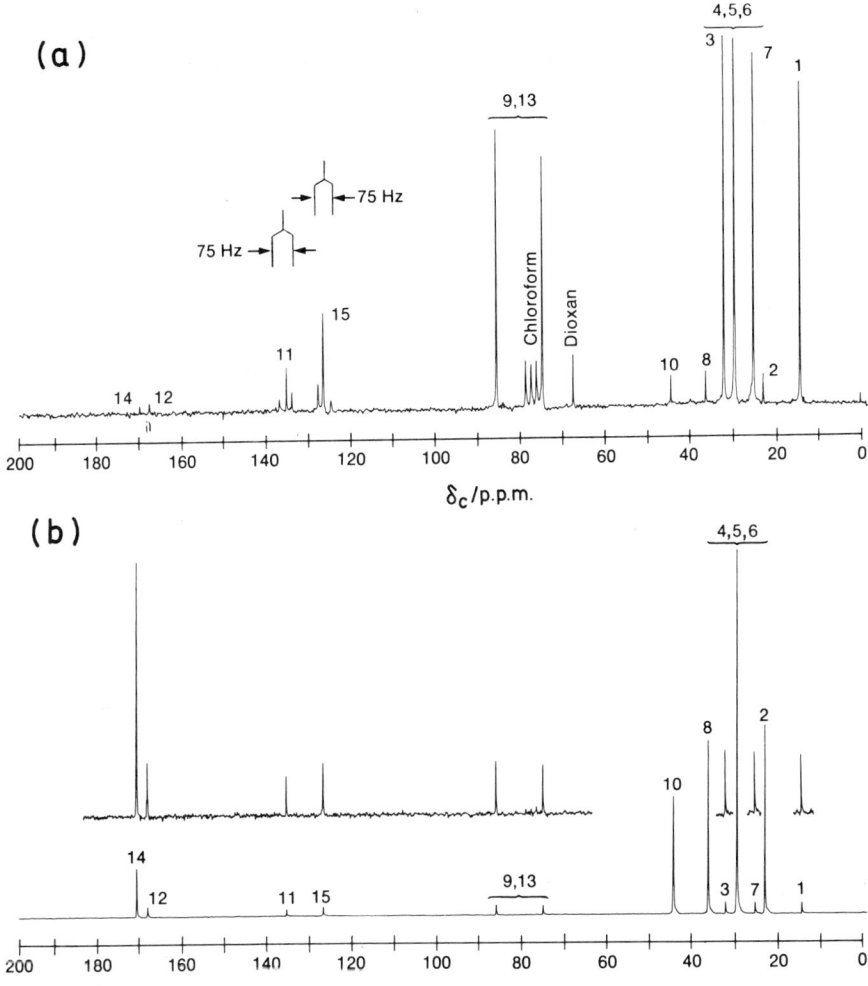

Figure 9 *FT ^{13}C n.m.r. spectra of avenaciolide*: (a) *from* $^{13}CH_3CO_2Na$; (b) *from* $CH_3{}^{13}CO_2Na$.

which would be expected if 2-^{13}C-labelled acetate was incorporated into this acid in the citric acid cycle.[43]

In the avenaciolide from sodium [1-^{13}C]acetate (Figure 9b), the alternate carbons (C-2, C-4, C-6, C-8, C-10, and C-14) are labelled. The shift data are

[43] Ref. 34, p. 620.

summarized in Table 12. These labelling studies are therefore in agreement with the biosynthetic pathway (Scheme 8) for avenaciolide that has 3-keto-dodecanoic acid (21) as an intermediate, which arises from the acetate–malonate pathway. The keto-acid condenses with succinyl-CoA (22), which is generated in the citric acid cycle to give a condensation product (23) that is then transformed into avenaciolide. The higher level of enrichment observed for the moiety derived from 3-ketododecanoic acid in avenaciolide indicates that acetate is converted more rapidly into this intermediate than into succinic acid. The lower intensity of C-12 compared with C-14 in Figure 9b is further evidence for the biosynthetic origin of C-12 from succinic acid that has been diluted in the citric acid cycle.

Table 12 ^{13}C N.m.r. data for avenaciolide, δ_C/p.p.m.^{a}

Carbon No.	Precursor	
	$CH_3{}^{13}CO_2Na$	$^{13}CH_3CO_2Na$
C-1	—	14.0
C-2	22.6	—
C-7	—	24.9
C-4	29.1	—
C-5	—	29.1
C-6	29.1	—
C-3	—	31.8
C-8	35.8	—
C-10	44.0	—
C-9	—	74.6^b
C-13	—	85.6^b
C-15^d	—	126.2^c
C-11^d	—	134.9^c
C-12^d	167.9	—
C-14	170.3	—

^{a} Downfield from TMS.
^{b} These values can be reversed.
^{c} Carbon–carbon coupling, $J = 75$ Hz.
^{d} These positions are labelled by both [1-^{13}C]- and [2-^{13}C]-acetate since succinic acid cycles through the citric acid cycle with dispersion of the label.

Ochratoxin A (24). The main features of the biosynthesis of this hepatic mycotoxin were established by feeding experiments conducted with L-[^{14}C]phenyl-alanine and sodium [2-^{14}C]acetate. Incorporation of these precursors into ochratoxin A (24) is in accord with the biosynthetic Scheme 9.

The biosynthetic origin of the one-carbon unit in ochratoxin A was examined by administration of sodium [^{13}C]formate (65%) to the culture medium.[44] The ^{13}C-labelled ochratoxin was hydrolysed with hydrochloric acid and esterified with diazomethane to yield the methyl methoxyisocoumarin-carboxylate (25). The 15 MHz swept CW ^{13}C n.m.r. spectrum of this methyl

[44] Y. Maebayashi, K. Miyaki, and M. Yamazaki, *Chem. and Pharm. Bull. (Japan)*, 1972, **20**, 2172.

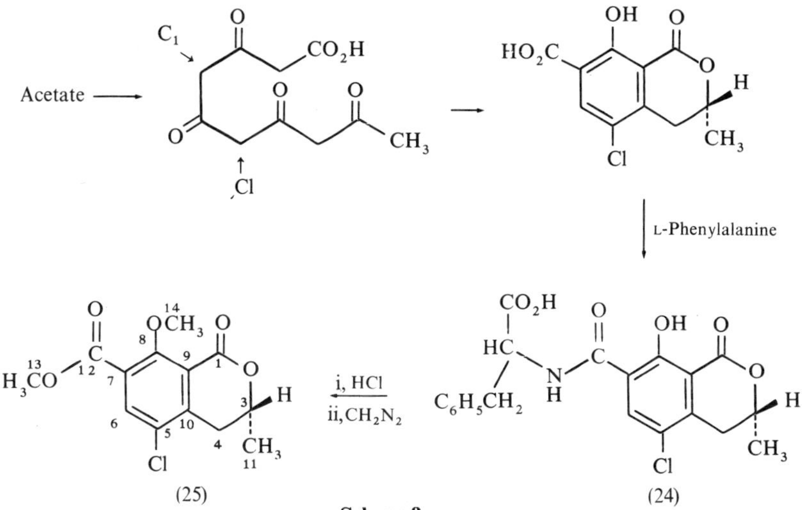

Stable Isotopes in Biosynthetic Studies 271

Scheme 9

ester in deuteriochloroform solution fully confirms the formate origin of the C-12 carboxy-group of ochratoxin A, since of the two low-field carbonyl signals at 161.24 and 165.15 p.p.m. found in natural-abundance material, only the signal at 165.15 p.p.m. was greatly enhanced in the ^{13}C-enriched ester.

An off-resonance-decoupled spectrum of natural-abundance ochratoxin A aided in the shift assignment of some of the resonances of (25). These are compiled in Table 13.

Table 13 ^{13}C N.m.r. data for methyl methoxyisocoumarincarboxylate in CDCl$_3$

Carbon No.	Multiplicity[a]	δ_C/p.p.m.[b]	
		Natural abundance	H^{13}CO$_2$Na
C-1	s	161.24	—
C-3	d	74.01	—
C-4	t	34.14	—
C-5			
C-7			
C-8	—	121.72—143.68	—
C-9			
C-10			
C-6	d	~135	—
C-11	q	20.85	—
C-12	s	165.15	165.15
C-13	q	53.00[c]	—
C-14	q	64.61[c]	—

[a] Singlet (s), doublet (d), triplet (t), quartet (q).
[b] Relative to ^{13}C resonance of TMS calculated using δ_C(TMS) = 192.8 − δ_C(CS$_2$).
[c] Values within any vertical column may be transposed.

Aureothin (26). The familiar problem of the biological origin of 'extra' methyl groups appended to aureothin (26) was investigated by using $^{13}CH_3CH_2CO_2Na$ as a label.[45]

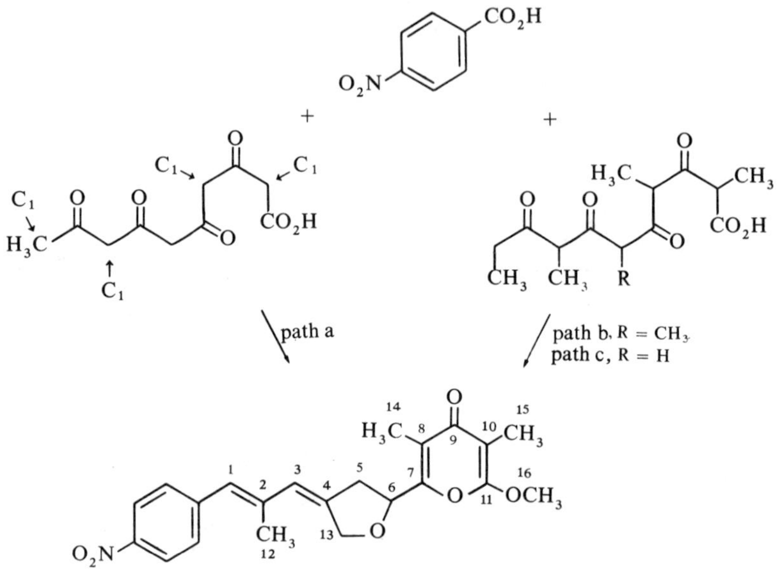

Scheme 10

Two plausible biosynthetic schemes can be considered. In pathway a (Scheme 10), methionine as the one-carbon unit can be introduced into a linear polyketide derived from acetate and *p*-nitrobenzoate. In pathway b (Scheme 10), a linear polyketide formed from propionate and *p*-nitrobenzoate can cyclize to yield aureothin. Preliminary studies showing the incorporation of ^{14}C-labelled propionate and the very minimal incorporation of [14C]-methionine indicated the operation of pathway b or c (Scheme 10). Conclusive evidence for these routes was established with the 3-^{13}C-labelled propionate. The FT ^{13}C n.m.r. spectrum of the isolated ^{13}C-labelled metabolite in CDCl$_3$ solution showed strongly enhanced signals for the four carbons derived from propionate at C-12, C-13, C-14, and C-15. The natural-abundance FT ^{13}C n.m.r. spectrum of aureothin shows the up-field aliphatic carbon resonances of C-5, C-6, C-12, C-13, C-14, C-15, and C-16 to be well resolved; off-resonance decoupling was employed for these carbon shift assignments.

Since [^{14}C]acetate was also incorporated into aureothin, pathway c (Scheme 10), incorporating an acetate unit into the polyketide, is the more probable route.

[45] M. Yamazaki, F. Katoh, J. Ohishi, and Y. Koyama, *Tetrahedron Letters*, 1972, 2701.

Stable Isotopes in Biosynthetic Studies

Antibiotic X-537A (27). The biosynthetic origin of the three unique C-ethyl groups of this antibiotic (27) is of interest, and this problem was investigated with the aid of ^{13}C-labelled substrates.

[Structure (27)]

Prior ^{14}C-incorporation studies suggested that the carbon skeleton of the antibiotic was derived from a linear polyketide composed of seven acetates, four propionates, and two butyrates.[46] Reversible intraconversion of acetate and butyrate by the organism also resulted in some randomization of these labels. The four C-methyl groups at C-4, C-10, C-12, and C-16 are derived from propionate, and the C-23-methyl group comes from the starter acetate unit of the polyketide. The ^{14}C study was unable to determine if one, two, or three of the C-ethyl groups were derived from butyrate.

The use of sodium [1-^{13}C]butyrate in X-537A fermentations and FT ^{13}C n.m.r. spectroscopy on the labelled antibiotic revealed the butyrate origin of the three ethyl groups. Additional experiments also confirmed the origin of the C-methyl groups in the antibiotic from [1-^{13}C]propionate.[47]

The proton-noise-decoupled, natural-abundance, FT ^{13}C n.m.r. spectrum at 22.63 MHz for the C_{34} compound in chloroform showed 33 resonances. Twenty of the signals were assigned either by comparsion with model compounds and degradation products or by biosynthetic considerations. The FT ^{13}C n.m.r. spectrum of the [1-^{13}C]butyrate-labelled antibiotic indicated increased signal intensities for C-13, C-17, and C-21. Because of the known difficulties in peak height or area comparisons, the criterion chosen for signal enhancements that were caused by ^{13}C-enrichments was at least a doubling in the height of the natural-abundance peaks. Similar intensity considerations for the [1-^{13}C]propionate-derived antibiotic in the FT ^{13}C n.m.r. spectrum showed enriched carbons at C-3, C-9, C-11, and C-15.

Slight randomization of the ^{13}C label from the [1-^{13}C]butyrate-derived antibiotic was observed by apparent enrichments at C-5 and C-23, which are formally carbons derived from C-1 of acetate. This indicates that the microorganism is capable of conducting the β-oxidation of [1-^{13}C]butyrate to [1-^{13}C]acetate, and this agrees with the ^{14}C results.

The chemical-shift assignments and the enrichment levels of these carbons in X-537A are presented in Table 14.

---

[46] J. W. Westley, R. H. Evans, jun., D. L. Pruess, and A. Stempel, *Chem. Comm.*, 1970, 1467.
[47] J. W. Westley, D. L. Pruess, and R. G. Pitcher, *J.C.S. Chem. Comm.*, 1972, 161.

Table 14 ^{13}C N.m.r. data for antibiotic X-537 A^a

Position	Assignment^b		δ_C/p.p.m.	Abundance/%	
	In (27)	Function		CH$_3$CH$_2$CH$_2$^{13}CO$_2$Na	CH$_3$CH$_2$^{13}CO$_2$Na
C-13	C=O	ketone	218.4	4	1
C-1	C=O	CO$_2$H	177.0	1	1
C-3	aromatic	C—OH	166.0	1	4.5
C-7	aromatic	C—CH$_2$	143.4	1	1
C-5	aromatic	CH	131.3	1.5	1
C-4	aromatic	C—Me	123.2	1	1
C-6	aromatic	CH	119.8	1	1
C-2	aromatic	C—CO$_2$H	118.7	1	1
C-18	C—O	C—O	87.7	1	1
C-15	C—O	CH—O	83.4	1.5	4
C-23	C—O	CH(Me)—O	77.3	1.5	1
C-22	C—O	C(Et)—OH	71.3	1	1
C-11	C—O	CH—OH	70.7	1.5	4
C-19	C—O	CH—O	68.7	1	1
C-12	CH, CH$_2$	CH	56.0	1	1
C-8	CH, CH$_2$	CH$_2$	49.0	1	1
C-21	CH, CH$_2$	CH$_2$	38.5	5	1
C-9	CH, CH$_2$	CH$_2$	37.8	1	4
C-17	CH, CH$_2$	CH$_2$	29.5	4	1

^{a} Shifts are δ_C/p.p.m. downfield from TMS.
^{b} The atoms assigned are those that are underlined.

Virescenosides. The virescenosides are a group of diterpene glycosides formed by the mushroom *Oospora virescens* (Link) Wallr. Structural analysis indicates that they are isopimaridiene aglycones conjugated to altrose. The biosynthetic pathway to two of the aglycones, (28a) and (28b), has been clarified by FT ^{13}C n.m.r. spectroscopy.[48] This was facilitated by the known carbon chemical shifts of two model isopimaridienes, (29a) and (29b). These shift assignments were made by the application of carbon chemical-shift theory and previously recognized trends in shifts of alicyclic compounds. Use of the paramagnetic shift reagent Pr(dpm)$_3$ on a ring A alcohol derivative aided in the ring A assignments.

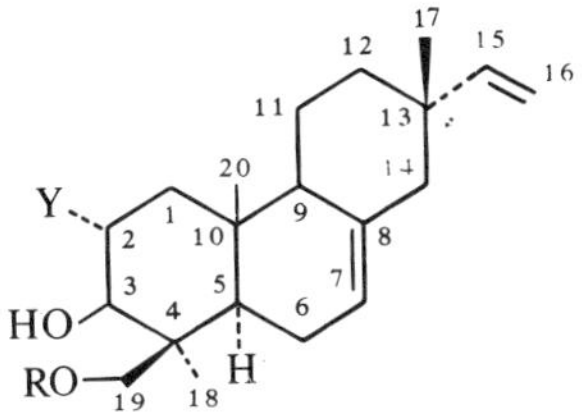

(28) a; R = H, Y = OH
b; R = β-D-altropyranosyl
Y = OH

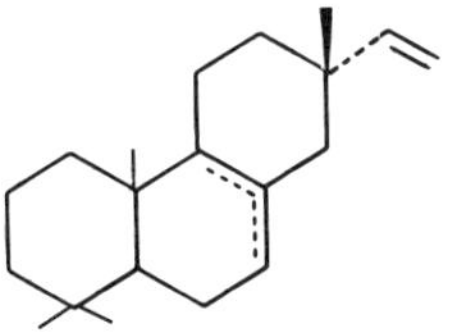

(29) a; Δ$^{7(8)}$
b; Δ$^{8(9)}$

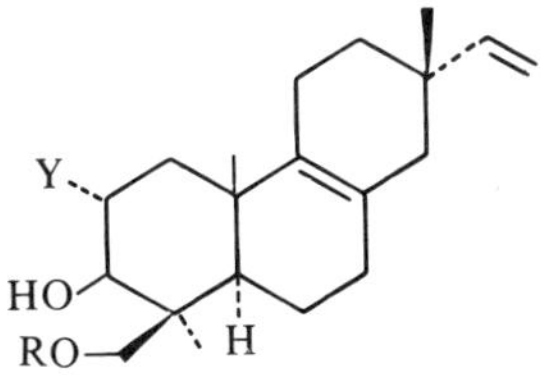

(30) R = H, Y = OH

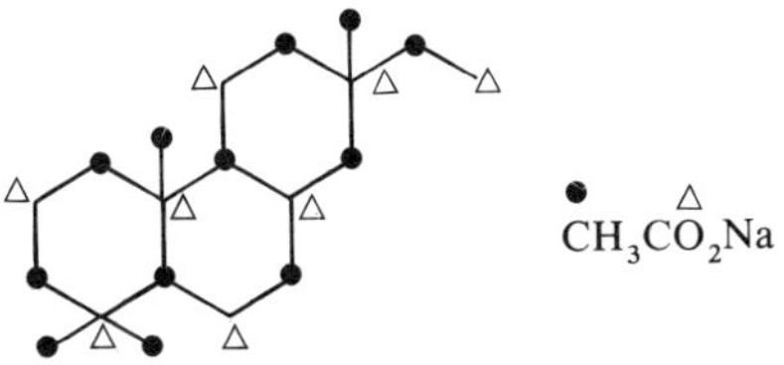

(31)

CH$_3$CO$_2$Na

Treatment of mushroom cultures with sodium [1-^{13}C]acetate and sodium [2-^{13}C]acetate yielded the two differently labelled glycosides (28b). Acidic hydrolysis of the glycoside resulted in partial double bond migration, and two alygcones, (28a) and (30), were isolated.

FT ^{13}C n.m.r. spectra at 15.077 MHz of the labelled materials yielded the predictable labelling pattern shown in (31). The shift assignments for (28a) and (30) are given in Table 15. Quantitative ^{13}C-incorporation data were promised in a later publication in conjunction with a parallel ^{14}C study.

Pyrrolnitrin (32). A novel application of FT ^{13}C n.m.r. spectroscopy involved a subtle mechanistic problem in the biosynthesis of the antifungal agent pyrrolnitrin (32) from *Pseudomonas aureofaciens*.[49] Earlier work had established that

[48] J. Polonsky, Z. Baskevitch, N. Cagnoli-Bellavita, P. Ceccherelli, B. L. Buckwalter, and E. Wenkert, *J. Amer. Chem. Soc.*, 1972, **94**, 4369.
[49] L. L. Martin, C.-J. Chang, H. G. Floss, J. A. Mabe, E. W. Hagaman, and E. Wenkert, *J. Amer. Chem. Soc.*, 1972, **94**, 8942.

Table 15 ^{13}C N.m.r. data for virescenosides, δ_C/p.p.m.^{a}

Position	(28a)	(30)
C-1	43.3	42.6
C-2	69.1	69.6
C-3	85.8	85.4
C-4	43.3	43.2
C-5	51.8^b	51.9
C-6	23.6	21.6
C-7	121.8	32.8
C-8	135.9	125.0
C-9	52.5^b	134.9
C-10	37.4^c	38.4
C-11	21.2	19.6
C-12	36.7	35.0
C-13	37.0^c	35.3
C-14	46.4	42.0
C-15	150.6	146.1
C-16	110.2	111.3
C-17	22.2	28.1
C-18	24.0	23.3
C-19	66.0	65.5
C-20	17.8	21.6

^{a} Spectra in chloroform at 15.077 MHz, chemical shifts downfield from TMS.
^{b,c} Values within any vertical column may be transposed.

D-tryptophan was a direct precursor. The use of various isotopically labelled D-tryptophans had demonstrated that the carbon C-2 of the indole ring of this precursor is converted into the carbon C-2 of the pyrrole ring in the product, as the amino-nitrogen is then oxidized to the nitro-group. Tritium from C-2 of the side-chain of L-tryptophan, but not D-tryptophan, is retained during the biosynthesis.[50]

These studies support the proposal that the enzyme chloroperoxidase initiates the biosynthesis by chlorinating the β-indole or C-4 carbon of tryptophan, followed by formation of a new pyrrole ring and loss of the chlorine atom (Scheme 11). This route (pathway a) includes the penultimate step of chlorination of both aromatic rings before the oxidative conversion of the aromatic amino-group into a nitro-group. This biosynthetic proposal requires electrophilic chlorination of the 3-position of the pyrrole ring, which normally reacts with electrophiles at C-2 or C-5.

An alternative pathway, b, was also considered possible. The 3-chloro-substituent in the pyrrolenine ring is retained, and pyrrolnitrin formation proceeds by the 1,2 migration of the aminophenyl group from C-4 to C-3. Subsequent aromatization to the pyrrole then occurs by dehydrogenation rather than dehydrochlorination. Operation of pathway b in the organism

[50] D. H. Lively, M. Gorman, M. E. Haney, and J. A. Mabe, *Antimicrobial Ag. Chemotherapy*, 1966, 462; R. Hamil, R. Elander, J. A. Mabe, and M. Gorman, *ibid.*, 1967, 388; H. G. Floss, P. E. Manni, R. L. Hamill, and J. A. Mabe, *Biochem. Biophys. Res. Comm.*, 1971, **45**, 781.

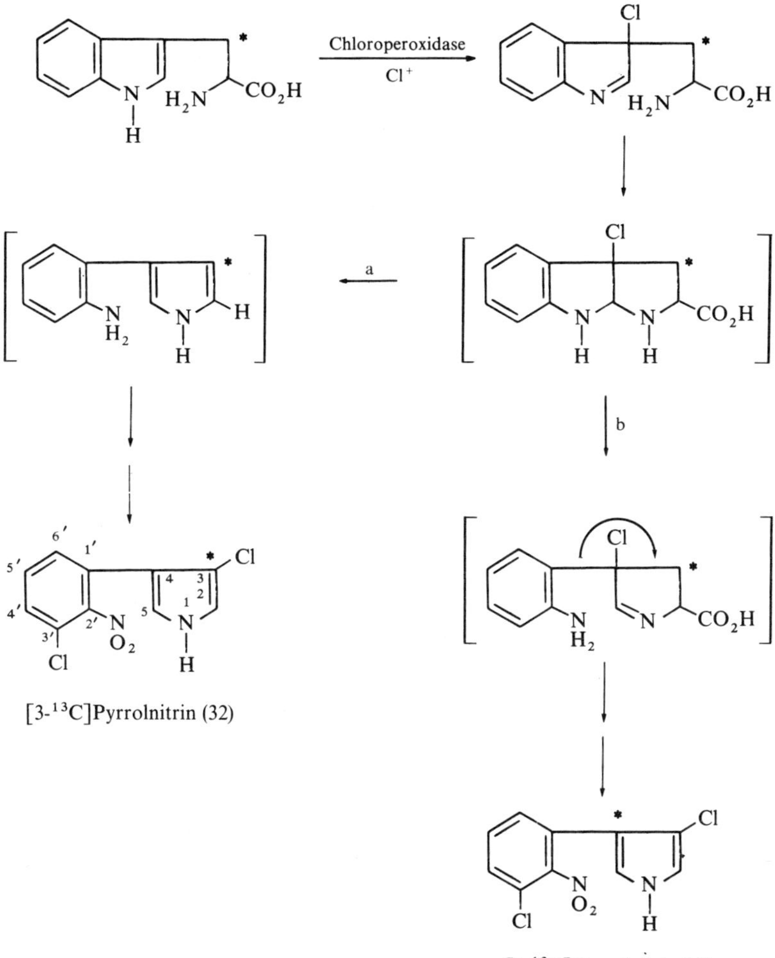

Stable Isotopes in Biosynthetic Studies

Scheme 11

[3-^{13}C]Pyrrolnitrin (32)

[4-^{13}C]Pyrrolnitrin (32)

requires that C-3 of tryptophan becomes C-4 of the pyrrole ring in the antibiotic, and pathway a requires that C-3 of tryptophan becomes C-3 of the pyrrole ring.

This question was examined with the aid of DL-tryptophan-[3-^{13}C]alanine prepared from Ba^{13}CO$_3$ (60% isotopic excess). The ^{13}C-labelled amino-acid was administered to the cultures of the organism, which then yielded 20 mg of labelled pyrrolnitrin containing 28.5 atom % excess ^{13}C.

The assignment of all ten carbon signals of natural-abundance material in the FT ^{13}C n.m.r. 15.077 MHz spectrum in chloroform solution was achieved

Table 16 ^{13}C N.m.r. data for substituted phenylpyrroles (32)—(35), δ_C/p.p.m.^{a}

Position	(32)	(33)	(34)	(35)
C-2	117.5^b	118.8	118.1	117.8^b
C-3	111.7	105.6	106.1	111.5
C-4	115.3	123.8	122.3	118.7^c
C-5	116.7^b	114.5	114.9	117.2^b
C-1'	127.8	135.5	135.9	120.2
C-2'	148.3	124.9	126.4	141.4
C-3'	124.8	128.4	132.2	119.3^c
C-4'	130.2	124.5	128.4	130.2
C-5'	130.2	128.4	130.2	116.0
C-6'	128.5	124.9	123.1	128.2

^{a} Chemical shifts relative to TMS in chloroform.
^{b,c} Values within any vertical column may be transposed.

with the aid of model compounds, including nitrobenzene, pyrrole, chlorobenzene, 4-phenylpyrrole (33), 4-(3',4'-dichlorophenyl)pyrrole (34), and aminopyrrolnitrin (35). The resonances of these C_{10} compounds are given in Table 16.

The most important assignments are for the C-3 (111.3 p.p.m.) signals of the pyrrole ring. These assignments were confirmed by reduction of the nitrogroup of the antibiotic. The C-3 signal remained unchanged, whereas that of the C-4 signal was shifted to 118.7 p.p.m. in aminopyrrolnitrin (35). The

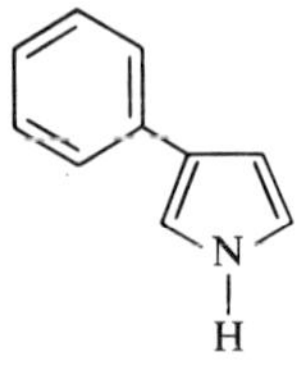

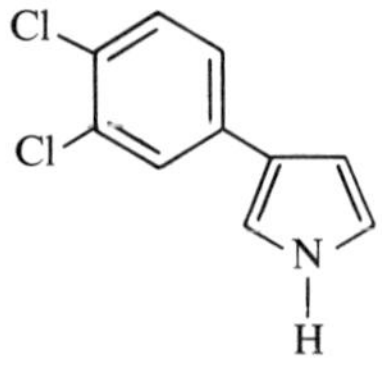

labelled pyrrolnitrin shows significant enhancement of the 111.7 p.p.m. signal, which established that C-3 of the pyrrole ring originated from C-3 of the tryptophan side-chain, and therefore 1,2 aryl migration did not occur during pyrrolnitrin biosynthesis.

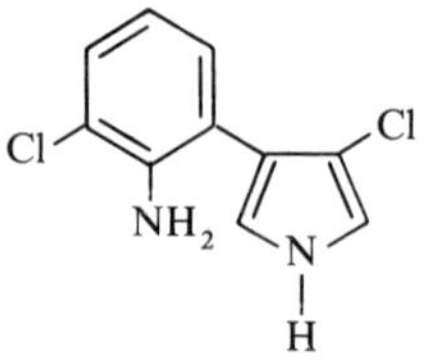

Protoporphyrin-IX (36). FT ^{13}C n.m.r. was used to good advantage to investigate the biosynthetic origin of the meso-carbons of protoporphyrin-IX methyl ester (36).[51] Initially, the ^{13}C chemical shifts of a series of porphyrins at natural abundance were determined. These resonances are tabulated in Table 17. The assignment of the sharp side-chain signals was made by comparison within the spectra of the four model porphyrins (36), (37), (38), and (39)

[51] A. R. Battersby, J. Moron, E. McDonald, and J. Feeney, *J.C.S. Chem. Comm.*, 1972, 920.

Table 17 ^{13}C N.m.r. data for porphyrins at 25.5 MHz, δ_C/p.p.m.^{a}

Porphyrin	Ar—$\underline{C}$H$_2$—$\underline{C}$H$_3{}^a$	Macrocyclic carbons	Ar—$\underline{C}$H$_2$—$\underline{C}$H$_2$—$\underline{C}$O—O$\underline{C}$H$_3$	Meso-carbons^b
(39)	19.3, 18.5	141.2—143.5^c	—	96.2
(37)	19.7(t), 17.6(q)	135—147^c	21.9, 37.0, 173.5, 51.5	96.4
(38)	—	133.5—145.5^c, 128.0^b	21.5, 36.8, 173.2, 51.5	95.5, 96.7, 99.0, 99.8
(36)^d	—	136—145^c	21.7, 36.8, 173.2, 51.6	95.7, 96.7, 97.0, 97.6
[meso-^{13}C]-(36)	—	—	—	96.0, 97.0, 97.3, 97.9

^{a} Chemical shifts relative to TMS in chloroform; d, t, q refer to multiplicity.
^{b} Strong signals.
^{c} Weak signals.
^{d} Also shows Ar—$\underline{C}$H=$\underline{C}$H$_2$ at 130.0 (d), 120.2 (t), and Ar—$\underline{C}$H$_3$ at 11.5 and 12.5 p.p.m.

and by off-resonance decoupling. The carbons of the macrocyclic ring gave two sets of signals near 135 p.p.m. and 140 p.p.m. in the FT ^{13}C n.m.r. spectra. The weakness of the set of signals in the 140 p.p.m. region was attributed to saturation effects on those carbons not directly bonded to protons. The four sharp strong resonances at ~97 p.p.m. were then assigned to the α, β, γ, and δ meso-methine carbons. These assignments agree with an earlier 15.08 MHz ^{13}C n.m.r. study of protoporphyrin IX.[52]

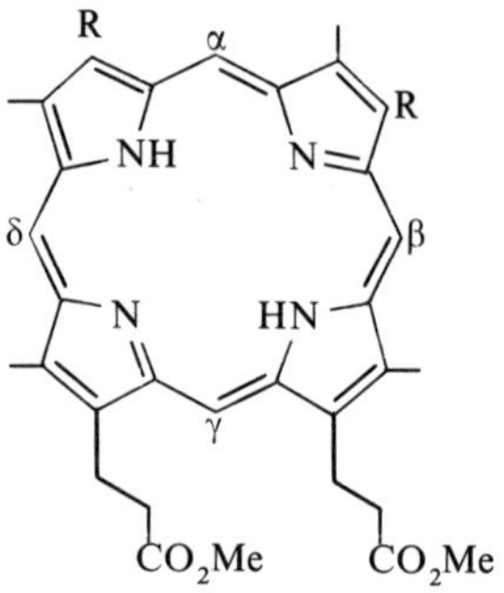

(36) R = CH:CH$_2$
(37) R = Et
(38) R = H

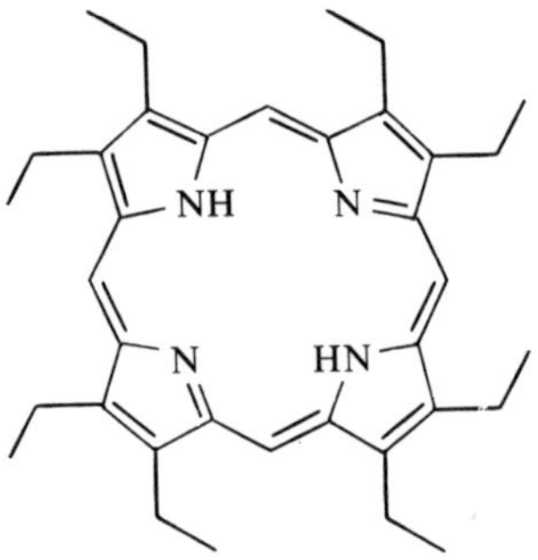

(39)

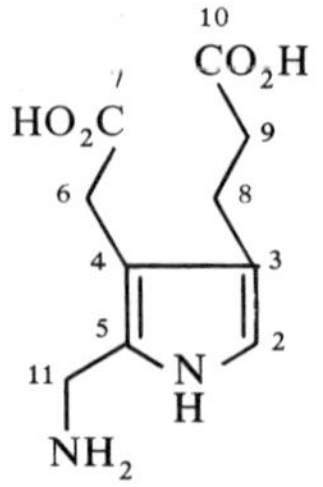

(40)

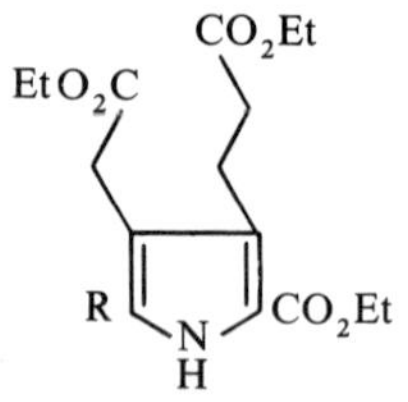

(41) R = H
(42) R = ^{13}CH$_3$
(43) R = ^{13}CH$_2$Cl
(44) R = ^{13}CH$_2$N$_3$

The precursor employed for the tracer experiment, [11-^{13}C]porphobilinogen, PBG (40), was prepared by reductive methylation of (41) with 60% ^{13}C-enriched formaldehyde to give (42), which was chlorinated to (43). The chloro-compound was converted into the azide (44) with sodium azide. The azide (44) was hydrogenated to the amine, which produced an intermediate lactam, then (40).

Incubation of (40) with an enzyme system from *Euglena gracilis* produced protoporphyrin-IX (35), which was isolated as its methyl ester. The FT

[52] D. Doddrell and W. S. Caughey, *J. Amer. Chem. Soc.*, 1972, **94**, 2510.

^{13}C n.m.r. spectrum of the ester in CDCl$_3$ showed only four sharp signals of equal intensity near δ 97 p.p.m., which were assigned to the meso-carbons. Insufficient sample prevented the observation of the natural-abundance signals.

This study established that the meso-carbons of protoporphyrins are derived from C-11 of PBG (40) and that these positions are equally labelled.

Vitamin B_{12} (45). The corrin ring of vitamin B_{12} (45), although structurally related to the porphyrins, is characterized by the absence of a bridge carbon between rings A and D. It is more highly reduced and also contains, in addition to the *gem*-dimethyl group at C-12, six other 'extra' methyl groups at C-1,

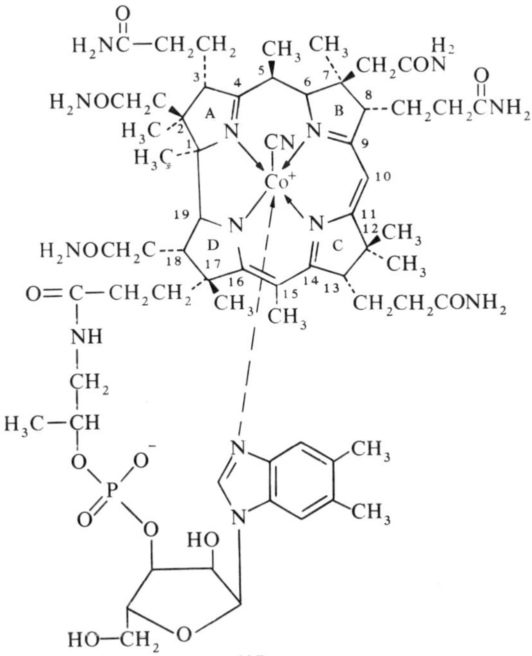

(45)

C-2, C-5, C-7, C-15, and C-17. One C-12-methyl group arises by decarboxylation of an acetic acid residue. ^{14}C-Tracer studies have demonstrated that six of the methyl groups in vitamin B_{12} are derived by C-alkylation with [*Me*-^{14}C]methionine.[53] Labelling studies with δ-amino[5-^{14}C]levulinic acid (ALA) on the origin of the C-1-methyl group gave equivocal results. Incorporation of this labelled ALA at C-4, C-9, C-14, and C-16 and the bridge carbon at C-5, C-10, and C-15 was established. If the C-1-methyl group were also labelled by

---

[53] R. C. Bray and D. Shemin, *Biochim. Biophys. Acta*, 1958, **30**, 647; *J. Biol. Chem.*, 1963, **238**, 1501.

[5-^{14}C]ALA, Kuhn–Roth oxidation should have yielded acetic acid labelled in the methyl group.[53] Instead, the acid isolated contained only 8—9% of the expected label. Therefore, the origin of the C-methyl group from either the δ-methine bridge of the A–D porphyrin rings or from methionine remained unestablished.

Since degradative procedures for vitamin B_{12} were not adequate to determine the origin of the C-1-methyl group, two groups of workers simultaneously investigated this problem with the aid of FT ^{13}C n.m.r. spectroscopy and ^{13}C-labelled precursors.

Shemin's group employed δ-[5-^{13}C]ALA, prepared from 75% enriched [2-^{13}C]malonic acid, as the vitamin B_{12} precursor.[54] Administration of the labelled ALA (50 mg l^{-1}) to cultures of *Propionibacterium shermanii* yielded 5 mg of the B_{12} ester, which was isolated by ion-exchange chromatography. From the known pathways of porphyrin biosynthesis, the [5-^{13}C]ALA should yield a vitamin molecule labelled at C-4, C-5, C-9, C-10, C-14, C-15, and C-16, and possibly at C-1. The FT ^{13}C n.m.r. spectra, which were recorded on a 12 mmol l^{-1} solution of the labelled vitamin in D_2O with a total of 52 600 pulses, showed signals from seven labelled carbon nuclei. Under the same instrumental conditions, no signals from vitamin B_{12} at natural abundance were observed. An examination of the FT ^{13}C n.m.r. spectrum showed carbon–carbon coupling in many of the resonances, which was in accord with that expected since ALA is known to label some adjacent sites in vitamin B_{12}. Mass spectrometry on the ^{13}C-labelled vitamin indicated ~40% ^{13}C at each labelled site. Singlets were found at 181.0 p.p.m. and 180.0 p.p.m., and two very closely spaced peaks at 166.2 p.p.m., 107.3 p.p.m., and 92.5 p.p.m. The doublets had $J_{^{13}C-^{13}C}$ of 60—70 Hz. A more complex splitting pattern was found, centred at 104.8 p.p.m., and no signals were observed in the 38.8—30.8 p.p.m. region, which is the expected region for the C-1-methyl group resonance. No ^{13}C-satellite peak was observed in the ^{1}H n.m.r. absorption region of the C-1-methyl proton signal; this further confirmed the absence of ^{13}C-enrichment at the C-1 position. These results established that the 5-carbon of ALA is not the precursor of the methyl group at C-1 of vitamin B_{12}.

At the same time, Scott's group[55] had administered 60% enriched [2-^{13}C]-ALA to *P. shermanii*, which yielded a labelled B_{12} that had a FT ^{13}C n.m.r. spectrum with eight high-field signals in the CH_2 and CH_3 regions due to the seven CH_2CONH_2 groups and the C-12-methyl group. This methyl group arises by decarboxylation of the original acetate side-chain.

Scott's results were identical with Shemin's findings on administration of [5-^{13}C]ALA. The spectrum of the enriched B_{12} had only seven signals, with the expected carbon–carbon splitting patterns at 187.2, 186.7, 183.6, 180.7, 113.2, 110.3, and 100.4 p.p.m. shifts downfield from hexamethyldisilazane

---

[54] C. E. Brown, J. J. Katz, and D. Shemin, *Proc. Nat. Acad. Sci. U.S.A.*, 1972, **69**, 2585.
[55] A. I. Scott, C. A. Townsend, K. Okada, M. Kajiwara, P. J. Whitman, and R. J. Cushley, *J. Amer. Chem. Soc.*, 1972, **94**, 8267.

(HMDS) for the four C≡N resonances and three C=C resonances. No enrichment at the C-1-methyl group from [5-^{13}C]ALA was detected.

The results of both groups therefore indicated that a ^{13}CH$_2$NH$_2$ terminus from one of the eight ALA units is lost during vitamin B$_{12}$ formation.

The origin of the C-1-methyl group was demonstrated by Scott to be methionine. Although the FT ^{13}C n.m.r. spectrum of cyanocobalamin obtained by feeding [Me-^{13}C]methionine showed only six signals enriched above natural abundance, the dicyanocobalamin derivative revealed seven well-resolved signals, which are at 26.53, 23.78, 23.56, 22.27, 21.35, 20.11, and 19.57 p.p.m., downfield from HMDS, in the region of the ^{13}C n.m.r. spectrum expected for methyl groups.

Any pathway to vitamin B$_{12}$ biosynthesis therefore requires that (i) one animomethyl terminus of the eight ALA units involved in the biosynthesis be lost, and (ii) that seven C-methyl groups be introduced by methionine – two in ring A, one in each of rings B, C, and D, and one in each of the two meso-positions (C-5 and C-15).

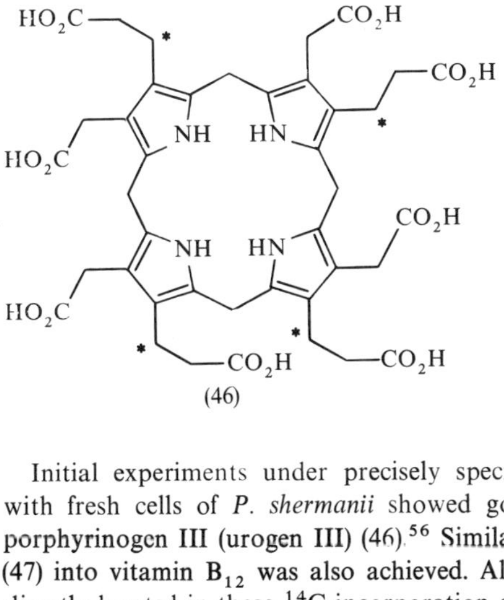

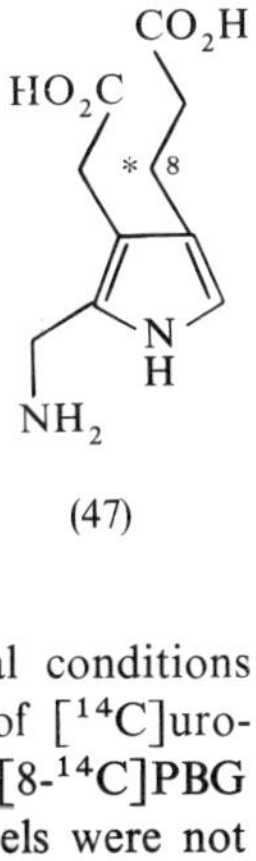

Initial experiments under precisely specified microbiological conditions with fresh cells of *P. shermanii* showed good incorporation of [^{14}C]uroporphyrinogen III (urogen III) (46).[56] Similar incorporation of [8-^{14}C]PBG (47) into vitamin B$_{12}$ was also achieved. Although the ^{14}C labels were not directly located in these ^{14}C-incorporation experiments, it was assumed that no randomization had taken place, and the result was interpreted as being in favour of specific and intact incorporation of urogen III. Parallel feeding experiments with [^{14}C]urogen I showed zero incorporation of this precursor.

To confirm these ^{14}C results and avoid degradation to locate the labels, Scott turned to the use of [8-^{13}C]PBG (47) and [^{13}C]urogen I—IV isomers labelled with ^{13}C, both precursors having 90% enrichment at each labelled site. The [^{13}C]urogen I—IV isomers were prepared from [8-^{13}C]PBG (47).

---

[56] A. I. Scott, C. A. Townsend, K. Okada, M. Kajiwara, and R. J. Cushley, *J. Amer. Chem. Soc.*, 1972, **94**, 8269.

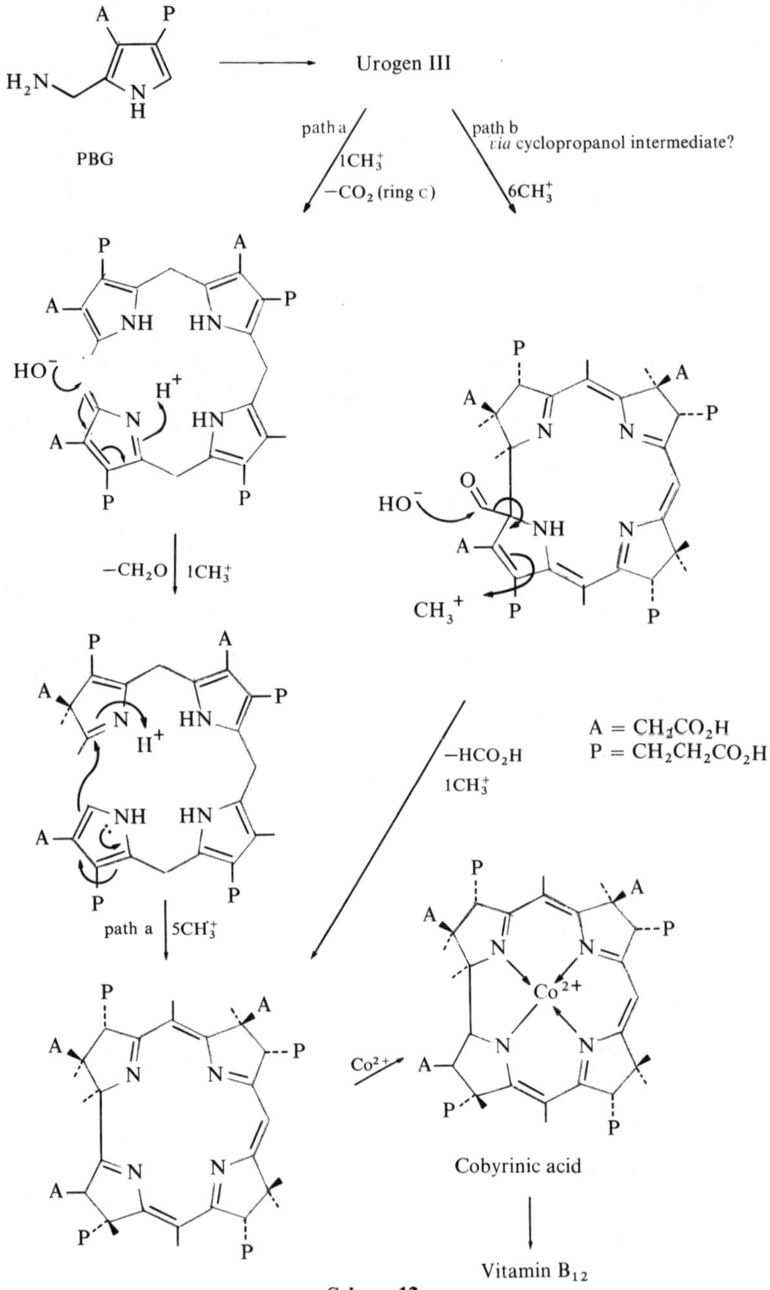
A
P
H2N
N
H
PBG
Urogen III
path a
1CH3+
−CO2 (ring c)
path b
via cyclopropanol intermediate?
6CH3+
HO−
H+
−CH2O
1CH3+
CH3+
−HCO2H
1CH3+
A = CH2CO2H
P = CH2CH2CO2H
path a
5CH3+
Co2+
Cobyrinic acid
Vitamin B12

Scheme 12

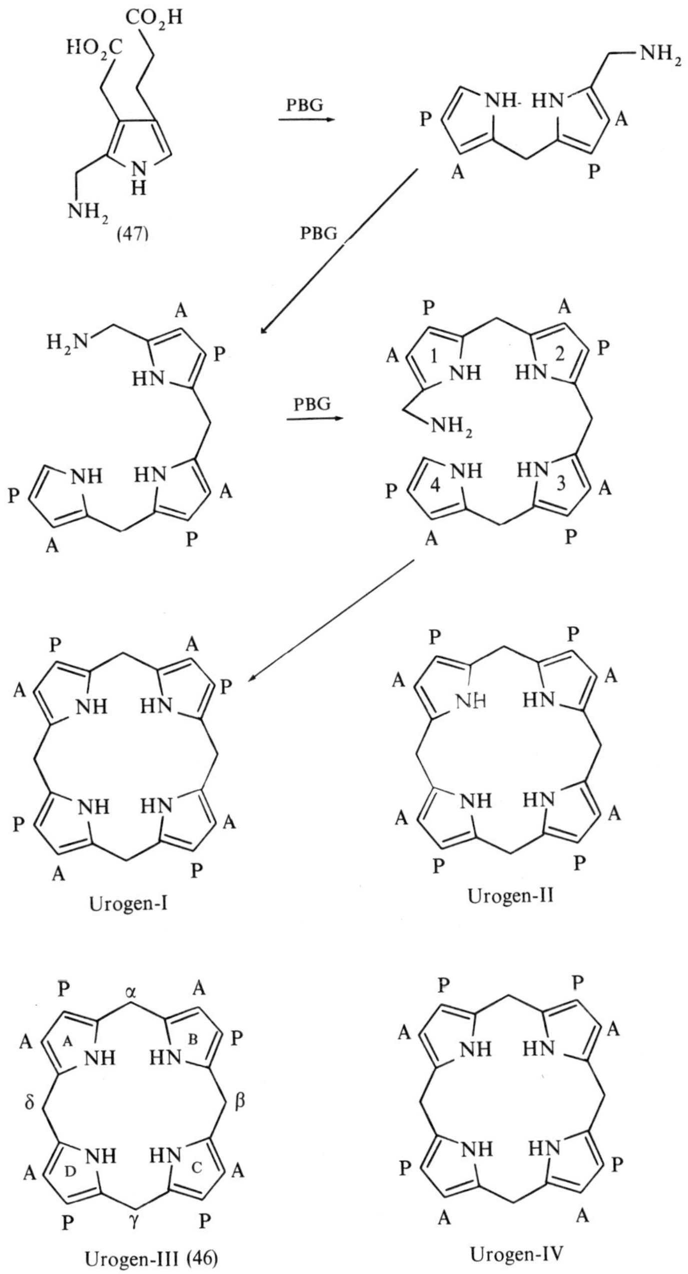
CO2H
HO2C
N
H
NH2
(47)
PBG
NH HN
NH2
P
A
A
P
PBG
H2N
A
P
HN
NH HN
P
A
A
P
PBG
P
A
A
1
NH
HN
2
P
NH2
NH
HN
P
4
3
A
A
P
P
A
A
NH HN
P
NH HN
P
A
A
P
Urogen-I
P
P
A
NH
HN
A
NH
HN
A
A
P
P
Urogen-II
P
α
A
A
A
NH
HN
B
P
δ
β
NH
HN
A
D
C
A
P
γ
P
Urogen-III (46)
P
P
A
NH
HN
A
NH
HN
P
P
A
A
Urogen-IV

Scheme 13

which had been prepared by known methods. The FT ^{13}C n.m.r. spectrum of labelled vitamin B_{12} obtained from [8-^{13}C]PBG contained three resonances at 37.75, 33.44, and 31.50 p.p.m. Conversion of the vitamin into the dicyano-form resolved the 31.50 p.p.m. signal into a pair of peaks, thereby demonstrating that [8-^{13}C]PBG specifically labels four sites in vitamin B_{12}.

Administration of the [^{13}C]urogen isomer mixture afforded a vitamin molecule, the FT ^{13}C n.m.r. spectrum of which again showed the same set of four enhanced methylene signals. These results established that the biosynthetic route to B_{12} is

$$\text{ALA} \rightarrow \text{PBG} \rightarrow \text{urogen III} \rightarrow \text{vitamin } B_{12}$$

Two plausible mechanisms were advanced for the transformation of urogen III to B_{12}, Scheme 12, which accounted for the one-carbon contraction of urogen III at the δ-methine bridge. Pathway a is initiated by methylation at C-1 in ring A of urogen, which results in scission of the bond between the δ-meso-carbon and C-1, with subsequent loss of the δ-meso-carbon as form-aldehyde. In mechanism b, the joining of rings A and D proceeds through a cyclopropanol intermediate and loss of the δ-meso-carbon as formic acid.

The mechanism of urogen III biosynthesis from PBG remains unclarified.[57] Its biosynthesis does not proceed through urogen I (Scheme 13). Urogen II and urogen IV are biologically inert isomers.

These ^{13}C-labelling studies added to and helped clarify some unknown ^{13}C-shift assignments of vitamin B_{12} that were reported in an earlier study.[58]

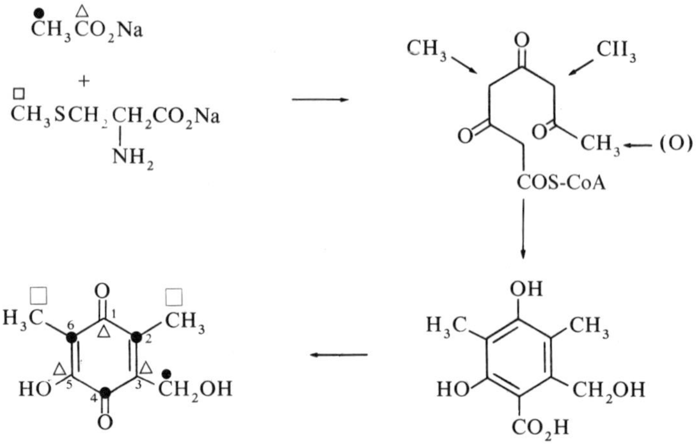

Scheme 14

[57] A. R. Battersby, *Pure Appl. Chem.*, 1971, **5**, 8.
[58] D. Doddrell and A. Allerhand, *Proc. Nat. Acad. Sci. U.S.A.*, 1971, **68**, 1083.

Shanorellin (48). Previous studies with ^{14}C-labelled precursors indicated that acetate and the *S*-methyl group of methionine were efficiently incorporated into shanorellin (48),[59] which is a benzoquinone pigment obtained from *Shanorella spirotricha* Benjamin. Oxidation with chromic acid of shanorellin labelled by [1-^{14}C]acetate yielded an inactive acetic acid, whereas quinone samples labelled by [*Me*-^{14}C]methionine and [2-^{14}C]acetate gave, after oxidation, acetic acid labelled at the methyl and carboxy-groups, respectively. On this basis, the biosynthetic route in Scheme 14 was proposed.

Table 18 [13]C N.m.r. data for shanorellin, δ_C/p.p.m.

Position	Precursor[a]		
	$CH_3{}^{13}CO_2Na$	$^{13}CH_3CO_2Na$	$H^{13}CO_2Na$
C-1	187.7	—	—
C-2	—	146.8	—
C-3	137.8	—	—
C-4	—	183.3	—
C-5	152.0	—	—
C-6	—	117.4	—
CH$_2$OH	—	54.8	—
CH$_3$	—	—	12.0
CH$_3$	—	—	7.8

[a] ~ 5% enrichments with these precursors were observed.

By administration of $H^{13}CO_2Na$, $^{13}CH_3CO_2Na$, and $CH_3{}^{13}CO_2Na$ as labelled precursors, ^{13}C-enriched shanorellins were isolated. They were examined in dioxan solution by [13]C n.m.r. spectroscopy in the CW swept mode. The results obtained are given in Table 18.[60] The known ^{13}C chemical shift assignments of a naturally occurring quinone perezone (49) aided in making the shanorellin ^{13}C-shift assignments.[61] C-1, C-3, and C-5 are labelled by [1-^{13}C]acetate; C-2, C-4, C-6, and the hydroxymethyl group are labelled by [2-^{13}C]acetates; the two methyl groups are labelled by [^{13}C]methionine. These ^{13}C-results fully support the proposed biosynthetic route of Scheme 14, wherein a triacetyl unit is methylated by methionine at C-2 and C-6, and at some subsequent stage the terminal carboxy-group of the polyketide is lost.

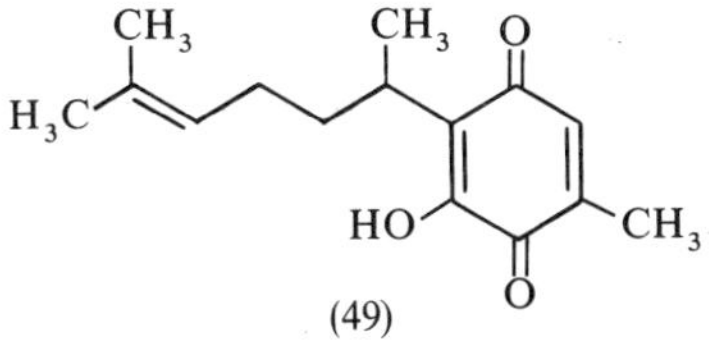

(49)

[59] C.-K. Wat and G. H. N. Towers, *Phytochemistry*, 1971, **10**, 103; C.-K. Wat and G. H. N. Towers, *ibid.*, p. 1355.
[60] C.-K. Wat, A. G. McInnes, D. G. Smith, and L. C. Vining, *Canad. J. Biochem.*, 1972, **50**, 620.
[61] P. Joseph-Nathan, Ma. P. Gonzalez, L. F. Johnson, and J. N. Shoolery, *Org. Magn. Resonance*, 1971, **3**, 23.

Palmitoleic Acid (50). Carbon-13-labelled lipids have been obtained from *Saccharomyces cerevisiae*. The yeast was grown on an enriched complex medium that was supplemented with 0.11 % sodium [2-^{13}C]acetate (50—60 % enriched). The lipid fraction was isolated from the harvested cells, and the fatty acid methyl esters were then isolated and purified by gas chromatography.[62]

$$H_3C \cdots CO_2CH_3$$

(50)

The FT ^{13}C n.m.r. spectrum at 15.08 MHz of ^{13}C-enriched methyl palmitoleate (50), which was isolated from the lipid fraction, was recorded on 1.2 mg in CHCl$_3$ solution. The spectrum showed enhanced signal intensities for the eight alternate carbon atoms (C-2, C-4, C-6, C-8, C-10, C-12, C-14, and C-16). This is the expected labelling pattern for the fatty acid derived from ^{13}CH$_3$CO$_2$Na. Mass spectrometry of the ^{13}C-enriched ester indicated a 32 % enrichment for the eight alternate sites.

3 Mass Spectrometry

General.—Since the electron-impact fragmentation patterns of many classes of natural products are now known, mass spectrometry can be used in conjunction with heavy isotopes in biosynthetic studies.[63] A study of the fragmentation pattern of a metabolite with and without isotopic labelling yields direct information on the position of the labelled atom in the molecule without recourse to wet chemical degradation. Essential requirements for the biosynthetic application of mass spectrometry are (i) thermal stability of the metabolite at the probe inlet temperatures and (ii) a reasonably simple fragmentation pattern.

Recent practical applications of mass spectrometry of metabolites labelled with the stable isotopes ^{2}H, ^{13}C, ^{15}N, and ^{18}O will be reviewed.

Patulin (52). Although the conversion of 6-methylsalicylic acid (51) into patulin (52) by *Penicillium patulum* was known,[64] the nature of the aromatic intermediates on the pathway to the lactone was obscure. The discovery of several phenolic co-metabolites,[65] *m*-cresol (53a; R = H), *m*-hydroxybenzyl alcohol (54), and toluquinol (55) in this organism suggested that (53) is a likely intermediate, as shown in Scheme 15. Administration of 3-methyl[2,4,6-^{2}H$_3$]phenol (53b; R = ^{2}H) to *P. patulum* afforded an enriched patulin.[66] The (*M* + 2)

---

[62] A. L. Burlingame, B. Balogh, J. Welch, S. Lewis, and D. Wilson, *J.C.S. Chem. Comm.*, 1972, 318.
[63] K. Biemann, 'Mass Spectrometry,' McGraw-Hill, New York, 1962; H. Budzikiewicz, C. Djerassi, and D. H. Williams, 'Mass Spectrometry of Organic Compounds,' Holden-Day, San Francisco, 1967; and ref. 2.
[64] S. W. Tannenbaum and E. W. Bassett, *J. Biol. Chem.*, 1959, **234**, 1961; *Biochim. Biophys. Acta*, 1960, **40**, 535.
[65] J. D. Bu'Lock, D. Hamilton, M. A. Hulme, A. J. Powell, H. M. Smalley, D. Shepherd, and G. N. Smith, *Canad. J. Microbiol.*, 1965, **11**, 765.
[66] A. I. Scott and M. Yalpani, *Chem. Comm.*, 1967, 945.

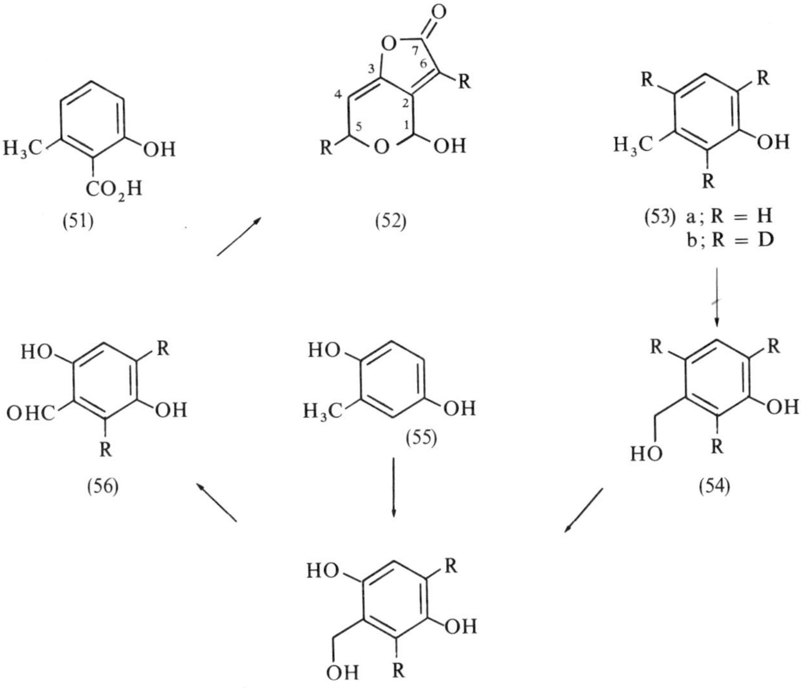

Scheme 15

peak corresponded to 30% incorporation of (53b; R = ²H). In a glucose-deficient growth medium, 57% incorporation was observed.

The relative deuterium abundance of seven principal fragment ions (52a—g) (Scheme 16) was determined (Table 19). The molecular composition of each fragment ion was supported by high-resolution mass measurements. These

Table 19 *Relative abundance of* $(M + 2)$ *and* $(M + 1)$ *peaks in deuteriopatulin* $(52; R = {}^2H)$

m/e	Fragment (52)	Ratio	
		$(M + 2)/M$	$(M + 1)/M$
156	M + 2	1.4	—
138	(a)	1.4	—
128	(b)	1.4	—
111	(c) (111.0061)	—	1.3
99	(d) (99.0392)	1.4	—
83	(e) (83.0115)	—	1.5
72	(f) (72.0197)	—	1.3
56	(g) (56.0241)	—	1.3

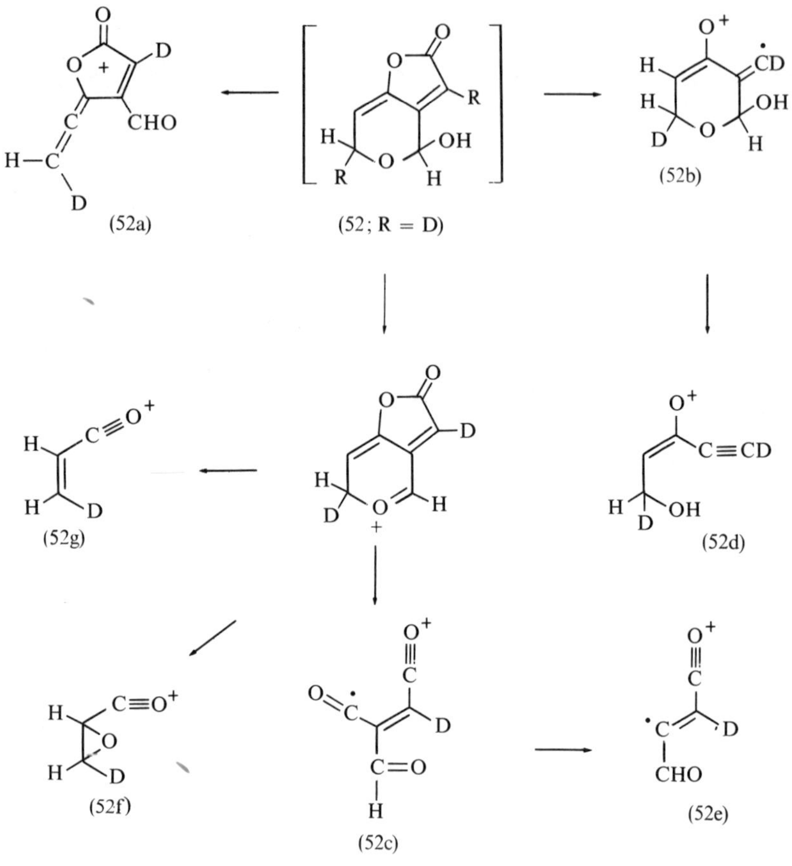

Scheme 16

results established that the observed labelling pattern agrees with the proposed biosynthetic route to patulin.

The formation of optically active deuteriopatulin (49; R = ^{2}H) was observed. This result showed that the intermediate formed from the ring fission of gentisaldehyde (53) underwent a stereospecific reduction during patulin formation. Natural patulin is optically inactive.

Deuterium Labelling.—*Vindoline* (57). Biosynthetic mapping by direct mass spectrometry was successfully achieved with the indole alkaloid vindoline (57; $R^1 = R^2 = H$).[67] A deuteriovindoline was prepared by the administration of

---

[67] E. S. Hall, F. McCapra, T. Money, K. Fukumoto, J. R. Hanson, B. S. Mootoo, G. T. Phillips, and A. I. Scott, *Chem. Comm.*, 1966, 348.

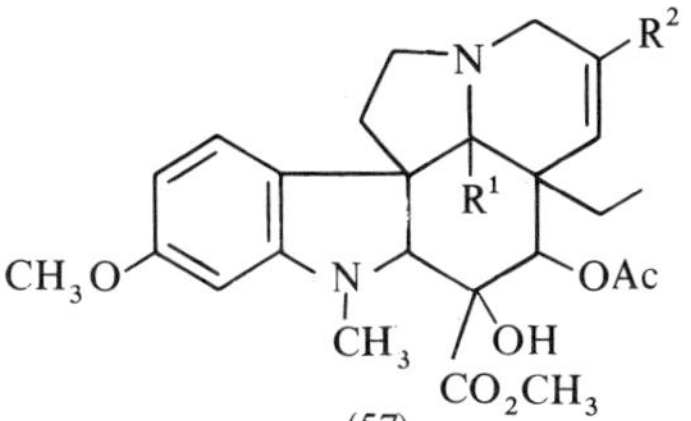

Stable Isotopes in Biosynthetic Studies

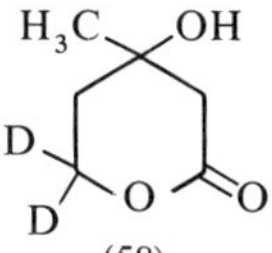

Table 20 *Incorporation data for deuteriovindolines*

Fragmentation	% Enrichment	
	$(57; R^1 = R^2 = {}^2H)$	$(57; R^1 = {}^2H, R^2 = {}^1H)$
$M + 1$	0.00	1.50
$M + 2$	1.55	0.00
(57a) + 1	0.30	1.90
(57a) + 2	1.55	0.00
(57b) + 1	0.00	0.00
(57b) + 2	0.00	0.00
(57c) + 2	0.00	0.00
(57c) + 1	0.00	0.00
(57d) + 2	1.50	0.00
(57e) + 1	0.30	1.90
(57e) + 2	1.90	0.00

[5-^{2}H$_2$]mevalonic acid lactone (58) to *Vinca rosea* plants. The deuterium-enrichment data for the deuteriovindoline were obtained by mass spectrometry and compared with those for unlabelled vindoline obtained under identical instrumental operating conditions. The data, as presented in Table 20, show that both deuteriums are present in M^+ and in the fragments (57a), (57d), and

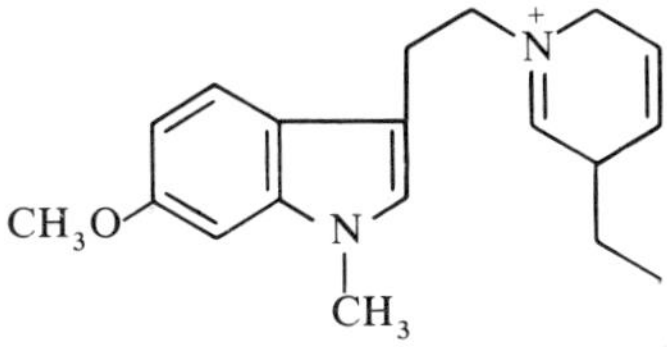

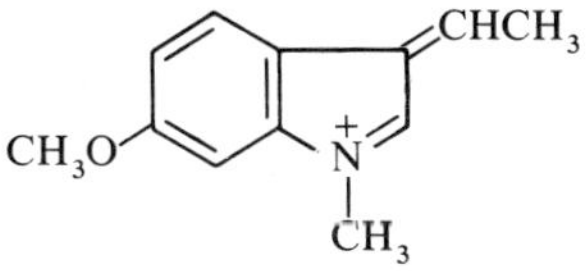

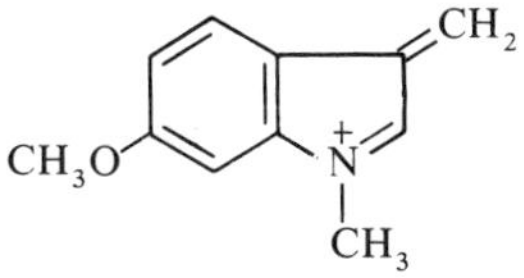

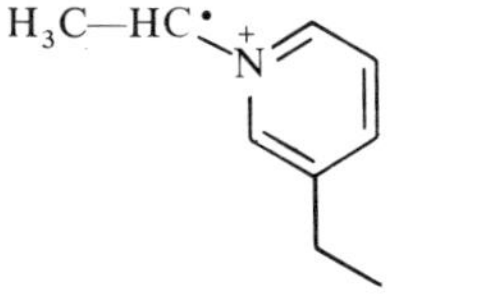

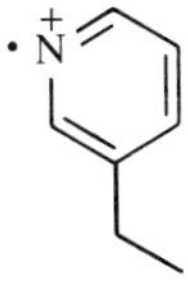

Scheme 17

(57e) (Scheme 17). The average enrichment of 1.6% in the (m/e + 2) peaks, namely M + 2, (57a) + 2, (57d) + 2, and (57e) + 2, corresponds to an incorporation of 0.2% of deuteriomevalonate into vindoline. Extensive hydrogen transfers during the fragmentation process can invalidate mass spectrometric results, and repetitive 'hot' and 'cold' runs under the same conditions should be performed to increase precision. These mass spectrometric results are compatible with the dimerization of mevalonate to geraniol (59), followed by cyclization to a cyclopentanoid system (60), cleavage to (61), rearrangement to (62), and their ultimate conversion into vindoline (57) (Scheme 18).[68]

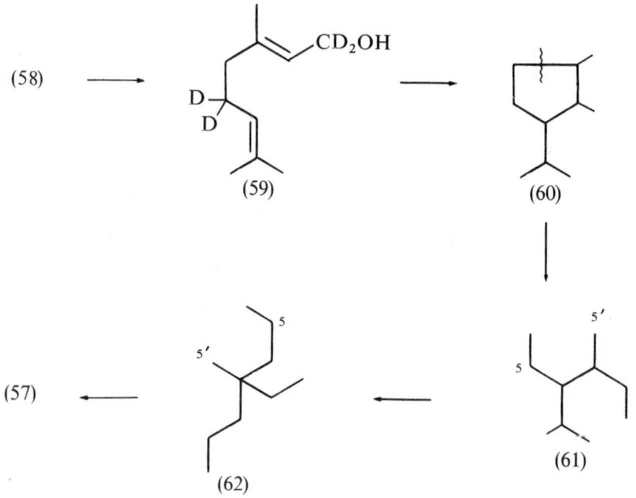

Scheme 18

Lincomycin (63). The bio-origin of the methyl groups of lincomycin (63) was investigated with L-[Me-$^2H_3$]methionine.[69] Addition of this labelled amino-acid to *Streptomyces lincolnensis* yielded a deuteriated antibiotic. The mass spectrum of this material showed molecular ions at m/e 406 (M +), 409 (M + 3), 412 (M + 6), and 415 (M + 9). This result, which agrees with a parallel ^{14}C study, indicated that a maximum of three CD_3 groups were incorporated.

Fragment ions at m/e 359 and 362 ($M - SCH_3$ and $M - SCD_3$) confirmed that the *S*-methyl group was labelled. Evidence for the location of the other two CD_3 groups came from the fragments (63a), m/e 263, and (63b), m/e 132.

---

[68] A. R. Battersby, in 'The Alkaloids,' ed. J. E. Saxton, (Specialist Periodical Reports), The Chemical Society, London, 1971, Vol. 1, p. 31.
[69] A. D. Argoudelis, T. E. Eble, J. A. Fox, and D. J. Mason, *Biochemistry*, 1969, **8**, 3408.

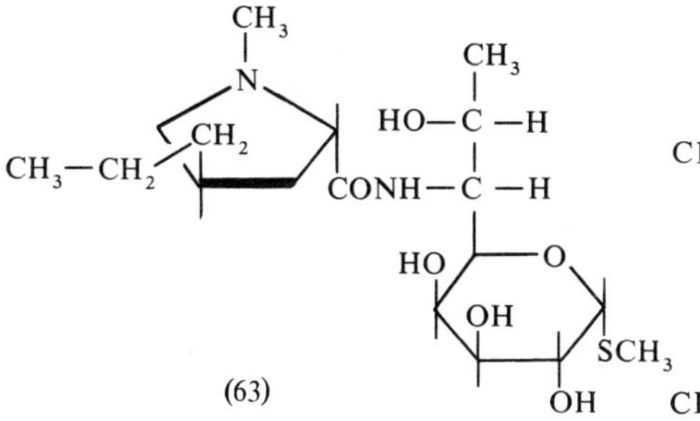

(63)

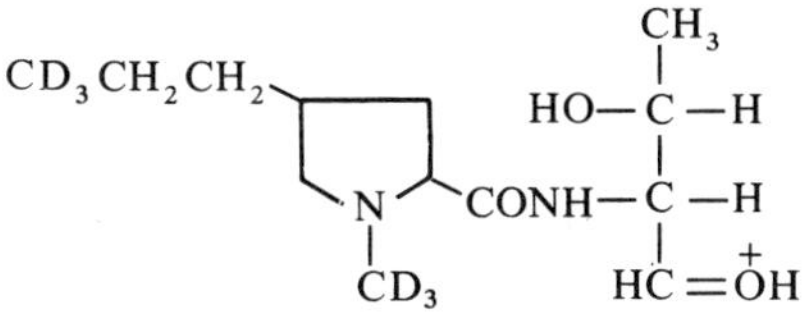

(63a) m/e 263

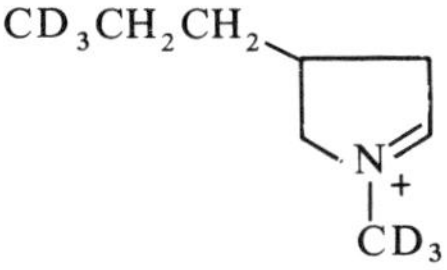

(63b) m/e 132

These results present a unique example of a biological methylation reaction on nitrogen, sulphur, and carbon occurring in the same system.

Mycarose (64). To establish whether transfer of an intact methyl group from methionine to the C-3 carbon of mycarose (64) occurs, DL-[*Me*-²H₃]methionine was used as a labelled precursor.[70]

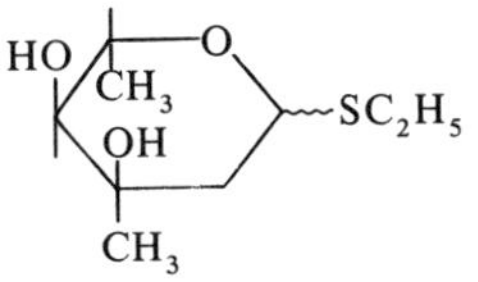

(64)

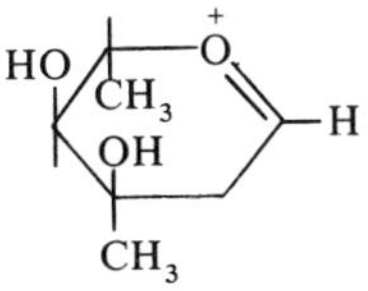

(64a)

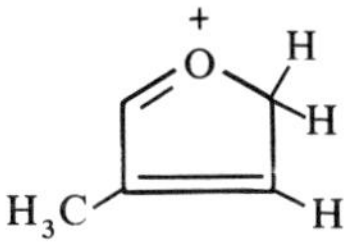

(64c)

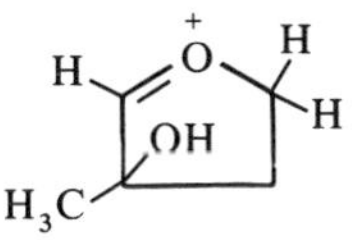

(64b)

The labelled mycarose isolated was converted into its methylthioglycoside. The mass spectrum of this glycoside had a molecular-ion peak (M^+) and fragment-ion peaks (64a), (64b), and (64c) that all retained the C-3-methyl group. All of these peaks were accompanied by additional peaks that were three mass units higher. The composition of the fragment ions was confirmed by high-resolution mass spectrometry.

[70] H. Pape, R. Schmid, H. Grisebach, and H. Achenbach, *European J. Biochem.*, 1969, **10**, 479.

These results proved that an intact methyl group is transferred to mycarose from methionine. This should be compared with the biological methylation reactions on carbon leading to tuberculosteric acid and ergosterol that resulted in the transfer of only two of the three protons of the methyl group of methionine.[71]

Vitamin K_2 (65) and Ubiquinone Q-8 (66). To help define the biological origin of the ring methyl groups in vitamin K_2 (65) and ubiquinone Q-8 (66), a

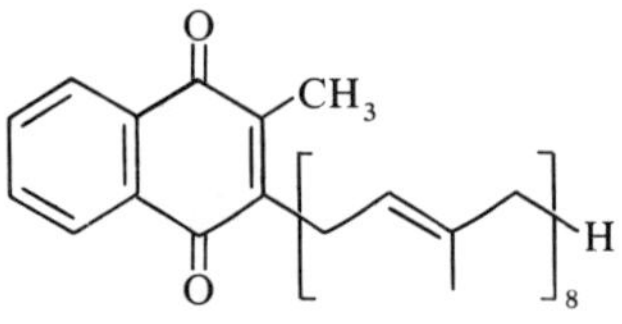

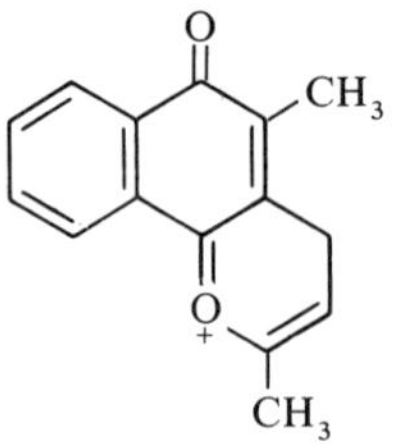

methionine auxotroph of *E. coli* was grown on [*Me*-$^2H_3$]methionine.[72] The deuteriated vitamin K_2 isolated had a mass spectrum with a strong molecular-ion peak at 719 ($M + 3$) with a relative abundance of 27%. The base peak for the unlabelled vitamin K_2 is at m/e 225 and is attributed to the ion a or a'. The deuteriated vitamin K_2 exhibited a comparable base peak at m/e 228.

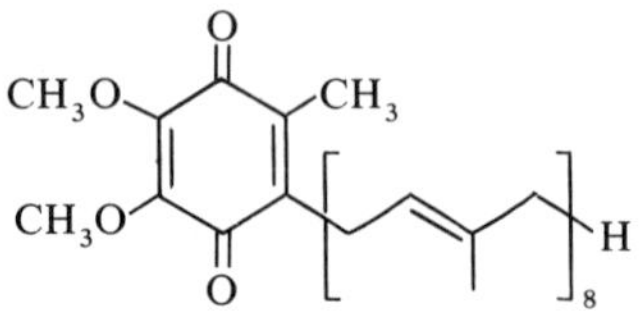

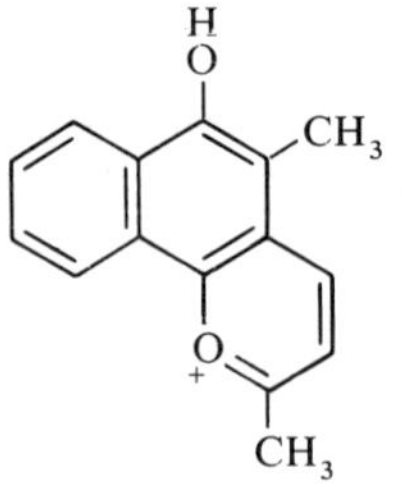

Further confirmation for a deuteriated methyl group in vitamin K_2 was achieved by ^1H n.m.r. spectroscopy. The signal for the unsaturated C-methyl group was absent at δ 2.17. This indicates a very high incorporation yield of labelled precursors in the methionine auxotroph.

Similar results were found with the deuteriated ubiquinone Q-8 (66), which showed a base peak in the mass spectrum at m/e 244, compared with m/e 235 for the unlabelled material. This fragment ion for the quinone is similar to a or a'. The molecular ion was at m/e 735 for the deuteriated material, compared with m/e 726 for the unlabelled quinone. Intensity comparisons showed an

---
[71] G. Jaureguiberry, J. H. Law, J. A. McCloskey, and E. Lederer, *Compt. rend.,* 1964, **258**, 3587; G. Jaureguiberry, J. H. Law, J. A. McCloskey, and E. Lederer, *Biochemistry,* 1965, **4**, 347.
[72] L. M. Jackman, I. G. O'Brien, G. B. Cox, and F. Gibson, *Biochim. Biophys. Acta,* 1967, **141**, 1.

incorporation of 99% for nine deuterium atoms. The absence of proton n.m.r. peaks at δ 3.90 (OCH_3) and δ 2.00 (C—CH_3) in the deuteriated ubiquinone confirmed this incorporation yield.

Protoporphyrin-IX (69a). Deuterium labelling helped to clarify the mechanism involved in the oxidative enzymatic conversion of coproporphyrinogen-III (67) into porphyrinogen-IX (68), which is the precursor of protoporphyrin-IX (69a).[73]

Porphobilinogens (70) labelled with two deuteriums at either position x or y in the propionic acid side-chain were prepared and incubated with an enzyme

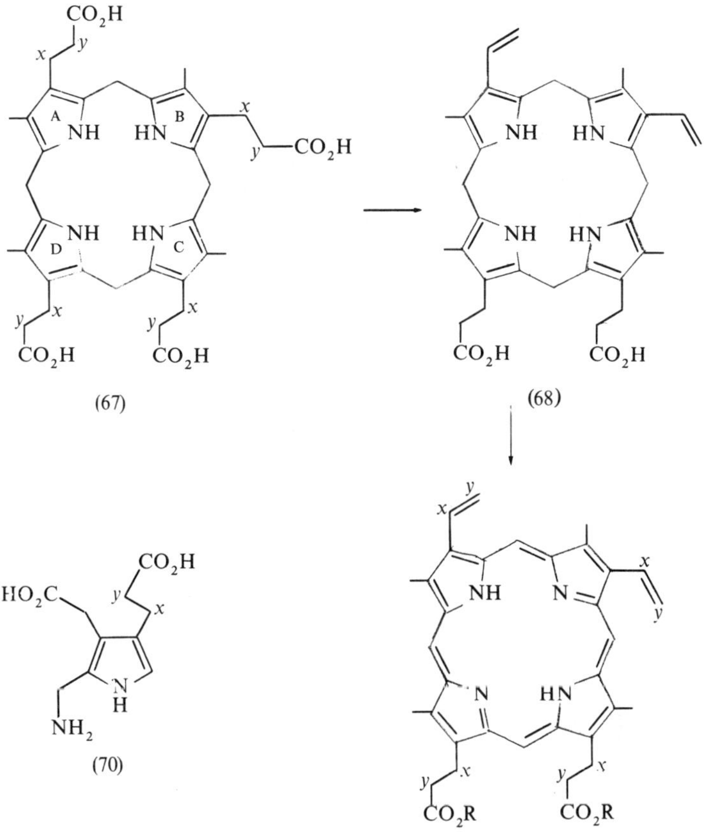

Scheme 19

[73] A. R. Battersby, J. Baldas, J. Collins, D. A. Grayson, K. J. James, and E. McDonald, *J.C.S. Chem. Comm.*, 1972, 1265.

system from *Euglena gracilis* to give mixtures of (67) and (68), which were converted photochemically into the protoporphyrins (69a) (Scheme 19). Their methyl esters (69b) were analysed by mass spectrometry. The porphobilinogen (70) labelled at *x* with two deuterium atoms yielded initially (67) with eight deuterium atoms, and conversion of (67) into (70) yielded a porphyrin with six deuterium atoms. Similarly, (70) labelled at *y* with two deuterium atoms yielded coproporphyrinogen-II (67) with eight deuterium atoms. Incubation of *y*-labelled (67) afforded the vinyl derivative (69a), which had retained most of the deuterium.

These results are consistent with an oxidative decarboxylation mechanism for the formation of the vinyl group in (69a), and thus ketone or acrylic acid intermediates are eliminated.

Nitrogen-15 Labelling.—*Gliotoxin* (71). Nitrogen metabolism in microbial systems can be conveniently studied by mass spectrometric analyses of compounds formed from ^{15}N-labelled precursors. Incorporation results with DL-[^{3}H]phenylalanine, DL-[*Me*-^{14}C]methionine, DL-[3-^{14}C]serine, DL-[1-^{14}C]-serine, and [2-^{14}C]glycine had determined that phenylalanine was the precursor of the indole moiety of the antibiotic gliotoxin (71), which is produced by *Trichoderma viride*.[74]

To establish whether the indole N-5 nitrogen of gliotoxin is derived intact from phenylalanine, incorporation studies with [^{15}N]phenylalanine (90—95%) were conducted. Parallel feeding studies with [^{15}N]glycine (90—95%),

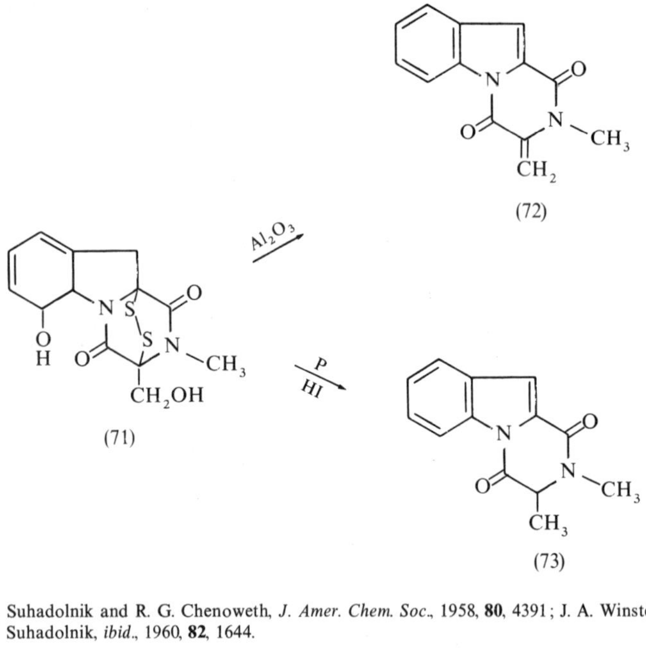

[74] R. J. Suhadolnik and R. G. Chenoweth, *J. Amer. Chem. Soc.*, 1958, **80**, 4391; J. A. Winstead and R. J. Suhadolnik, *ibid.*, 1960, **82**, 1644.

[2-^{13}C]glycine (60%), [1-^{13}C]glycine (60%), and [^{13}C]formate (60%) were also performed.[75]

Although gliotoxin (71) was found to decompose thermally to a $(M - S_2)^+$ peak in the heated inlet, two stable dethio-derivatives of gliotoxin, (72) and (73), were found to be suitable for the mass spectrometric determination of heavy isotope enrichment. Compound (72) was prepared by activated alumina treatment of gliotoxin, and (73) by red phosphorus and hydriodic acid reduction of the antibiotic. The fragment ions of (72) used to determine the isotope incorporation were m/e 226 (M^+), m/e 143 ($C_9H_3NO^+$ or $C_{10}H_9N^+$), which contains the indole moiety, and m/e 115. The fragment ions of (70) examined were m/e 228 (M^+), m/e 213 $(M - CH_3)^+$, and the m/e 143 and m/e 115 peaks. The data on the incorporation of the labelled precursors into (72) and (73) are presented in Tables 21 and 22.

Table 21 *Incorporation data on compound* (72)

	Δ^b			
Fragment P,^{a} m/e	[^{15}N]Phenylalanine	[^{15}N]Glycine	[2-^{13}C]Glycine	[1-^{13}C]Glycine
226 (M^+)	1.3	2.8 10.4	1.4	2.4
143	1.0	0.8 3.5	0.3	0.0
115	0.8	0.6 3.2	0.0	0.7

^{a} P = height of peak at m/e, $I_{rel} = [(P + 1)/P] (100\%)$.
^{b} Δ = labelled I_{rel} − unlabelled I_{rel}.

These results showed that [^{15}N]phenylalanine is incorporated into gliotoxin; however, more information is required before the quantity of incorporation of the intact amino-acid *versus* that occurring through a nitrogen-free intermediate can be determined. The incorporation of [^{15}N]glycine, [2-^{13}C]glycine, [1-^{13}C]glycine, and [^{13}C]formate into gliotoxin is in agreement with the ^{14}C study. The labelling of the N-2 and N-5 positions by [^{15}N]glycine indicated that the amino-group was released into the nitrogen pool during glycine metabolism. The carbons of the ^{13}C-labelled glycine were not incorporated into the indole ring of this antibiotic.

Table 22 *Incorporation data on compound* (73)

Fragment P,^{a} m/e	[^{15}N]Glycine, Δ^b (± 0.2), %
228 (M^+)	8.9
213 $(M - CH_3)^+$	8.9
143	2.3
115	2.3

^{a} P = height of peak at m/e, $I_{rel} = [(P + 1)/P] (100\%)$.
^{b} Δ = labelled I_{re} − unlabelled I_{rel}.

[75] A. K. Bose, K. G. Das, P. T. Funke, I. Kugajevsky, O. P. Shukla, K. S. Khanchandani, and R. J. Suhadolnik, *J. Amer. Chem. Soc.*, 1968, **90**, 1038.

Oxygen-18 Labelling.—β-*Nitropropionic Acid*. The biological precursors of the nitro-group of β-nitropropionic acid, a metabolite of *Penicillium atrovenetum*, were investigated by incorporation studies with ^{15}N- and ^{18}O-labelled substrates.[76] Mass spectral analyses of the methyl esters of the β-nitropropionic acid samples indicated that the ammonium ion, which was administered as ^{15}NH$_4$Cl, was incorporated in preference to the nitrate ion administered as K^{15}NO$_3$ and KN^{18}O$_3$. The molecular ion at m/e 133 (M^+) was not observed, but the fragment ion at m/e 102 (M − OCH$_3$) was shifted to m/e 103 for ^{15}N-enriched samples and to m/e 104 for ^{18}O-enriched samples. It was established that the amino-group of [^{15}N]aspartic acid was utilized in preference to the ammonium ion (2:1) for the synthesis of the nitro-group.

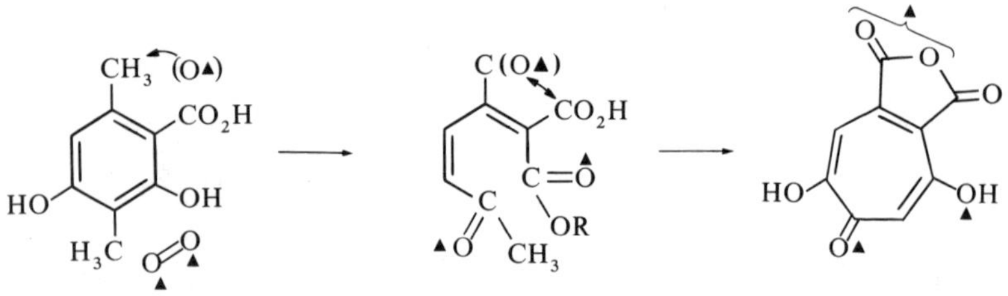

Scheme 20

Stipitatonic Acid (75). The mechanism for the oxidative cleavage of 3-methylorsellinic acid (74) to afford the tropolone stipitatonic acid (75) in *Penicillium stipitatum* was very elegantly demonstrated with ^{18}O labelling.[77] Two pathways were considered, the dioxygenase recyclization route, Scheme 20, and the mono-oxygenase ring-expansion route, Scheme 21.

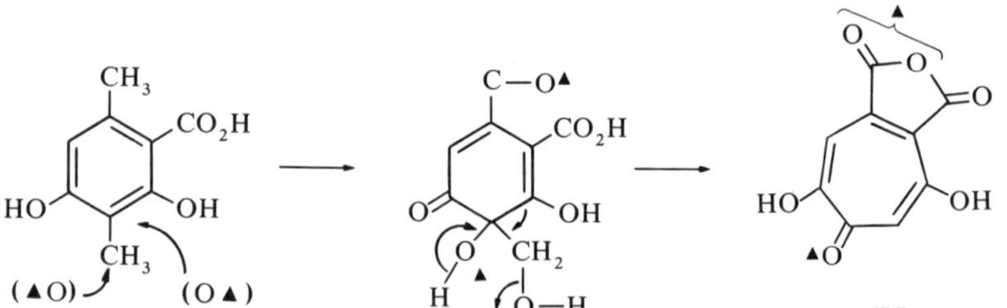

Scheme 21

[76] P. D. Shaw and J. A. McCloskey, *Biochemistry*, 1967, **6**, 2247.
[77] A. I. Scott, *Pure Appl. Chem.*, 1971, **5**, 21.

Since the mass spectra of the ^{18}O-enriched stipitatonic acid and other tropolone metabolites isolated showed only $M + 2$ peaks and no $M + 4$ peak, which would have been required by the dioxygenase mechanism, the monooxygenase pathway was favoured.

Carbon-13 Labelling.—*Phlebiarubrone* (76). Phenylalanine was established as a precursor of phlebiarubrone (76) as a result of the administration of DL-[3-^{13}C]phenylalanine (50% excess ^{13}C) to *Phlebia strigosozonata*.[78] The

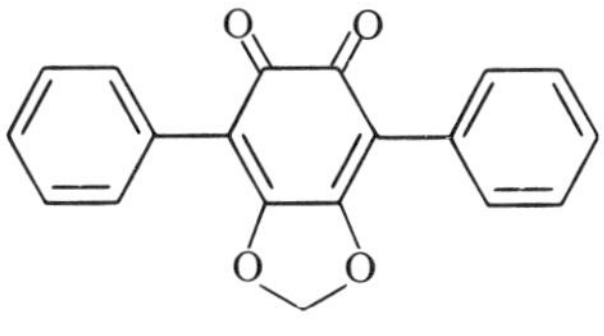

(76) 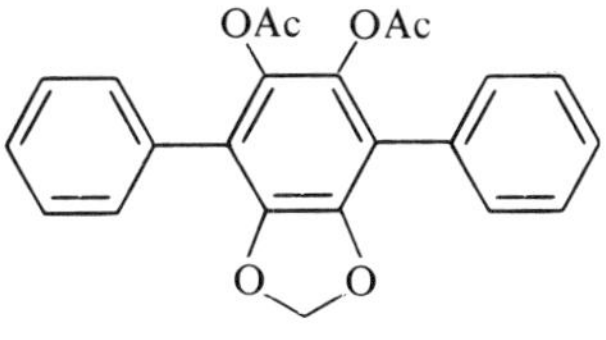   (77)

labelled metabolite was converted into the more soluble and volatile leucoacetate derivative (77), which exhibited a mass spectrum with 5.4% of the molecules enriched with one ^{13}C and 1.8% of the molecules doubly labelled with ^{13}C. Incorporation of H^{13}CO$_2$Na into the methylenedioxy-carbon was demonstrated by the ^{13}C-satellite peaks and mass spectrometry.

The Reporter would like to thank L. Cheney, M. I. Dawson, G. Detre, R. H. Peters, H. Seto, K. T. Suzuki, and D. Yasuda for their help in the preparation of this Report.

[78] A. K. Bose, K. S. Khanchandani, P. T. Funke, and M. Anchel, *Chem. Comm.*, 1969, 1347.

Author Index

Abe, F., 36
Abe, H., 90
Abou-chaar, C. I., 112
Abut-Hajj, Y. J., 36
Achari, R. G., 182
Achenbach, H., 293
Achilladelis, B. A., 10
Achiwa, K., 120
Acklin, W., 8
Adams, E., 86
Adams, J. B., 44
Adams, P. M., 10
Adesida, G. A., 239
Adesogan, E. K., 239
Aexel, R. T., 30
Agurelle, S., 117
Ahmad, A., 93
Akhtar, M., 20, 28, 29
Akiyama, M., 30
Alam, N. H., 48
Albersheim, 80
Aldag, R., 86
Alexander, K., 28
Alfsen, A., 35
Ali, A. A. E. R., 111
Alibert, G., 226
Allcock, C., 5
Allen, C. M., 3
Allerhand, A., 260, 286
Allison, M. J., 117
Altman, L. J., 65
Anchel, M., 299
Andersen, N. H., 11
Anderson, D. R., 212, 260
Anderson, R. J., 27, 61
Anderson, W. A., 252
Anding, C., 57
Andrewes, A. G., 65
Anjyo, T., 40
Aragón, M. C., 2, 22
Arcamone, F., 216
Arditti, J., 60, 232
Argoudelis, A. D., 292
Arigoni, D., 8
Arimasa, N., 36
Arnold, R., 200
Arnold, W. N., 78
Asen, S., 230

Ash, L., 65
Asworth, T. S., 75
Atherton, L., 55
Atkin, S. D., 27, 28
Avery, M. D., 44
Axelrod, L. R., 33, 40
Ayaki, Y., 44, 46

Baggaley, K. H., 27
Bailey, B. K., 195
Baillie, A. C., 217
Baisted, D. J., 25
Bakker, H. J., 14, 15
Balce, L. V., 90
Baldas, J., 295
Ball, C. D., 122
Ball, E. A., 232
Ball, P., 44
Ballantine, J. A., 187
Balogh, B., 288
Balsubramaniam, S., 46
Banesji, A., 59
Banthorpe, D. V., 1, 2, 4, 6, 23
Barbier, M., 64
Bardon, O., 64
Barlow, S. A., 12
Barnes, F. J., 65
Barth, C., 21
Barton, D. H. R., 28, 49, 55, 110
Barz, W., 206, 217, 234
Baskevitch, Z., 12, 275
Bassett, E. W., 288
Bassett, R. A., 4
Bates, R. B., 92
Bateson, J. H., 14
Batra, P. P., 67
Batterham, T. J., 87
Battersby, A. R., 5, 110, 182, 224, 248, 278, 283, 292, 295
Baulieu, E.-E., 35
Baumann, T. W., 182
Baur, A. H., 80
Bayne, C.-J., 63
Baxter, C., 124, 182
Beadle, A. S., 2
Beastall, G. H., 24

Bechtold, M. M., 29
Beck, A. B., 232
Bedford, C. T., 187
Beedle, A. S., 21
Beg, Z. H., 20
Beijersbergen, J. C. M., 83
Beimer, N., 78
Bell, E. A., 76, 83, 88, 89
Ben-Aziz, A., 67, 70
Benedict, C. R., 70
Benes, P., 44
Benjamin, W. A., 183
Benn, H. M., 115
Bentley, R., 22, 187, 192, 196, 246, 253
Benvensite, P., 24, 48, 57
Bergman, B. H. H., 83
Berlin, J., 206, 217
Bernhard, R. A., 78
Beytia, E., 20, 24
Bhat, J. V., 227
Bhavnani, B. R., 37
Bhavsar, G. C., 118
Bibb, P. C., 232
Biddlecombe, W., 35
Bidwell, R. G. S., 79
Biemann, K., 288
Billing, B. H., 45
Bimpson, T., 48
Binks, R., 110
Biollaz, M., 8, 266
Birch, A. J., 2, 183, 188, 202, 231, 242
Bissett, F. H., 91
Björkhem, I., 34, 35, 36
Björkmann, L. J., 65
Black, H. S., 20
Blanchet, M.-F., 64
Blaschke, G., 113
Bloch, K., 24, 25, 186
Bloxham, D. P., 20
Boar, R. B., 28
Bodansky, M., 85
Bodem, G. B., 110
Bohm, H., 182, 222
Bond, F. T., 55
Booth, R., 18
Booth, W. D., 40

300

Author Index

Bordas, E., 91
Borkowski, B., 96
Bose, A. K., 297, 299
Boudet, A., 226
Bouquet, A., 92
Boutry, M., 21
Bowen, D. H., 13
Boyd, G. S., 31, 44
Boyle, J. E., 85
Braekman, J.-C., 182
Brambow, H. J., 205
Bramley, P. M., 67
Brandt, E. V., 210, 230, 231
Brandt, K., 40
Brandt, R. D., 57
Braunstein, J. D., 120
Bray, R. C., 281, 282
Breuer, H., 44
Bricker, L. A., 20
Briedis, A., 18, 20
Briggs, M. H., 64
Britton, G., 67, 70
Brochmann-Hanssen, E., 182
Brodie, H. J., 40
Brooks, C. J. W., 10, 31
Brooksbank, W. L., 40
Brophy, P. J., 38
Brown, C. E., 282
Brown, E. G., 75
Brown, J. R., 200
Brown, R. O., 65
Broyer, T. C., 79
Brunengraber, H., 20
Budzikiewicz, H., 288
Büchi, G., 266
Buckwalter, B. L., 12, 275
Bugany, H., 63
Bu'Lock, J. D., 183, 202, 288
Burbolt, A. J., 4, 5
Burleson, D., 111
Burlingame, A. L., 288
Burnett, A. R., 196
Burstein, S., 31, 32
Butcher, D. N., 124, 125
Butler, G. W., 77, 92, 93, 94
Byerrum, R. U., 122

Caglioiti, L., 216
Cagnoli-Bellavita, N., 12, 275
Calimbas, T., 64
Cameron, E. H. D., 37
Campbell, I. M., 238
Campbell, J. M., 253
Canonica, L., 10, 59, 187
Caporale, G., 3
Capozzi, A., 3
Cardemil, E., 7, 22
Cardillo, R., 214
Carlson, K. D., 88
Caroll, M., 88
Carimir, J., 80, 83, 85

Carr, D. J., 13
Caspi, E., 26, 30, 59
Castric, P. A., 76
Caughey, W. S., 280
Cave, A., 238
Cawthorne, M. A., 28
Ceccherelli, P., 12, 275
Ceska, O., 233
Chadenson, M., 229
Chadha, M. S., 59
Chakraborty, J., 40
Chandler, J. L. R., 122
Chang, C.-J., 275
Chang, Y.-F., 86
Chapdelaine, A., 34
Chapman, P. J., 218
Charlwood, B. V., 1, 4, 6
Chaudhuri, N. K., 31
Chayet, L., 7, 22
Chedekel, M. R., 111
Chen, C.-H., 182
Chen, D. M., 79
Chenoweth, R. G., 296
Cherest, H., 80
Chiang, H.-C., 182
Chidester, C. G., 59
Chilton, W. S., 85
Chin, C. C., 37
Chisholm, M. D., 98, 100, 104
Chiyoda, Y., 61
Chopin, J., 229
Chow, C. M., 79
Chromiński, A., 60
Chu, M., 217
Ciranni, G., 216
Clapp, R. C., 91, 92
Clay, P. T., 202
Clements, J. B., 13
Clements, J. H., 182, 248
Clifford, K., 23
Clinkenbeard, K., 16
Co, N., 32
Coates, R. M., 24
Coburn, R. A., 91
Cochran, D. W., 260
Coggins, C. W. jun., 67
Cohen, C. F., 49
Collie, J. N., 183
Collins, D. C., 41
Collins, J., 295
Collins, J. F., 182
Comai, K., 29
Cone, C., 83, 88
Conn, E. E., 76, 92, 94
Conti, S. F., 70
Cook, I. F., 62
Coolbaugh, R. C., 12
Cooper, D. Y., 32
Cooper-Driver, G., 218
Corbella, A., 8, 9
Corbett, J. R., 217
Cori, O., 7, 22

Corcoran, J. W., 184
Cordes, E. H., 260
Corner-Zamodits, J. J., 218
Cornforth, J. W., 23, 26
Corse, J., 210
Coscia, C. J., 7
Cossins, E. A., 80
Couchman, R., 223
Court, W. E., 182
Cowan, R. A., 35
Cox, G. B., 294
Craig, M. C., 20
Crastes de Paulet, A., 37, 40
Creveling, R. K., 78
Crews, L., 78
Croizat, B., 44
Crombie, L., 6, 65, 206
Cronholm, T., 36, 37, 45
Cross, B. E., 13, 14
Croteau, R., 1, 4, 5, 9
Crout, D. H. G., 115, 183, 192, 233
Crowden, R. K., 231
Cumbus, B. J., 20
Curtis, R. F., 197, 200
Cushley, R. J., 212, 260, 282, 283

Dagley, S., 218
Dahm, K. H., 7, 61
Dain, J. G., 187
Dalis, A., 77
Dall'Acqua, F., 3
Danielak, R., 96
Danielsson, H., 34, 36, 46
Darby, F. J., 184
Dardenne, G. A., 83, 85
Das, K. G., 297
Davenport, H. E., 232
Davies, B. H., 67
Davies, D. D., 225
Davies, E. J., 55
Davies, M. E., 232
Davies, T., 31
Davis, B. P., 38
Davis, J. B., 65
Davis, N. B., 80
Davis, N. M., 115
Davison, A. N., 30
Daxenbichler, M. E., 88
Dean, P. D. G., 27
De Benedict, C., 67
Decker, K., 21
Delwiche, C. V., 29
Dempsey, M. E., 29, 33
Dennick, R. G., 16
de Robichon-Szulmaister, 80
Desaty, D., 243
Descomps, B., 37, 40
Detre, G., 243
Devon, T. K., 1

Devys, M., 64
Dewey, L. J., 122
Dewick, P. M., 206
Dhar, A. K., 29
Dickson, L. G., 49
Digenis, G. A., 112
Dimroth, P., 184
Dixon, W. R., 32
Djerassi, C., 288
Dobson, T. A., 110, 182
Dodd, W. A., 80
Doddrell, D., 260, 280, 286
Dönges, D., 124
Döpp, H., 63
Doisy, E. A., 46
Domagala, J., 120
Doney, R. C., 78
Donnelly, W. J., 182
Donninger, C., 21
Donovan, F. W., 183
Doonan, H. J., 6
Douglas, T. J., 27
Downing, M., 7
Dowsett, J. R., 217
Doyle, P. J., 30, 49, 85
Dreiding, A. S., 108
Dresler, S., 227
Drosdowsky, M. A., 64
Duchamp, D. J., 59
Dugan, R. E., 18, 20, 25
Duncan, J. M., 55
Dunham, D. J., 10
Dunkelblum, E., 108
Dunphy, P. J., 5, 23
Dupont, M. S., 232
Durley, R. C., 13
Durst, F., 57
Dutky, S. R., 49
Dvornik, D., 35

Eagles, J., 223
Ebel, J., 228
Ebersole, R. C., 26
Eble, T. E., 292
Edwards, J. M., 195, 239
Edwards, P. A., 18, 19
Egger, K., 229
Einarsson, K., 35, 36
Elander, R., 276
Elliott, M. C., 96, 101
Elliott, W. H., 44, 46
Ellis, P. E., 63
Ellyard, R. K., 44
El-Masry, S., 112
El-Olemy, M. M., 59, 60
Elze, H., 125
Endress, R., 228
Endo, K., 186
Engelsma, G., 225
English, P. D., 28
Epstein, W. W., 65

Eriksson, H., 36
Ernst, R., 60
Ernst, R. R., 252, 257
Escherich, W., 90
Esders, T. W., 57
Eto, Y., 30
Ettlinger, M., 91, 218
Ettlinger, M. G., 91, 92, 95
Evans, R., 12
Evans, R. H., jun., 273
Evans, W. C., 115, 118, 120, 235
Eyjólfsson, R., 91, 218
Eyolfson, J. L., 117

Faini, F., 22
Fairbairn, J. W., 111, 112, 197
Fairlie, J. C., 187
Falardeau, P., 34
Fales, H. M., 89, 182
Faraj, B. A., 112
Farnden, K. J. F., 76
Farrar, K. R., 2
Faulds, W. F., 243
Fayez, M. B. E., 89, 182
Fazekas, A. G., 34
Fazli, F. R. Y., 58
Feeney, J., 278
Feldbruegge, D. H., 20
Fellows, L. E., 89
Ferezou, J. P., 64
Ferreira, D., 210, 230, 231
Ferreira, N. P., 214
Ferrito, V., 187
Fiecchi, A., 29
Findlay, D. A. R., 65
Firth, P. A., 6
Fisch, M. H., 60
Fisher, N., 108
Flavin, M., 80
Flick, B. H., 60
Floss, H. G., 87, 88, 215, 275, 276
Forrester, J. M., 10
Forrester, P. I., 186, 187
Foulkes, D. M., 110, 182
Fourie, T. G., 234
Fowden, L., 74, 75, 76, 82, 83, 84, 85, 87
Fox, J. A., 292
Francis, M. J. O., 4, 23
Freeman, C. W., 29
Freeman-Cooper, 183
Frey, G., 235
Friedell, G. H., 30
Fritig, B., 220
Fritsch, H., 206
Frohofer, H., 108
Fryberg, M., 49
Fu, P., 122
Fuganti, C., 182, 214

Fujioka, S., 63
Fujiwasa, A., 78
Fukumoto, K., 290
Funke, P. T., 297, 299
Furfine, C. S., 37
Fushiki, H., 221, 222

Galbraith, M. N., 63
Gallay, J., 35
Galli, G., 29
Games, D. E., 3
Gander, J. E., 90
Garcia-Peregrin, E., 2, 22
Garg, A. K., 182
Gariboldi, P., 8, 9
Gatenbeck, S., 200, 202, 246
Gaucher, G. M., 186, 187
Gautheret, R. J., 125
Gautier, J., 238
Gaylor, J. L., 16, 29, 55
Gear, J. R., 182
Gefter, M. L., 3
Geissman, T. A., 115, 183, 192, 233
Gellert, E., 85
Gering, R. K., 77
Geynet, P., 35
Ghani, A., 120
Ghiringhelli, D., 214
Ghisalberti, E. L., 15
Gholson, R. K., 122
Ghraf, R., 44
Giangrasso, D., 214
Gibb, W., 37
Gibbons, G. F., 27, 44, 46
Gibson, C. A., 116
Gibson, F., 87, 294
Gibson, M. R., 182
Gilbert, J. D., 31
Gilbert, L. I., 61, 63
Gilbertson, T. J., 86, 107
Giles, C. A., 35
Giovanelli, J., 74, 80
Givner, M. L., 35
Glass, A. D. M., 222
Glotter, E., 60
Glover, J., 48
Gmelin, R., 91, 96, 99
Goad, L. J., 48, 55, 57, 64
Godtfredsen, W. O., 26
Goldfarb, S., 18, 19
Goldman, A. S., 36
Gonzalez, A. G., 11
Gonzalez, Ma. P., 287
Goodeve, A. M., 116
Goodfellow, R., 64
Goodfellow, R. D., 60
Goodwin, T. W., 16, 24, 48, 55, 57, 59, 62, 67, 70
Gorell, T. A., 63
Gorman, M., 276

Author Index

Gottlieb, O. R., 223
Gould, R. G., 18, 19
Gower, D. B., 38, 40
Graebe, J. E., 13
Grambow, H. J., 205
Granroth, B., 74
Grant, D. M., 250
Grant, D. R., 79, 80
Grant, J. K., 35
Grasselli, P., 214
Grayson, D. A., 295
Green, J., 27, 28
Green, T. R., 25
Greene, R. C., 80
Gregonis, D. E., 65
Gregory, K. W., 18
Greten, H., 30
Grey, J. C., 22
Griffith, G. R., 7
Griffiths, K., 37
Griffiths, L. A., 236
Grisebach, H., 205, 206, 228, 234, 293
Grobbelaar, N., 79
Groger, D., 106, 107
Gross, D., 115, 182
Grotzinger, E., 238
Grove, J. F., 255
Grundon, M. F., 182
Grutner, J. B., 259
Guarnaccia, R., 7
Guder, W., 18
Guilford, H., 192
Gupta, M. P., 182
Gupta, P., 210
Gupta, R. N., 85, 108, 182
Gustafsson, J.-Å., 35, 36, 40, 44
Gut, M., 31, 32, 33

Hackenschmidt, J., 21
Hadwiger, L. A., 226
Hagaman, E. W., 275
Hagen, C. W., 232
Haginiwa, J., 75, 76, 89, 182
Hahlbrock, K., 94, 206, 227, 228, 232
Haider, K., 217
Halpern, B., 85
Hall, D. A., 31
Hall, E. S., 290
Hall, S. F., 11
Hamasaki, T., 212, 256, 264, 267
Hamberg, M., 202
Hamill, R. L., 122, 276
Hamilton, D., 288
Hammer, C. F., 37
Hammond, A. L., 241
Hamon, N. W., 117
Hamprecht, B., 18, 19

Haney, M. E., 276
Hanka, L. J., 85
Hansbury, E., 30
Hanson, J. R., 1, 6, 10, 11, 12, 13, 290
Hanson, K. R., 182, 224
Harborne, J. B., 216, 220, 231, 232, 234, 235
Hardman, R., 58
Harland, W. A., 31
Harms, H., 217
Harrison, D. M., 182
Harrison, F. A., 40
Harrison, P. G., 195
Hartmann, M. A., 57
Hartmann, T., 90, 124
Hasegawa, M., 220
Hasegawa, S., 226
Haslam, E., 233
Hassall, C. H., 187, 197, 200
Hatanaka, H., 20
Hatanaka, S., 85
Hattori, H., 90
Haupt, O., 44
Hauser, S., 19, 46
Hauteville, M., 229
Havir, E. A., 224, 225
Hayakawa, S., 20
Hawker, J., 13
Heap, R. B., 40
Hefendehl, F. W., 6
Heftmann, E., 113
Hegarty, M. P., 223
Hegnauer, R., 91
Heidel, P., 30
Heintz, R., 24, 48, 57
Helmkamp, G. M., 186
Hemming, F. W., 22
Hendrickson, H. R., 76
Hendrickson, J. B., 183
Henning, G. L., 67
Henrick, C. A., 61
Herber, R., 65, 67
Herbert, R. B., 182, 248
Herriot, D., 37
Hickman, P. E., 20
Hieronimus, B., 41
Higgins, M., 18
Higuchi, T., 221, 222
Hikino, H., 63
Hill, R. K., 88
Hillis, W. E., 202
Hirata, K., 27
Hirth, L., 220
Hochberg, R. B., 32
Hockenhall, D. J. P., 243
Hodgkins, D. S., 225
Hodgkinson, A. J., 223
Holm, C. H., 250
Holmlund, C. E., 27, 64
Holtom, A. M., 12
Holton, R. A., 27

Honma, S., 44
Honwad, V. K., 236
Hopkins, R., 40
Hopla, R. E., 27
Horan, H., 182
Horie, Y., 61
Horn, D. H. S., 63
Horning, E. C., 46
Horning, M. G., 46
Horton, B. J., 20
Hosel, W., 235
Hosoda, H., 40
Houghton, R. P., 6
Howard, A. S., 223
Howard, F. D., 78
Howes, C. D., 67
Hoyer, G.-A., 63
Hsia, S. L., 38
Hsu, W. J., 67
Huber, J., 18, 19
Hudson, A. T., 187
Hughes, J. C., 220
Hulme, M. A., 288
Hunt, G. E., 82
Hutchinson, C. R., 83
Hyde, P. M., 44
Hyun, J., 31

Ikawa, S., 44
Ikekawa, N., 61
Ilert, H.-I., 124
Iljin, G. S., 112, 182
Imaseki, H., 115
Impellizzeri, G., 109
Innerarity, L. T., 221, 226
Inoue, T., 182
Ishikura, N., 202

Jackman, L. M., 65, 294
Jacob, G., 7, 22
Jacobsen, J. V., 78
Jacques, D., 233
Jadot, J., 80
James, D. H., 3
James, K. J., 295
Jankowski, W. C., 250
Jansen, F. H., 45
Jautelat, M., 250
Jaureguiberry, G., 294
Jayaram, M., 62
Jedlicki, E., 22
Jefcoate, C. R. E., 29
Jefferies, P. R., 14, 15
Jeffrey, A. M., 235
Jeffrey, G. A., 91
Jeffrey, J., 37
Jellinck, P. H., 44
Johansson, G., 34
Johne, S., 106, 107
Johnson, C. M., 79
Johnson, F. J., 77

Johnson, L., 193, 212, 252, 255, 256, 264
Johnson, L. F., 250, 267, 287
Jommi, G., 8, 9
Jones, A. J., 250
Jones, J. B., 35
Jones, J. P., 30
Josefsson, E., 100, 101, 104
Joseph-Nathan, P., 287
Jubier, M. F., 86
Jungalwala, F. B., 65

Kahnt, F. W., 34
Kajiwara, M., 282, 283
Kan, K. W., 33
Kanayama, K., 44
Kanazawa, A., 64
Kande, J., 44
Kandutsch, A. A., 30
Kaneko, K., 182
Kapadia, G. J., 89, 182
Kaplan, M. M., 80
Kaplanis, J. N., 61, 63, 64
Karlson, P., 63
Karlsson, K.-A., 65
Karr, D., 80
Kass, L. R., 186
Kates, M., 25
Katkov, T., 40
Katoh, F., 214, 272
Katsuki, H., 20
Katz, J. J., 282
Kawachi, T., 18
Kawaguchi, A., 20
Keates, R. B., 10
Keen, N. T., 236
Kefeli, V. I., 90
Keglević, D., 90
Kekwick, R. G. O., 22
Keller, H., 182
Kelly, R. B., 85
Kempe, U. M., 49, 55
Kenneally, M. F., 182
Khanchandani, K. S., 297, 299
Khatoon, N., 65
Kienle, M. G., 29
Kim, H. S., 91
Kim, W. K., 90
Kimball, H. L., 31
Kimura, K., 36
Kindl, H., 97, 100, 103
King, D. S., 63
Kirby, G. W., 109, 110
Kircheis, U., 184
Kirk, L., 85
Kirkland, D. F., 98, 102
Kirven, E. P., 120
Kisaki, T., 121
Kiss, P., 217
Kitagawa, T., 120
Kjaer, A., 95, 96, 99
Kleiman, R., 88

Klimentyeva, N. I., 182
Klyshew, L. K., 182
Knapp, F. F., 48, 55
Knight, M., 235
Knittel, P., 187
Knox, J. R., 14, 232
Knuppen, R., 44
Kobus, J., 122
Kochakian, C. D., 36
Koga, K., 120
Konishi, S., 182
Konz, W. E., 27
Kopcewicz, J., 60
Koreeda, M., 63
Kowerski, R. C., 65
Koyama, T., 2, 23, 24
Koyama, Y., 214, 272
Králová, M., 101
Kramer, J. K. G., 25
Kredich, N. M., 74
Kreuzaler, F., 227, 232
Krishnamurty, H. G., 235
Kroszezynski, W., 10, 187
Kruger, P. E. J., 223
Kruglychina, G. K., 182
Kubota, T., 195
Kugajevsky, I., 297
Kuhlmann, K. F., 250
Kunesch, G., 238
Kuramoto, H., 75, 76, 89
Kuschmusadov, J. K., 182
Kushwaha, S. C., 25
Kutáček, M., 90, 101
Kuttan, R., 182

Ladésić, B., 90
Laing, D. G., 5
Laird, W. M., 223
Lamers-Stahlhofen, G. J. M., 38
Lambiotte, M., 44
Lane, M. D., 16
Lanzaraini, G., 227
Lapar, V., 19
Larsen, B. R., 65
Larsen, P. O., 87, 89, 215
Larson, R. L., 231
Lavie, D., 60
Law, J. H., 294
Lawton, R. G. 10
Layne, D. S., 41
Leblanc, H., 34
Lederer, E., 294
Lee, C.-J., 105
Lee, E., 192, 193
Lee, H. J., 182
Lee, J. L.-C., 111
Leete, E., 83, 89, 106, 110, 111, 113, 115, 116, 117, 120, 121, 182, 202
Lehnen, B., 44

Lehoux, J.-G., 34
Leistner, E., 196, 197
Lejeune, B., 86
Lem, B., 182
Lemke, P. A., 259
Lenton, J. L., 55
Le Patourel, G. N. J., 23
Leung, K., 36, 44
Levenberg, B., 85
Lever, M., 77
Levey, G. S., 20
Levy, G. C., 250
Lewis, B. G., 79
Lewis, J. R., 210
Lewis, S., 288
Liaaen-Jensen, S., 70
Lieberman, S., 32
Liebisch, H. W., 118, 182
Light, R. J., 57, 184, 186
Liljegren, D. L., 106
Lindner, H. R., 36
Lindquist, E. L., 63, 64
Lipsky, S. R., 212, 260
Lisboa, B. P., 44
Liston, A. J., 34
Liu, G. C. K., 60
Lively, D. H., 276
Lloyd-Jones, J. G., 30, 59, 62
Lo, W. B., 20
Loeber, D. E., 65
Loke, K. H., 38
Lonergan, C. M., 231
Long, L., jun., 91, 92
Loomis, W. D., 1, 4, 5, 9
Losel, W., 200
Louden, M. L., 115, 116
Loveys, B. R., 13
Lovkova, M. Y., 112, 182
Lowe, D. A., 217
Lowenstein, J. M., 21
Lubet, P., 64
Lukacs, G., 242
Lundström, J., 182
Lunenfeld, B., 38
Lupien, P. J., 20
Luttrell, B., 32
Lux, S. E., 29
Lynen, F., 184
Lyttle, C. R., 44

Mabe, J. A., 275, 276
Mabry, T. J., 92, 108
McCapra, F., 290
McCarthy, J. L., 31
McCloskey, J. A., 294, 298
McCloskey, P., 217
McConnell, W. B., 77, 79
McCorkindale, N. J., 187
McCreadie, T., 11
McDermott, J. C. B., 67
McDonald, E., 182, 193, 248, 278, 295

Author Index

Macdonald, J. C., 3
McDonald, P. D., 32
McFarlane, I. J., 182
McGarrity, J. F., 55
McGhie, J. F., 28
MacGibbon, C. M., 85
Machida, Y., 10
McInnes, A. G., 193, 243, 247 255, 287
MacLean, D. B., 182
MacMillan, J., 13
McMurtrey, K., 182
McNamara, D. J., 18, 19
MacSweeney, D. A., 40
Madyastha, K. M., 7
Magalhaes, M. T., 223
Mae, T., 78
Maebayashi, Y., 270
Maghuin, G., 80
Mahadevan, S., 102
Mahler, H. R., 260
Maier, V. P., 226
Makino, I., 45
Malathi, K., 76
Malhotra, H. C., 48, 70
Mallaby, R., 23
Malmstrom, L., 200
Mandinas, B., 65, 67
Mann, L. K., 78
Manni, P. E., 276
Mansell, R. L., 223
Manson, M. E., 44
Manuel, M. F., 110
Marcenko, E., 90
Marciani, S., 3
Marsh, H. V., 225
Marshal, B. J., 216
Martin, C. W., 46
Martin, D. G., 85
Martin, J. D., 11
Martin, L. L., 275
Marumo, S., 7, 23, 90
Mason, D. J., 292
Massey, S. R., 109
Massey-Westropp, R. A., 242
Matsui, M., 36
Matsuo, M., 98, 99, 103, 104
Mayer, D., 46
Mayor, F., 2, 22
Mazelis, M., 75, 78, 80, 84
Mazza, M., 182
Mead, R. J., 75
Mecker, M. A., 74
Meigs, R. A., 40
Mellon, F. A., 85
Mentzer, C., 267
Merkatz, J., 44
Merwe, J. P., 231
Messe, M. T., 250
Metzler, M., 7, 61
Meusy, J.-J., 64
Meyer, D., 7, 61

Middleton, B., 18, 217
Middleton, E. J., 63
Miersch, J., 80
Mikolajczak, K. L., 92
Milewich, L., 33, 40
Millborrow, B. V., 11
Miller, H. E., 108
Millington, D. S., 85
Mills, S. E., 90
Milne, G., 266
Minari, O., 30
Minato, H., 55, 59
Minghetti, A., 216
Misiti, D., 216
Mitchell, D. J., 79
Mitchell, E. D., 7
Mitropoulos, K. A., 27, 44, 46
Mitsuhashi, H., 182
Miura, G. A., 90
Miyaki, K., 270
Mizusaki, S., 121
Moir, N. J., 29
Money, T., 10, 187, 202, 290
Moore, D. P., 74
Moore, R. J., 36
Moore, T. C., 12
Mootoo, B. S., 290
Moran, J., 278
Morgan, B., 27, 28
Morgan, E. D., 63
Morisaki, M., 61
Moriyama, H., 63
Morris, C. J., 74, 86, 89
Mosbach, E. H., 19, 46
Mosbach, K., 196, 202
Moss, G. P., 1
Moss, J., 16
Mothes, K., 106
Mudd, S. H., 74, 80
Mühlenberg, C., 104
Müller, E., 126
Müller, O.-A., 18
Müller, P., 75, 88, 89
Muhtadi, F. J., 197
Mulder, E., 38
Mulheirn, L. J., 16
Muller, N., 242
Munakata, K., 90
Munday, K. A., 2, 21
Murakoshi, I., 75, 76, 89, 182
Muroya, H., 19
Murphy, G. M., 45
Murray, B. G., 231
Murray, M. J., 6
Murrill, J. B., 115
Muscio, F., 65
Myant, N. B., 44

Nadeau, R., 13
Nagasaki, T., 59
Nagel, D. W., 214

Nakamura, E., 63
Nakanishi, K., 63
Nakashima, R., 3
Nakasone, S., 61
Nall, B. W., 46
Nambara, T., 40, 44
Nambudiri, H. M. D., 227
Nancarrow, C. D., 37
Nash, C. H., 259
Neher, R., 34
Neish, A. C., 90
Nelson, G. L., 250
Nes, W. R., 48
Neumann, D., 126
Neuss, N., 259
Newcombe, F., 182
Newkome, G. R., 88
Ngo, T. T., 74
Nicholas, H. J., 30, 46
Nicholson, R., 31
Nickolson, R. C., 33
Nicolas, J. C., 40
Nielsen, B., 182
Nigam, S. N., 77, 79
Niimura, Y., 85
Nilsson, K., 65
Nimrod, A., 36
Nishimoto, T., 24
Nishino, T., 2
Nocke-Fink, L., 44
Noguchi, M., 121
Norton, K., 13
Norum, K. R.,
Nowacki, E., 182
Nozoe, S., 7, 10, 191
Nulu, J. R., 83, 88
Nurakow, D., 182

Oakey, R. E., 40, 41
O'Brien, I. G., 294
O'Donovan, D., 113
O'Donovan, D. G., 182
Oehlschlager, A. C., 11, 49
Oertel, G. W., 44
Ogura, K., 2, 23, 24
Ogura, M., 44
Ohira, K., 78
Ohishi, J., 214, 272
Ohizumi, Y., 63
Ohmiya, S., 89, 182
Ohtaka, H., 61
Oka, T., 235
Okada, K., 282, 283
Okada, M., 36
Okauchi, T., 63
Okubayashi, M., 61
Okuda, K., 46
Olesch, B., 182
Olson, J. O., 111, 202
Onderka, D. K., 87, 88, 215
Oriente, G., 182

Ourisson, G., 57, 220
Overeem, J. C., 83
Overton, K. H., 11
Ozon, R., 64

Pace-Ascak, C., 202
Padmanaban, G., 76
Paleg, L. G., 27
Palmer, E. D., 28
Panke, D., 91
Paoletti, E. G., 29
Paoletti, R., 29
Pape, H., 293
Paris, M., 92
Paris, R.-R., 92
Parke, D. V., 40
Parker, H. I., 113
Parker, T., 2, 24
Parkhurst, J. R., 225
Parks, L. W., 55
Parry, D. R., 197, 200
Patterson, G. W., 49
Peisker, K., 182
Pennington, D., 70
Percy-Robb, I. W., 44
Perlman, D., 85
Perold, G. W., 223
Peterson, P. J., 82
Petrosilius, U., 46
Phillips, D. O., 48
Phillips, G. T., 23, 184, 187, 290
Piatelli, M., 109, 182
Pifferi, P. G., 227
Pilgrim, J., 59
Pinder, A. R., 182
Pitcher, R. G., 202, 273
Pitot, H. C., 19
Plasse, J.-C., 44
Plimmer, J. R., 230
Pokorny, M., 90
Polito, A., 2, 24
Pollard, J. K., 83
Pollow, B., 41
Pollow, K., 41
Polonsky, J., 12, 238, 275
Pons, M., 40
Pont-Lezica, R., 7, 22
Popják, G., 2, 21, 24
Popplestone, C. R., 86
Porter, J. W., 18, 20, 24, 25, 65
Posner, B. I., 80
Poulter, C. D., 6
Pousette, Å., 36
Powell, A. J., 288
Praez, J. C., 234
Prather, C. W., 122
Price, R. J., 57
Pritchard, D. E., 242
Pruess, D. L., 202, 273
Pryce, R. J., 205
Przybylska, J., 79

Pugh, E. L., 25

Quackenbush, F. W., 19
Qureshi, A. A., 20, 24, 65

Radhakrishnan, A. N., 182
Radwan, A. S., 182
Rahimtula, A. D., 29
Railton, I. D., 13
Ramage, R., 5, 182, 248
Raman, T. S., 62
Ramasarma, T., 20, 21
Ramm, P. J., 16
Rampini, E., 38
Ramsey, R. B., 30
Ramstad, E., 117
Ramuz, H., 110
Randall, P. J., 57, 62
Ranjiva, R., 226
Ranzi, B. M., 10, 187
Rao, G. S., 20, 44, 89, 182
Rao, M. L., 44
Rao, P. V. S., 227
Rapoport, H., 110, 113
Rappaport, L., 13
Ratledge, C., 216
Ratner, S., 241
Reay, P. F., 92
Rees, A. F., 67
Rees, H. H., 16, 24, 57, 59, 62
Reeves, B. E. A., 46
Reich, H. J., 250
Reid, D. M., 13
Reid, P. D., 225
Reinbothe, H., 80
Reininger, W., 200
Richards, J. H., 183
Richards, R. W., 242
Rilling, H. C., 26, 65
Rindone, B., 10, 187
Rios, A., 30
Ritter, M. C., 29, 33
Robbins, W. E., 61, 62, 63, 64
Roberts, J. D., 250
Roberts, F. M., 75
Roberts, M. F., 111
Robins, D. J., 196
Robinson, I. M., 117
Robinson, R., 183
Robinson, T., 122
Robinson, W. H., 24
Roddick, J. G., 124, 125
Rodwell, V. W., 18, 19
Röller, H., 7, 61
Rönsch, H., 182
Rösler, H., 108
Rohmer, M., 57
Rohringer, R., 90
Ronchetti, F., 59
Rosa, N., 90
Rosenbaum, N., 3
Rosenberg, H. 182

Rosenthal, G. A., 81, 82
Rosenthal, O., 32
Rother, A., 182
Roughley, P. J., 212
Roux, D. G., 210, 230, 234
Rowan, M. G., 27
Roy, D. N., 76
Rudney, H., 18
Rudzats, R., 85
Rueppel, M. L., 110
Russell, G. B., 92
Russkov, Yu. A., 182
Russo, G., 59
Ryan, K. J., 40
Ryvarden, L., 70

Saat, Y. A., 40
Sabine, J. R., 20, 21
Sacher, J. A., 225
Safe, S., 55
Saintot, M., 37
Saito, T., 63
Sakagami, T., 30
Sakamoto, Y., 27
Sakazaki, R., 82
Saleemuddin, M., 20
Salisbury, P., 187
Samuelsson, B., 202
Sandor, H., 113
Sandermann, W., 4
Sander, T., 34
Santaniallo, E., 10, 187
Sarma, P. S., 76
Sastry, S. D., 182
Sato, F., 37, 220
Saucier, S. F., 30
Sauer, L. A., 34
Savin, M. A., 80
Scala, A., 29
Scallen, T. J., 29, 30
Schaller, H. J., 118
Schechter, J., 24
Scherf, H., 90
Schiefer, S., 100, 103
Schleyer, H., 32
Schmid, R., 293
Schmitt, J. H., 90
Schmitt, R. C., 195, 239
Schmitz, M., 232
Schmotzer, L. A., 105
Schriefers, H., 44
Schrift, A., 79
Schüler, M., 91
Schütte, H. R., 75, 88, 89, 115, 118, 182
Schultz, Ch., 182
Schumacher, R., 44
Schuster, A., 96
Schuster, M. W., 29
Schwarting, A. E., 182
Schwimmer, S., 78, 113
Sciuto, S., 182

Author Index

Scolastico, C., 8, 10, 187
Scott, A. I., 184, 187, 192, 193, 248, 282, 283, 288, 290, 298
Seaman, F., 92
Seamark, R. F., 37
Segal, W., 75
Seifried, H. E., 29
Seigler, D., 92
Seigler, D. S., 92
Seiler, M. P., 27
Seitz, V., 232
Self, R., 223
Serif, G. S., 105
Seshadri, T. R., 238
Seto, H., 242
Seto, S., 2, 23, 24, 193, 245, 246, 252, 264
Shackleton, C. H. L., 44
Shah, S. N., 30
Shapiro, D. J., 19
Shapiro, M. J., 35
Shargool, P. D., 74
Sharples, D., 91, 94
Shaw, P. D., 298
Shefer, S., 19, 46
Shemin, D., 281, 282
Sheperd, D., 288
Shepherd, K. S., 18
Sheppard, R. C., 85
Shibata, S., 196
Shikita, M., 37
Shimada, M., 221, 222
Shimaoka, A., 59
Shiori, T., 120
Ship, S., 35
Shoji, J., 82
Shoolery, J. N., 242, 287
Shukla, O. P., 297
Siddall, J. B., 7, 61
Siddiqui, M., 20
Siddons, P. T., 65
Sijpensteijn, A. K., 83
Simonet-Thierry, N., 44
Simpson, E. R., 31
Simpson, F. J., 235
Sims, J. J., 236
Singh, R. K., 67, 70
Sipahimalani, A. T., 59
Sipe, J. D., 27, 64
Siperstein, M. D., 16
Sisler, E. C., 122
Sjövall, J., 45
Skinner, S. J. M., 29
Skrdlant, H. B., 29, 30
Skursky, L., 111
Slakey, L. L., 18, 20
Slater, G. P., 3
Slaytor, M., 124, 182
Smalley, H. M., 288
Smith, A. G., 64
Smith, A. R. H., 57
Smith, C. R., jun., 92

Smith, C. Z., 18
Smith, D. G., 193, 243, 247, 255, 287
Smith, E. C., 221, 226
Smith, E. H., 115
Smith, G. E., 236
Smith, G. N., 202, 288
Smith, H., 233, 242
Smith, I. K., 74
Smith, J. D., 20
Smith, T. A., 124
Sørensen, H., 87, 89
Sørup, P., 89
Sofowora, E. A., 58
Sokolowski, G., 41
Solomon, S., 36, 37, 44
Somell, A., 36
Spenser, I. D., 85, 93, 108, 182
Splittstoesser, W. E., 80
Spohn, M., 30
Spring, M. S., 91, 94, 200
Spyropoulos, C. G., 182
Srikantaiah, M. V., 30
Srinivasan, V., 182
Stafford, H. A., 227
Starr, P. R., 55
Statham, C. M., 231
Staunton, J., 182, 184, 193, 202, 224
Steck, W., 195
Steglich, W., 110, 200
Steiner, M., 124
Steinreich, P., 109
Steinstra, T. M., 125
Stempel, A., 273
Steward, F. C., 79, 83
Stewart, J. C., 13
Steyn, P. S., 214
Stickland, R. G., 232
Stockingt, J., 223
Stockmann, H., 4
Stoddart, J. L., 14
Stoffel, W., 30
Stohs, S. J., 59, 60
Stoker, J. R., 91, 94, 200
Stokke, K. T., 30, 31
Stone, K. J., 25
Stothers, J. B., 250
Stowe, B. B., 96, 101, 102
Strange, P. G., 224
Strobel, G. A., 76
Styles, E. D., 233
Suárez, M. D., 2, 22
Subbarayan, C., 20
Sugii, M., 82
Sugiyama, T., 16
Suhadolnik, R. J., 296, 297
Suketa, Y., 82
Sulimovici, S., 38
Sunderland, N., 232
Sung, M.-L., 85, 87
Suwal, P. N., 111

Suzuki, N., 191
Suzuki, K., 30
Suzuki, K. T., 7, 191
Suzuki, T., 82, 182
Suzuki, Y., 7, 23, 267
Svoboda, J. A., 61, 62, 63
Swain, T., 218, 220
Sykes, R. J., 212, 260
Synge, R. L. M., 223
Syrdal, D. D., 11
Szymanski, E. S., 37

Tabata, M., 125
Tack, B. F., 218
Tahler-Dao, A., 37
Tai, H.-H., 25
Taira, S., 182
Tait, A. D., 33
Takagi, Y., 37
Takahashi, E., 182
Takahashi, V., 245
Takao, I., 182
Takeda, K., 59
Takemoto, T., 63
Takenaka, S., 193
Takeuchi, Y., 120
Tamaki, E., 121
Tamura, S., 245
Tan, L., 34
Tanabe, M., 212, 243, 245, 246, 252, 256, 264, 266, 267
Tanabe, Y., 121
Taniguchi, K., 85
Tannenbaum, S. W., 288
Tanner, R. J. N., 233
Tapper, B. A., 93, 94
Tash, J., 63
Teale, J. D., 31
Telle, J., 125
Téllez, R., 7, 22
Tenfel, E., 235
Teshima, S., 64
Teuscher, E., 125
Than, A., 67
Thomas, D., 212, 256
Thomas, G. M., 110
Thomas, R., 4, 195
Thompson, E. D., 55
Thompson, J. F., 74, 77, 78, 86, 89
Thompson, M. J., 49, 61, 62, 63, 64
Thompson, R. S., 233
Thomson, J. A., 63
Thomson, R. H., 195
Threlfall, D. R., 90
Thurman, D. A., 81
Tikonav, Yu. B., 182
Tissut, M., 229
Tjarks, L. W., 92
Toepfer, R., 80

Toft, P., 34
Tokoroyama, T., 195
Tomita, Y., 55
Tomkins, G. M., 74
Topham, R. W., 55
Toube, T. P., 65
Towers, G. H. N., 192, 225, 287
Townsend, C. A., 282, 283
Townsley, J. D., 40
Treadwell, C. R., 31
Trzeciak, W. H., 31
Tschesche, R., 58
Tsou, G., 85
Tubbs, P. K., 18, 217
Tuning, M. S., 13
Turnbull, K. W., 8
Turner, W. B., 183, 187, 267
Tweto, J., 80

Ullmann, H., 21
Underhill, E. W., 96, 97, 98, 99, 102, 115
Ungar, F., 33
Unrau, A. M., 49, 86
Uomosi, A., 55

Vahouny, G. V., 31
Vaishnav, Y. N., 89, 182
Valadon, L. R. G., 67
Van Cantfort, J., 46
Van der Horst, D. J., 64
van der Merwe, J. P., 210
Van Der Molen, H. J., 38
Van Etten, C. H., 88
Vangedal, S., 26
van Tamelen, E. E., 27
Veliky, I. A., 126
Vervier, R., 85
Verzar-Petsi, G., 182
Villoutreix, J., 65, 67
Vincent, F., 35
Vining, L. C., 193, 243, 247, 255
Virtanen, A. I., 78
Voelkert, E., 79, 80
Voges, A., 46
Volk, W. A., 70
von Bahr, C., 40
Voogt, P. A., 64

Wada, F., 27
Wagnières, M., 96

Waiblinger, K., 107
Waiss, A. C., 210
Walker, J. E., 85
Waller, G. R., 111, 182
Walsh, L. B., 46
Walter, H., 184
Wang, H. M., 34
Wanner, H., 182
Wareing, P. F., 13
Warren, J. C., 37
Warren, R. P., 82
Wasserman, H. A., 264
Wasserman, H. H., 212
Wat, C. K., 192, 287
Watanabe, M., 182
Weedon, B. C. L., 65
Weeks, O. B., 65
Weigert, F. J., 250
Weintraub, H., 35
Weisleder, D., 88
Weisner, K. J., 193
Weiss, U., 195, 239
Welch, J., 288
Wemple, J., 120
Wender, S. H., 221, 226
Wenkert, E., 12, 275, 276
Wenzel, M., 41
Westlake, D. W. S., 193, 217, 247
Westley, J. W., 202, 273
Wetter, L. R., 98
Whistance, G. R., 90
Whitaker, E. M., 41
White, A. F., 11, 13
White, I. H., 37
White, L. W., 18
Whitehouse, D., 115
Whiteside, J. A., 81
Whiting, D. A., 6, 65, 206, 212
Whitman, P. J., 282
Whitmore, D., jun., 61
Whitmore, E., 86
Wickramasinghe, J. A. F., 59
Widdowson, D. A., 49, 55
Widholm, J. M., 217
Wierenga, W., 27
Wiermann, R., 228
Wigfield, D. C., 182
Wightman, R. H., 182
Wilcox, M. E., 108
Wilkinson, C. F., 64
Williams, C. A., 231, 232
Williams, D. H., 288

Williamson, A. H., 229
Williamson, D. G., 41
Wilson, D., 288
Wilson, D. A. A., 40
Wilson, J. D., 36
Wiltshire, H. R., 224
Winstead, J. A., 296
Wirz-Justice, A., 2, 6
Witton, D. C., 2, 21, 29
Wohlport, A., 108
Wojciechowski, Z., 57
Wolfe, L. S., 202
Wolff, I. A., 92
Wollenweber, E., 229
Woodbridge, A. P., 63
Woods, D. K., 6
Woolley, J. G., 115, 120
Woolley, V. A., 118, 120
Wong, E., 230
Wright, J. L. C., 193, 247
Wrightman, R. H., 224
Wu, C. C., 120
Wyler, H., 108

Yagen, B., 30, 59
Yalpani, M., 187, 288
Yamada, S., 120
Yamaguchi, Y., 78
Yamasaki, G., 46
Yamasaki, H., 44
Yamasaki, K., 44, 46
Yamazaki, M., 103, 214, 270, 272
Yang, R. S. H., 64
Yang, S. F., 80
Yates, J., 40
Yazaki, Y., 202
Yokoyama, H., 67
Young, I. G., 87
Young, J. L., 86
Young, M. R., 4
Youngken, H. W., 115, 116

Zacharius, R. M., 83
Zaki, A. S., 236
Zamorani, A., 227
Zamoscianyk, H., 31
Zaprometov, M. T., 234
Zenk, M. H., 90, 196, 223
Zilg, H., 94